Proceedings in Life Sciences

Cellular Biology of Ectoenzymes

Proceedings of the International
Erwin-Riesch-Symposium on Ectoenzymes May 1984

Edited by
Georg W. Kreutzberg, Martin Reddington,
and Herbert Zimmermann

With 115 Figures and 46 Tables

Springer-Verlag
Berlin Heidelberg New York Tokyo

Professor GEORG W. KREUTZBERG
Dr. MARTIN REDDINGTON
Max-Planck-Institut für Psychiatrie
Abt. Neuromorphologie
Am Klopferspitz 18a
8033 Martinsried, FRG

Professor HERBERT ZIMMERMANN
AK Neurochemie
Zoologisches Institut der Universität
Siesmayerstraße 70
6000 Frankfurt/M., FRG

ISBN-13:978-3-642-70666-0 e-ISBN-13:978-3-642-70664-6
DOI: 10.1007/978-3-642-70664-6

Library of Congress Cataloging in Publication Data. Main entry under title: Cellular biology of ectozymes. (Proceedings in life sciences). Includes index. 1. Cytochemistry. 2. Enzymes. 3. Plasma membrane. 4. Extracellular space. I. Kreutzberg, Georg W. II. Reddington, Martin, 1947– . III. Zimmermann, Herbert, 1944– . IV. Title: Ectozymes. V. Series. QH611.C43 1985 574.87'6042 85-17317

Softcover reprint of the hardcover 1st edition 1986

2131/3130-543210

Preface

Cells do not normally live as single entities but are grouped together in specific functional and structural configurations in various tissues. Intracellular mechanisms maintain cellular viability and provide the means necessary for their specific cellular functions. The interaction between cells is maintained by mechanisms involving extracellular signalling. Such extracellular mechanisms may include special properties of the cell surface which involve immediate cell contact, but may also represent mechanisms which act at a distance and are mediated via special secretions and/or receptors. Recent studies on cell-cell contact have tended to stress cell surface components directly mediating cellular interactions; the extracellular medium as a metabolically active compartment has been rather neglected. However, it represents a vital medium through which cells communicate, being important in, for example, chemotaxis in primitive organisms, and in development and in the coordination of multiple functions in multicellular organisms. It is not surprising, therefore, that a number of molecular mechanisms have developed together with increasing biological complexity during evolution. Two aspects of the extracellular space have received increasing attention in the last few years. First, several macromolecules such as collagen, laminin and fibronectin have been identified as components of an extracellular matrix giving a structural dimension to the extracellular compartment. Secondly, the discovery of an increasing number of ectoenzymes which, by definition, are part of the plasma membrane or attached to it, and which have their catalytic site facing the extracellular space, has shed new light on the metabolic capabilities of the extracellular, interstitial medium. Clearly, this medium is not only a passive compartment involved in maintaining the extracellular ionic composition, in the transport of metabolites or of hormones and transmitter molecules; it is also a place of intensive metabolism involved in extracellular mechanisms.

Studies on ectoenzymes are underway in many laboratories and have been the concern of workers in disciplines as diverse as cancer research, cell biology, vascular physiology and cardiology, neurochemistry, zoology, and botany. This has inevitably led to a lack of close communication amongst scientists working on ectoenzymes and for this reason it was considered timely to bring researchers in this field together for the first time for an exchange of ideas and experimental data. A meeting was therefore held from May 6th–9th, 1984 under the auspices of the German Society of Cell Biology and the Max Planck Society at Ringberg Castle, situated above Lake Tegernsee in the Bavarian Alps. Its aim was to discuss various problems relevant to the function of ectoenzymes, from their biosynthesis, intracellular transport, membrane incorporation and membrane topography to questions of their regulation and their role in intercellular communication. Among the enzymes discussed were ATPases and ADPases, 5′nucleotidase, protein kinase, peptidases, acetylcholinesterase, and enzymes involved in glycoprotein metabolism. Despite the variety of topics covered, a good deal of common ground was evident. The need became apparent to draw up a set of criteria for establishing the "ecto" nature of an enzyme. Several criteria must be applied including action on non-permeable substrates, inhibition of enzyme activity by non-permeable inhibitors and inactivation by limited proteolytic digestion or covalent inhibitors. Electron microscopical techniques have also proved to be invaluable in localising reaction products to the extracellular medium. These approaches have been applied in various degrees to the enzymes discussed, as will be evident in the ensuing chapters, and have been treated in more detail in Appendix 1. It is clear, however, that several of the enzymes which can be found in an "ecto" position are not exclusively localised in this way. Many, even the almost classical ectoenzyme, 5′nucleotidase, also occur intracellularly, at least in some cell types. Others, such as protein kinase and acetylcholinesterase, can be released directly into the extracellular space. Acetylcholinesterase is even found intracellularly, as a membrane-bound ectoenzyme, and in a secreted form as an integral component of the interstitial space, where it could have direct modulatory effects on nerve cell activity. With most enzymes it is unclear whether the various localisations reflect the presence of isoenzymes or a diversity of cellular processing routes. The term "ectoenzyme" should therefore be considered at present as referring to a localisation rather than an intrinsic enzymatic function.

The variety of ectoenzymes considered in this volume indicates the richness of metabolism in the extracellular space. The nucleotidases, for instance, are thought to act together in a cascade to break down ATP via ADP and AMP to adenosine. Since in this case both substrates and products are known to mediate a variety of physiological responses, these metabolic interactions may have a profound significance for the functions of complex tissues. Extracellular ATP may also act as a substrate for ecto-protein kinases which could modulate the function of proteins on the cell surface or in the extracellular medium. Membrane components such as glycoproteins and glycolipids may also be modified by ecto-glycosyltransferases and glycosidases with important consequences for cell recognition phenomena, from simple sponges to neurons and glial cells. Peptides, which have assumed an important place as neuronal agents in recent years, are inactivated by ectopeptidases. Clearly, metabolic interactions in the extracellular space serve several vital biological functions.

It can be expected that the number of enzymes fulfilling the "ecto" criteria will increase in the next few years. We hope that this book will serve to focus the interest of cell biologists and biochemists on the metabolic activity of the cell surface and on the extracellular space as an important metabolically active compartment.

The Editors

Contents

Ectonucleotidases: 5′Nucleotidase

Ecto-Protein Kinase

Ectoglycosylation

Ectoacetylcholinesterase

Ectopeptidases

List of Contributors

You will find the addresses at the beginning of the respective contributions

Perspectives

Ectoenzymes: Their Modulation and Similarity to Certain Enzymes of Intracellular Membranes

M. L. KARNOVSKY [1]

1 Introduction

About a decade ago an attempt was made to lay down a definition of ectoenzymes, and to establish criteria by which one might judge whether an enzyme belongs to this category or not (DePierre and Karnovsky 1974a). The statement of definition read "The plasma membrane of cells may contain enzymes whose active sites face the external medium rather than the cytoplasm, and these enzymes are reffered to as ectoenzymes." The criteria for ectoenzymes given below are slightly altered from the original wording, but they are, in principle, unchanged from those earlier stated (DePierre and Karnovsky 1973, 1974a, b). They may be outlined as follows, and it is realized that not all criteria have been tested for every enzyme considered to be "ecto":

1) Detection of the enzyme should be possible using intact cells.
2) The question must be answered in the negative as to whether the enzyme under examination was released from the cells into the external medium.
3) The substrate employed should preferably be one that does not enter the cells. However, if criterion 2 above has not been definitively statisfied, one should guard against a circular argument in which an enzymatic activity is revealed with a substrate that is unable to penetrate into the cells, and this fact is then interpreted to indicate an ectoenzyme based solely on the nature of the substrate.
4) When the cells are broken, one should be able to account for any increase in enzymatic activity observed.
5) The products of the enzyme reaction should be localized to the exterior of the cell.
6) One should be able to inhibit the enzyme, selectively, with a nonpenetrating, covalently linking, reagent.

[1] Department of Biological Chemistry, Harvard Medical School, Boston, Massachusetts 02115, USA

2 Checking Criteria for Ectoenzymes

Our original studies (DePierre and Karnovsky 1974a, b) were with guinea pig granulocytes, which turned out to be an excellent choice, for a particular reason cited below. The enzymes that were examined were ATPase, AMPase (5′-nucleotidase), and p-nitrophenylphosphatase. These could all be detected and measured using intact cells and it was demonstrated that they were not "leached" from the cells. In the case of the first and third mentioned enzymes, breaking the cell by homogenization, sonication, or treatment with saponin, caused about a doubling of enzymatic activity. This suggested, but did not prove, that since the substrates used were all negatively charged, and entered the cells with difficulty, the activity seen with intact cells was presumably *ecto,* while the *increased* activity revealed upon the breakage of the cell was *endo,* i.e., located within the cytoplasmic compartment, including organelles. In the case of the second mentioned enzyme, the 5′-nucleotidase, and with these particular cells, it was observed that there was virtually *no increase* in activity on disruption of the cells by homogenization or sonication, but about a 50% increase with the saponin treatment. This was interpreted to mean that virtually all of this enzyme was "ecto" in these cells, i.e., in the plasmalemma, and that saponin may have released it from that locale, or otherwise activated it (Emmelot et al. 1964) resulting in increased activity in this particular case (DePierre and Karnovsky 1974b). In these experiments, the first four desiderata or criteria, mentioned above, were satisfactorily attained.

In the case of the fifth criterion, we undertook experiments in which cells were loaded with inorganic orthophosphate [^{33}P] (DePierre and Karnovsky 1974a). They were then exposed to [^{32}P]-labeled substrate. [^{14}C]-Inulin was included in the medium to allow measurement of residual medium present in the cell pellet that was ultimately examined. Using short incubation periods, it was possible to demonstrate that though most of the preloaded phosphate remained intracellular, almost none of the phosphate released from the substrate was cell associated, i.e., all the [^{32}P] was outside. Thus, the substrate was presumably cleaved by an ecto-

Table 1. Localization of inorganic [^{32}P]phosphate formed by hydrolysis of ATP, AMP, and p-nitrophenyl phosphate after loading intact cells with inorganic [^{33}P]phosphate

Substrate	Length of incubation	Volume of incubation mixture which was inside cells	$^{33}P_i$ inside cells	Product ($^{32}P_i$) inside cells
	min	%	%	%
None	7.0	50.2	89.9	–
ATP	7.0	48.7	85.6	0
AMP	2.0	55.1	93.6	1.2
p-nitrophenyl phosphate	2.0	53.1	94.8	0.8

Table 2. Comparison of activities of ATPase, AMPase, and p-nitrophenyl phosphatase in sonicates prepared from cells treated with the diazonium salt of sulfanilic acid and from untreated cells

	Activity[a] in:			
	Intact cells		Sonicates prepared from:	
Enzyme	Before treatment with reagent	After reagent	Untreated cells	Cells treated with reagent
ATPase	100	16.3	228	125
AMPase	100	11.6	105	17.7
p-nitrophenyl phosphatase	100	18.9	190	92.2

[a] All activities are expressed as a percentage of the activity of untreated cells (Column 1)

Monolayers were incubated with 3.5 m*M* diazonium salt for 30 min at 37°. The reaction was terminated as described in the text. Control monolayers were of two sorts: one set of controls received no reagent, while the other set was washed immediately after the diazonium salt was pipetted onto them. Some of the monolayers subsequently were assayed for the ATPase, AMPase, and p-nitrophenyl phosphatase of intact cells. Cells were scraped off other monolayers and sonicate d; these sonicates were also assayed for ATPase, AMPase and p-nitrophenyl phosphatase

enzyme. There was ample evidence that the movement of phosphate from inside to outside the cell was very slow indeed so the likelihood that the phosphate was released internally and conveyed to the outside, was minimal. The data are reproduced in Table 1.

To meet the last criterion cited above the Berg reagent (diazotized sulfanilic acid) was used as a nonspecific reagent that would attack exposed proteins, and was virtually unable to enter cells at appropriate concentrations, temperatures, and times of incubation (Berg 1969; DePierre and Karnovsky 1974a, b). When it was applied to guinea pig granulocytes with these strictures in mind, it may be observed from Table 2 to have eliminated more than 80% of each of the three enzymes studied. Virtually the whole *internal enzyme component* of ATPase and p-nitrophenylphosphatase, however, remained untouched, as was revealed when the cells were broken after this treatment (DePierre and Karnovsky 1974a).

In the case of AMPase, confirming our belief that virtually all the enzyme was ecto in these particular cells, it was noted that only a very small residual enzyme activity was evident when the cells treated with the reagent were subsequently rendered permeable. It could be shown that when monolayers of these cells were treated with the Berg reagent (0.7 m*M*) most of the ectoenzymes were gone in 30 min, while lactate dehydrogenase was unaffected (DePierre and Karnovsky 1974a).

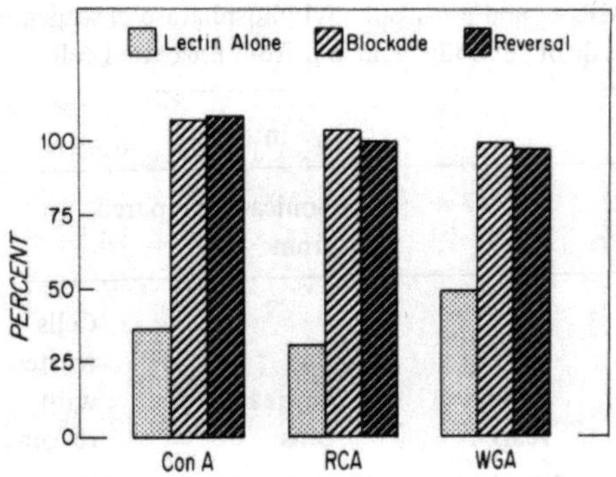

Fig. 1. Blockade and reversal of inhibition: 5′nucleotidase of a plasma membrane fraction. Samples of plasma membrane were incubated with 15 μg of ConA per ml; 15 μg of RCA per ml, or 100 μg of WGA per ml for 15 min at 4°C. For blocking studies, incubation with lectin was performed in the presence of a hapten sugar. For reversal studies, lectin incubation was followed by exposure to a hapten sugar for 15 min at 4°C. In both cases, the seccond 15 min incubation was initiated by adding 0.1 ml of concentrated stock to the suspension; washes were not performed between incubations (compare whole cells). The competing haptens were 50 m*M* α-methyl mannoside for ConA, 100 m*M* galactose for RCA, and 100 m*M* N-acetyl glucosamine for WGA. Very similar results were obtained with whole cells (guinea pig granulocytes)

2.1 Lectins and Ectoenzymes

In work performed later with the same cells, it was noted that lectins could provide excellent reagents for the detection of ectoenzymes by virtue of their size which prevented entry into cells, and the glycoprotein nature of membrane proteins (Smolen and Karnovsky 1980a). In Fig. 1 data illustrating the action of three different lectins that inhibit granulocyte plasmalemma ecto-5′-nucleotidase are presented. All three lectins are prevented from inhibiting that enzyme when the appropriate small molecule soluble hapten is present with the lectin. The same haptens are capable of *reversing* the inhibition when they are applied after exposure of cells to the lectin (Smolen and Karnovsky 1980a).

3 Application of Ectoenzymes

Ectoenzymes provide good markers for the plasmalemma (DePierre and Karnovsky 1973), and it has been the objective of several investigators to utilize studies of the disposition of plasmalemma markers to follow the process of phagocytosis. This is an endocytic process and the internalization of vesicles (phagosomes) is well established. In addition to quantitative studies of the amount of plasmalemma internalized during phagocytosis, done by following the entry of an ectoenzyme into the cell (Reviewed by Smolen and Karnovsky 1980b; also Werb and Cohn 1972; Tsan and Berlin 1971a, b), it was also possible to demonstrate that the vacuoles that one observes after the treatment of cells with phorbol myristate acetate (PMA) are indeed of plasmalemma origin. This was particularly applicable in the case of guinea pig granulocytes because the 5′-nucleotidase is, as pointed out above, virtually totally in the plasmalemma of these cells. Further, the situation was clear-cut because a specially sensitive cytochemical method for 5′-nucleotidase was available. This relies on cerium to precipate released inorganic phosphate (Badwey et al. 1984; Robinson and M.J. Karnovsky 1983).

An additional matter of some concern has been to demonstrate the origin of the superoxide and peroxide that is released during phagocytic stimulation of granulocytes or that by PMA. This function, too, appeared from studies done to detect the locale of release of peroxide, this time using cerium as a trapping agent for H_2O_2, to be located in the plasmalemma and the vacuoles that are formed from that structure (Briggs et al. 1975). Thus, we must believe that the enzymes involved in the release of active oxygen radicals when such cells are stimulated is an ectoenzyme complex. One might have inferred that this was simply a "plasma membrane localized" system, except for the fact that the phenomenon of H_2O_2 release was shown to be sensitive to the nonpenetrating reagent referred to above (diazotized sulfanilic acid) (Briggs et al. 1975).

4 Changes in the Activity of an Ectoenzyme During Modulation of Cell State

In macrophages an interesting fact is that cellular behavior may be modulated by products of T-lymphocytes. Thus, peritoneal macrophages may exist as "control" or "resting" macrophages, "activated" macrophages which are formed during long-term infections with bacteria (e.g., *Bacillus Calmette Guerin,* (BCG), or *Listeria monocytogenes*). Peritoneal macrophages may also be "elicited", or drawn into a sterile abscess in the peritoneum (("inflammatory macrophages"). The ecto-5′-nucleotidase diminishes from the control (resident) values with such treatments of the cells (Karnovsky et al. 1975) (Table 3). A question raised was the following: If indeed the adenosine, cleaved from the 5′-nucleotide on the exterior surface of the cell is of importance to the cell [especially since there is a specific transport protein for adenosine in this cell type (Tsan and Berlin 1971a, b)], would the cell respond to the treatments that diminish the ectoenzyme activity by *raising* the levels of intracellular enzymes that utilize the product of the ectoenzyme? Would the enzymes responsible for the further metabolism of adenosine, be modulated in their levels to meet a new situation resultant upon activation, or elicitation, of the cells? As Table 3 indicates, there is indeed a reciprocal change in the most immediate enzyme in this context, adenosine deaminase, and perhaps in the nucleoside phosphorylase and phosphoribosyl transferase, which is somewhat removed from the nucleotidase. Interestingly

Table 3. Activity[a] of the enzymes of purine metabolism of mouse peritoneal macrophage populations

Macrophage	Ecto-5′-nucleotidase	Adenosine Deaminase	Kinase	Nucleoside phosphorylase	Phosphoribosyltransferase
Resident	45.7 ± 2.1	38.1 ± 8.1	28.7 ± 6.0	6.5 ± 0.3	0.6 ± 0.3
Activated	19.1 ± 1.1	49.8 ± 10.0	32.4 ± 3.6	58.9 ± 11.6	4.0 ± 1.0
Elicited	2.7 ± 0.3	70.6 ± 27.3	19.7 ± 1.0	48.9 ± 18.3	4.4 ± 2.5

[a] Activity expressed as nmol/min per mg protein; n = 3–5

enough, the hypothesis was only half satisfied in the case of adenosine kinase (an immediately available reaction for the product of the ectoenzyme) where a reciprocal relationship is shown for the resident and activated cells between the levels of ecto-5′-núcleotidase and that of the kinase, but not in the case of the elicited cells (Philippeaux and Karnovsky, unpublished data).

Although the observations above deal with the *diminution* of ecto-AMPase in a physiological situation, cirumstances do exist under which this enzyme *increases* markedly. For example, Lazdins and Karnovsky (1978) observed that the presence of AMP, ADP, and ATP in the medium quadrupled the levels of the enzyme of cultured mouse macrophages. Such substances as adenosine, cyclic-AMP, and UMP also had some effect. The original cells cultured here were elicited peritoneal cells; it would have been interesting to note whether the same effects would pertain with resident cells under the same conditions, i.e., with starting material not already at a very low level of ecto-AMPase. The question of actual synthesis and degradation of an ectoenzyme has been attacked by Edelson and Erbs (1978) in the case of alkaline phosphodiesterase I.

5 An Enzyme of an Internal Cellular Membrane Studied by Ectoenzyme Methodology

An interesting enzyme, now recognized as an enzyme *system*, has been studied for many years (Nordlie 1974). This is, for example, the glucose-6-phosphatase of the endoplasmic reticulum of liver. This "enzyme" is now known to consist of a hydrolase on the lumenal side of the endoplasmic reticulum, and a putative protein transporter facing the cytoplasmic side (Arion et al. 1975). The latter ferries glucose-6-phosphate across the membrane to the hydrolase. Specificity with respect to the substrates of the system resides with the transporter (the translocase) and not with the hydrolase (Arion et al. 1975). The endoplasmic reticulum may be isolated as microsomes, for which this enzyme ist usually regarded as diagnostic, and the microsomes retain their configuration, with the cytoplasmic side facing the outside. (The lumenal side of the original endoplasmic reticulum, is now "endo", i.e., on the inside of the microsome). Ectoenzyme techniques could be used to probe the situation in intact microsomes. Thus, the Berg reagent was shown to inhibit the transport of glucose-6-phosphate into the lumen of the microsomes. The action of the hydrolase located there served as the indicator (Nilsson et al. 1978). Similarly, a more sophisticated reagent, 4,4′-diisothiocyanostilbene-2-2′-di-sulfonic acid (DIDS), which by virtue of its negative charge is nonpenetrating, and by virtue of its isocyano group forms covalent links with amino groups of protein, has been a successful probe of the situation (Zoccoli and Karnovsky 1980; Zoccoli and Karnovsky 1982). The reagent was previously used in a more conventional sense to probe Band 3 of erythrocytes (Cabantchik et al. 1978). Figure 2 illustrates the fact that using intact microsomes, DIDS or H_2DIDS (the hydrogenated form of the reagent) with an ethane rather an ethylene moiety between the two phenyl groups, abolishes activity to the extent of about 80%. However, activity can be totally

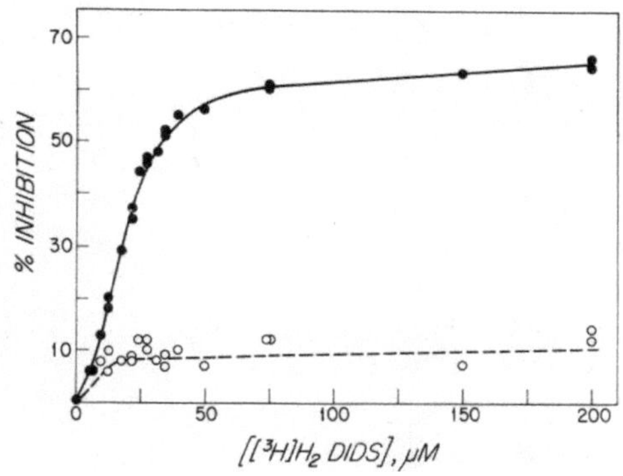

Fig. 2. The concentration dependence of inhibition of microsomal glucose-6-phosphatase activity by [^{3}H]H_2DIDS. Microsomes (5 mg of protein-ml^{-1}) were treated with inhibitor for 10 min at 30°C. Glucose-6-phosphatase activity was determined in control and inhibitor-treated microsomes in the absence (•) and the presence (○) of 0.03% (w/v) sodium deoxycholate

restored when treated microsomes are disrupted. Of even greater interest perhaps is the fact that *glucose*-6-phosphate, but not *mannose*-6-phosphate, will protect the system against the reagent, suggesting that the latter attacks the active site of the translocase. A great virtue of this reagent is that it may be prepared in a tritiated form. When this is reacted with intact microsomes, not only do these become inhibited with respect to glucose-6-phosphatase activity in the fashion just outlined, but a specific polypeptide, demonstrable by polyacrylamide gel electrophoresis becomes labeled (Zoccoli and Karnovsky 1982). This polypeptide has an M_r = 54.000, and the degree of attachment of the tritiated [^{3}H]H_2DIDS reagent conicides exactly with the functional inhibition observed, as shown in Fig. 3. This is an unique behavior on the part of the particular microsomal polypeptide mentioned, and leads one to believe that it is this polypeptide that forms at least a part of the translocase protein responsible for ferrying the substrate to the hydrolase in a totally specific fashion. Through such an approach, the polypeptide may be isolated and the system reconstructed.

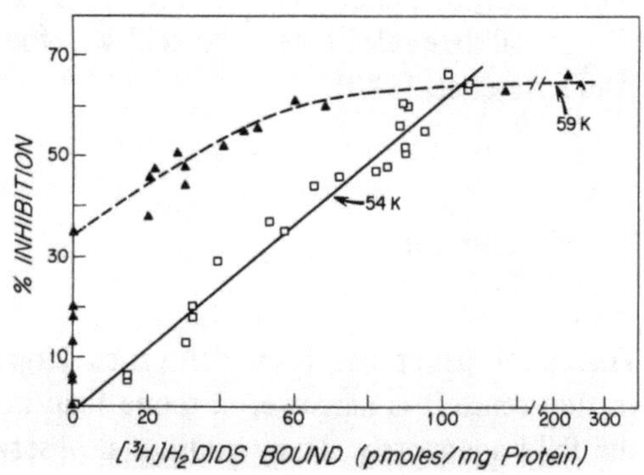

Fig. 3. Correlation between the inhibition of glucose-6-phosphatase activity and labeling by [^{3}H]H_2DIDS of microsomal polypeptides. The level of inhibition of glucose-6-phosphatase activity in the absence of deoxycholate was plotted as a function of the amount of [^{3}H]H_2DIDS bound to either the 54,000 dalton (□) or 59,000 dalton (▲) polypeptides. The *line* drawn to fit the data for the 54,000 dalton polypeptide was obtained by least squares analysis and had a correlation coeeficient of 0.98

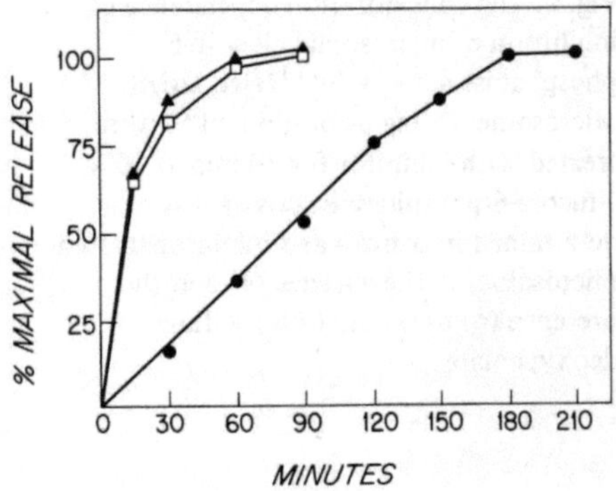

Fig. 4. Time course for removal of sialic acid from intact granulocytes by neuraminidase. Conditions for the soluble enzyme were: 10^8 intact cells were treated with 1.99 units (Neuraminyl-lactose) of soluble enzyme, in Krebs-Ringer phosphate (KRP), at 37°C. Those for the Sepharose-bound enzyme, washed on a Millipore filter with 100 vol KRP buffer immediately before use, were: 1.63 units (neuraminyl-lactose method) of bound enzyme per 10^8 cells, at 37°C in 0.5 ml of KRP buffer. Control tubes with cells and an equivalent amount of untreated Sepharose beads were subjected to the same treatment without any observable release of sialate. All samples were shaken vigorously throughout the incubation. *Symbols:* □, intact cells, soluble neuraminidase; ▲, sonicated material, soluble neuraminidase; ●, intact cells, Sepharose-bound neuraminidase

6 Models of Ectoenzymes

Models of ectoenzymes on the surface of cells do not abound but have been widely constructed for surface receptors, glycosylated proteins of membranes, etc. It may be remarked that enzymes attached to large objects, such as liposomes or plastic granules, may indeed serve to some extent as models of the situation. In one simple example, sialidase has been attached to a Sepharose B granule. The sialidase so attached has been permitted to make contact with guinea pig granulocytes, and as is seen in Fig. 4, sialic acid was released from the surface of those cells at approximately half the rate manifested by the soluble enzyme. Furthermore, all the sialidase-releasable sialic acid was found in this case to be an "ecto-substrate" (DePierre et al. 1980).

7 Conclusion

This brief paper has been concerned largely with work from a single laboratory. In that sense it is narrower in scope than might be, and has not attempted to review the field adequately. Many additional observations of great interest and importance could have been cited. Further, it must be evident that 5′-nucleotidase has received major, perhaps excessive, attention. It is, in truth, the ectoenzyme par excellence. It has been studied, as mentioned, in attempts to follow the fate of cellular membrane after endocytosis (e.g., Werb and Cohn 1972; Edelson and Cohn 1976; Smolen and Karnovsky 1980a, b; Widnell 1982); in studies of cells of the nervous system (e.g., Kreutzberg and Hussain 1982); in comparisons of the membrane of normal and malignant cells (e.g., Weiss and Sachs 1977; Carraway et al. 1976; Raz et al. 1978); in the functions of lymphocytes (e.g., Fleit et al. 1975) and even in rat heart muscle (e.g. Frick and Lowenstein 1978).

The true functional role of ectoenzymes in biology has not yet been defined to any extent. It has been suggested, for example, that these enzymes might be involved as sensors, or to control concentrations of external effectors. In this chapter it has been suggested that they may provide important substrates, which may be transported, as in the case of adenosine, for intracellular processing. In the case of the release of oxygen radicals, one can comprehend the utility of that release on what becomes the inside of the internalized endocytic vesicles, and thus, the importance of an ecto-oxido-reductase system, that produces the radicals. This locale is very close to, or even in contact with, the internalized (phagocytized) microbe, for example. Thus, the killing process is facilitated. On the other hand, one is puzzled at the fact that release of oxygen radicals also occurs on the outside of the cells, into the medium in which they are suspended, or onto a surface with which they may be in contact. Such a process could well be important in anti-tumor action, i.e., when a tumor cell and macrophage make contact. It could also have deleterious (pro-tumor) sequelae to the host (Badwey and Karnovsky 1980; Fridovich 1983) if the released radicals attack normal host cells. In general, though ectoenzymes are of inestimable utility to the investigator as markers of the plasmalemma, their usefulness to the cell itself and the entire organism remains a question to titillate us further.

Acknowledgments. The following Figures are published with the kind permission of the Journals indicated, to whom the author is grateful: Fig. 1: *Infection and Immunity;* Fig. 2 and 3: the *Journal of Biological Chemistry;* Fig. 4: the *Biochemical Journal* (London).

References

Arion WJ, Wallin BK, Lange AJ, Ballas LM (1975) On the involvement of a glucose-6-phosphate transport system in the function of microsomal glucose-6-phosphatase. Mol Cell Biochem 6:75–83

Badwey JA, Karnovsky ML (1980) Active oxygen species and the functions of phagocytic leukocytes. Annu Rev Biochem 49:695–726

Badwey JA, Robinson JM, Karnovsky MJ, Karnovsky ML (1984) Involvement of protein kinase C and calmodulin in superoxide production by neutrophils. Fed Proc 43:1521

Berg HC (1969) Sulfanilic acid diazonium salt: A label for the outside of the human erythrocyte membrane. Biochim Biophys Acta 183:65–78

Briggs RT, Drath DB, Karnovsky ML, Karnovsky MJ (1975) Localization of NADH oxidase on the surface of human polymophonuclear leukocytes by a new cytochemical method. J Cell Biol 67:566–586

Cabantchik ZI, Knauf PA, Rothskin A (1978) The anion transport system of the red blood cell: the role of membrane protein evaluated by the use of "probes". Biophys Acta 515:239–302

Carraway KL, Fogle DD, Chesnut RW, Huggins JW, Carraway CAC (1976) Ectoenzymes of mammary gland and its tumors. Lectin inhibition of 5′-nucleotidase of the 13.762 rat mammary ascites carcinoma. J Biol Chem 251:6173–6178

DePierre JW, Karnovsky ML (1973) Plasma membranes of mammalian cells: a review of methods for their characterization and isolation. J Cell Biol 56:275–303
DePierre JW, Karnovsky ML (1974a) Ecto-enzyme of the guinea-pig leukocyte I. Evidence for an ecto-adenosine monophosphatase, adenosine tri-phosphatase, and p-nitrophenyl phosphatase. J Biol Chem 249:7111–7120
DePierre JW, Karnovsky ML (1974b) Ecto-enzymes of the guinea-pig polymorphonuclear leukocyte II. Properties and suitability as makers for the plasma membrane. J Biol Chem 249: 7121–7129
DePierre JW. Lazdins J, Karnovsky ML (1980) The determination and localization of sialic acid in guinea pig granulocytes. Biochem J 192:543–550
Edelson PJ, Cohn ZA (1976) 5′-nucleotidase activity of mouse peritoneal macrophages II. Cellular distribution and effects of phagocytosis. J Exp Med 144:1596–1608
Edelson PJ, Erbs C (1978) Plasma membrane localization and metabolism of alkaline phosphodiesterase I in mouse peritoneal macrophages. J Exp Med 147:77–86
Emmelot P, Bos CJ, Benedetti EL. Rümke PH (1964) Studies on plasma membranes I. Chemical composition and enzyme content of plasma membranes isolated from rat liver. Biochim Biophys Acta 90:126–145
Fleit H, Conklyn M, Stebbins RD, Silber R (1975) Function of 5′-nucleotidase in the uptake of adenosine from AMP by human lymphocytes. J Biol Chem 250:8889–8892
Frick GP, Lowenstein JM (1978) Vectorial production of adenosine by 5′-nucleotidase in the perfused rat heart. J Biol Chem 253:1240–1244
Fridovich I (1983) Superoxide radical: an endogenous toxicant. Annu Rev Pharmacol Toxicol 23:239–257
Karnovsky ML, Lazdins J, Drath D, Harper A (1975) Biochemical characteristics of activated macrophages. Ann NY Acad Sci 256:266–274
Kreutzberg GW, Hussain ST (1982) Cytochemical heterogeneity of the glial plasma membrane: 5′-nucleotidase in retinal Müller cells. Neurocytol 11:53–64
Lazdins J, Karnovsky ML (1978) Effect of phosphate esters, nucleotides and nucleosides on 5′-nucleotidase of cultured mouse macrophages. J Cell Physiol 96:115–121
Nilsson OS, Arion WJ, DePierre JW, Dallner G, Ernster L (1978) Evidence for the involvement of a glucose-6-phosphate carrier in microsomal glucose-6-phosphatase activity. Eur J Biochem 82:627–634
Nordlie RC (1974) In:Horecker BL, Stadtman ER (eds) Current topics in cellular regulation, vol 8. Academic, New York, pp 33–117
Raz A, Collard JG, Inbar M (1978) Decrease in 5′-nucleotidase activity in malignant transformed and normal stimulated cells. Cancer Res 38:1258–1262
Robinson JM, Karnovsky MJ (1983) Ultrastructural localization of several phosphatases with cerium. J Histochem Cytochem 31:1197–1208
Smolen JE, Karnovsky ML (1980a) Effect of surface modifiers on an ecto-enzyme: granulocyte 5′-nucleotidase. Infect Immun 28:475–485
Smolen JE, Karnovsky ML (1980b) Effect of phagocytosis on guinea pig granulocyte membrane markers. J Cell Physiol 102:71–80
Soberman JR, Karnovsky ML (1980) Metabolism of purines in macrophages. J Exp Med 152: 241–246
Tsan M-F, Berlin RD (1971a) Effect of phagocytosis on membrane transport of non-electrolytes. J Exp Med 134:1016–1035
Tsan M-F, Berlin RD (1971b) Membrane transport in rabbit alveolar macrophages. The specificity and characteristics of amino acid transport systems. Biochim Biophys Acta 241:155–169
Weiss B, Sachs L (1977) Differences in surface membrane ecto-ATPase and ecto-AMPase in normal and malignant cells. J Cell Physiol 93:183–188
Werb Z, Cohn ZA (1972) Plasma membrane synthesis in the macrophage following phagocytosis of polystyrene latex particles. J Biol Chem 247:2439–2446
Widnell CC (1982) Evidence for a continual exchange of 5′-nucleotide between the cell surface and cytoplasmia membranes in cultured rat fibroblasts. Cell 28:61–70

Zoccoli MA, Karnovsky ML (1980) Effect of two inhibitors of anion transport on the hydrolysis of glucose-6-phosphate by rat liver microsomes: covalent modifications of the glucose-6-P transport component. J Biol Chem 255:1113–1119
Zoccoli MA, Karnovsky ML (1982) Identification of a rat liver microsomal polypeptide involved in the transport of glucose-6-phosphate: labeling with 4,4′-diisothiocyano-1,2-diphenyl-ethane 2,2′disulfonic acid. J Biol Chem 257:3919–3924

Ectonucleotidases: General Aspects

Ectonucleotidases of Vascular Endothelial Cells: Characterisation and Possible Physiological Roles

J. D. PEARSON [1]

1 Introduction

Vascular endothelial cells form a monolayer that is the biological interface between the bloodstream and extravascular tissue. Although classically regarded as a passive, albeit selective, permeability barrier, it is now clear that endothelial cells actively regulate many aspects of vascular homeostasis, including vascular permeability (both to small molecules and to blood cells), vessel tone, and the intravascular processes of thrombosis and blood coagulation. The recognition of the importance of endothelial metabolism in these diverse activities has been promoted substantially by the availability of endothelial cells as homogeneous cultures in vitro, with which to investigate endothelial properties in detail without the difficulty, encountered in other approaches, of possible interfering effects of other blood or vascular cells. The interested reader is referred to two recent books that document the current state of endothelial cell biology (Nossel and Vogel 1983; Jaffe 1984).

One important role of the endothelial cell is to limit the locally induced response of the vessel wall to vasoactive mediators. This can be achieved in several ways. For certain classes of vasoactive agents, endothelial cells possess efficient uptake mechanisms that remove the compound from the circulation, and subsequently metabolise it within the cell: in the case of prostaglandins and serotonin the inactive product is then released to the circulation (for review see Gillis and Roth 1976). Thrombin, which is vasoactive and a stimulator of blood platelet aggregation in addition to its procoagulant activity, is also inactivated by endothelial cells. This is due to anti-thrombin III, synthesised by endothelium and presented on its luminal surface, which tightly binds the thrombin and is then released to the circulation as an inactive protease-antiprotease complex (Lollar and Owen 1980; Chan and Chan 1981). A third manner in which endothelial cells regulate the levels of circulating vasoactive mediators is by possessing ectoenzymes. One of the most extensively investigated of these enzymes is angiotensin converting enzyme which, together with other endothelial ectoenzyme peptidases, controls the action of vasoactive peptides; it produces the arterial constrictor angiotensin II from its decapetide precursor angiotensin I, and also inactivates the nonapeptide dilator bradykinin (Ryan and Ryan 1984).

[1] Section of Vascular Biology, MRC Clinical Research Centre, Harrow, HA1 3UJ, UK

2 Biological Actions of Circulating Adenosine Derivatives

Adenosine derivatives are released into the circulation following tissue damage, hypoxia, blood platelet aggregation or the interaction of endothelial cells themselves with certain vasoactive mediators (Green and Stoner 1950; Pearson and Gordon 1979; Forrester 1981). These purines powerfully affect vascular tone and platelet function, and their activities, mediated via distinct purinoceptors, can often have opposing results: for example, adenosine inhibits, while ADP induces, platelet aggregation. Adenosine is a vasodilator in many tissues, acting at receptors on smooth muscle cells (Burnstock 1981). ATP and ADP are also potent dilators of most peripheral vessels, but in contrast to adenosine their action is via specific receptors on endothelial cells (Furchgott 1983; Gordon and Martin 1983). Because of these important biological effects of circulating adenosine derivatives it is apparent that, in common with other vasoactive mediators, mechanisms to regulate their production and inactivation must exist. In the following sections I have outlined the significance of endothelial ectonucleotidases in this process and reviewed our progress in characterising the enzymes. A more general review of ectonucleotidases and their possible physiological roles has recently been published (Pearson 1985).

3 Presence of Endothelial Ectoenzymes Metabolising Adenine Nucleotides

The disappearance of injected ATP on a single passage through a capillary bed (the lung) was first described by Binet and Burstein (1950). Since then, the use of organ perfusion and tracer dilution studies has demonstrated that circulating nucleotides are metabolised by endothelial ectoenzymes and the localisation of nucleotidase activities to the endothelial plasma membrane has been confirmed cytochemically (see Pearson 1985; Ryan and Ryan 1984 and references therein). These endothelial ectoenzymes are responsible for degrading nucleotides in the bloodstream in vivo: the half-life of ATP in cell-free plasma or whole blood ex vivo is several minutes, while in vivo its half-life is less than time to pass through one capillary bed, i.e. a few seconds (see references in Pearson and Gordon 1985).

4 Characterisation of Endothelial Ectonucleotidases

We first investigated the ectonucleotidases of endothelial cells using intact pig aortic cells in culture (Pearson et al. 1980). Addition of $1 \mu M$ ATP, ADP or AMP to the medium above 16 mm culture wells, each containing $\simeq 10^5$ cells, led to the breakdown of each nucleotide to adenosine. The half-lives of ATP, ADP and AMP were $\simeq 5$, 10 and 30 min respectively, and kinetic analysis showed sequential catabolism of ATP → ADP → AMP → adenosine (Fig. 1). Nucleotide catabolism was not accompanied by the release of any detectable cytoplasmic enzyme (lactate dehydrogenase)

Fig. 1. Time course of product formation from catabolism of exogenous ATP by pig aortic endothelial cells in culture. Monolayers of endothelial cells (= 10^5 cells per well) were incubated with 1μM [^{3}H]-ATP, and the products formed in the supernatant analysed by t.l.c. Based on Fig. 2 (Pearson et al. 1980; The Biochemical Society, London; reprinted by permission)

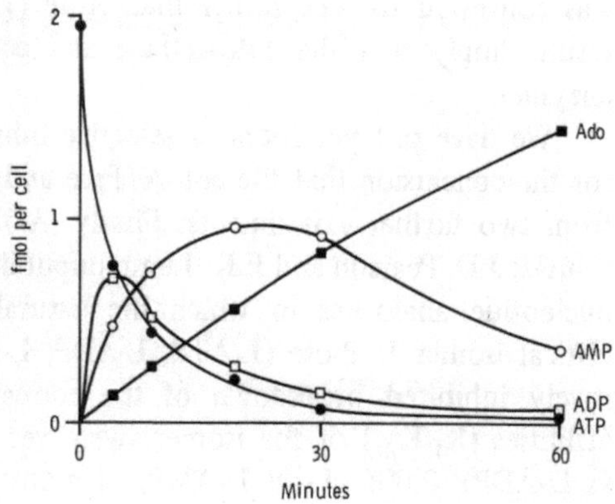

or nucleotidase activity into the supernatant. Breakdown of each nucleotide was inhibited by chelation of Mg^{2+}, but not of Ca^{2+}. The ectoenzymatic activity was not inhibitable by excess inorganic phosphate, β-glycerophosphate or *p*-nitrophenylphosphate. Estimates of the k_m and V_{max} values for ATP, ADP and AMP catabolism, derived from experiments with at least six separate cell lines in each case, were 475, 155 and 24μM and 67, 9 and 0.8 fmol min^{-1} $cell^{-1}$ respectively (Cusack et al. 1983).

Without devoting considerable effort to scaling up the culture techniques, only relatively small amounts of protein are available from in vitro cell studies. We therefore decided it would initially be more productive to determine whether the ectonucleotidase activities on endothelial cells were due to separate enzymes by the use of selective substrates or inhibitors, rather than to attempt to purify the enzymes.

5 Use of Selective Inhibitors and Substrates for Ectonucleotidases

AMP catabolism was power fully and selectively inhibited by adenosine-α, β-methylenediphosphate (k_i = 0.1 μM), which had no effect on the breakdown of ADP or ATP at concentrations up to 1mM, strongly suggesting that AMP breakdown is catalysed by 5′-nucleotidase, well characterised as an ectoenzyme in other cell types (see Pearson 1984, and other chapters in this volume).

ADP catabolism was inhibited by a wide range of ATP analogues (e.g. adenosine-β, γ-methylene-triphosphate; adenosine-β, γ-imido-triphosphate; adenosine-(γ-thio)-triphosphate (ATPγS), while ATP catabolism was much less affected by these ATP analogues; e.g. the K_i for adenosine-β, γ-imido- triphosphate as an inhibitor of ADPase activity was 30 μM, whereas it was 500 μM against ATPase activity, the pattern of inhibition being apparently competitive in each case (J.D. Pearson and E.K. Lund, unpublished data). We were not able to test the effectiveness of ATP itself as an ADPase inhibitor, because of ecto-(nucleoside diphosphate kinase) activity on the endothelial cells – in the presence of a nucleoside triphosphate, ADP was converted to ATP rather than AMP (Pearson et al. 1980, 1983) – but these results imply that the ectoADPase and ectoATPase activity are due to separate enzymes.

We have not yet found a selective inhibitor for the ectoATPase, but support for the conclusion that the ectoATPase and ectoADPase are distinct enzymes comes from two further experiments. Firstly, ADP is a poor inhibitor of the ATPase (K_i $>$ 3m*M*; J.D. Pearson and E.K. Lund, unpublished data). Secondly, we used a series of nucleotide analogues in which the natural D-ribose moiety was replaced by its optical isomer L-ribose (L-ATP, L-ADP, L-AMP). The L-nucleotides each competitively inhibited breakdown of the corresponding D-nucleotide, but the relative affinities (K_i/K_m) of the isomers were very different: 0.8 (L- vs D-ATP); 0.15 (L- vs D-ADP); 0.005 (L- vs D-AMP), demonstrating that each enzyme has a different tolerance to L-sugar substitution (Fig. 2). Correspondingly, the ability of each L-nucleotide to act as a substrate for endothelial ectonucleotidases was markedly different; L-ATP was catabolised to L-ADP moderately well, L-ADP to L-AMP poorly, and L-AMP catabolism to adenosine was not detectable (Fig. 3).

From the results of this series of inhibitor- and substrate-specificity experiments, we therefore concluded that the endothelial plasmalemma contains three distinct ectoenzymes catabolising nucleotides: nucleoside triphosphatase, nucleoside diphosphatase and 5′-nucleotidase.

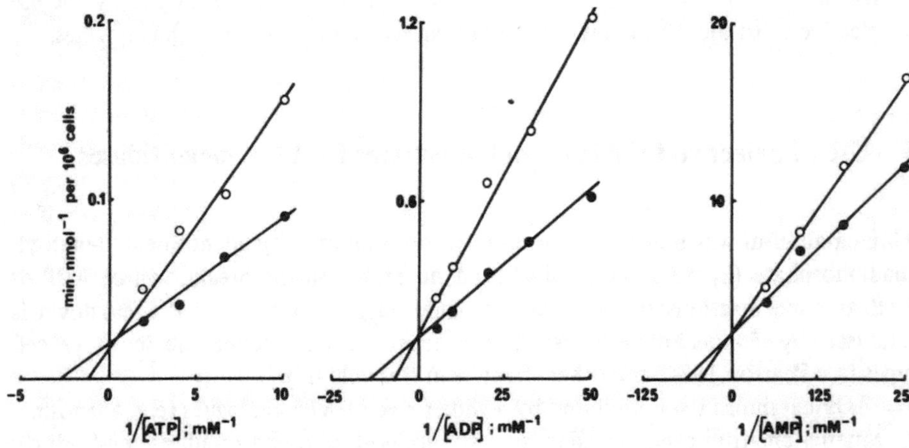

Fig. 2. Inhibition of catabolism of D-nucleotides by the corresponding L-enantiomer. Confluent monolayers of pig aortic endothelial cells (= 10^5 cells per well) were used. Double reciprocal plots are shown; •, in the absence of L-nucleotide; ○, in the presence of 1m*M* L-nucleotide (Cusack et al. 1983; The Biochemical Society, London; reprinted by permission)

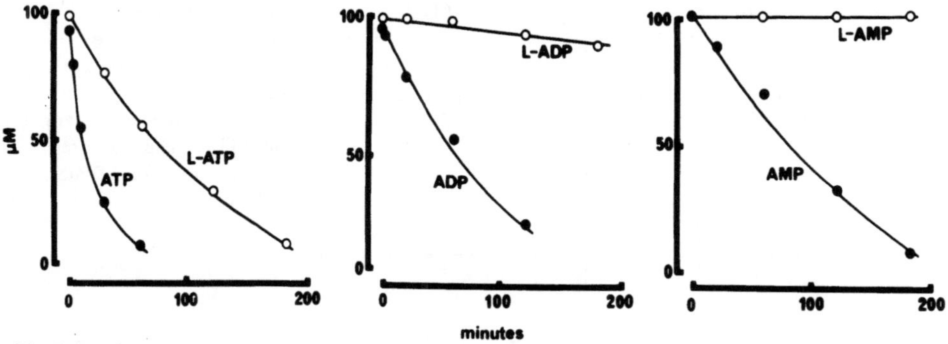

Fig. 3. Rates of catabolism of D- and L-enantiomers of ATP, ADP and AMP. Confluent endothelial monolayers (= 10^5 cells per well) were incubated with 100 μM nucleotides (Cusack et al. 1983; The Biochemical Society, London; reprinted by permission)

6 Use of Phosphorothioate Nucleotide Analogues to Study Ectonucleotidases

The synthesis and use of the phosphorothioate nucleotide analogues was first promoted by Eckstein (for review, see Eckstein 1979). In these analogues a sulphur atom is substituted for a non-bridging oxygen on one of the phosphate moieties, and they have proved extremely powerful tools with which to analyse the stereochemical course of phosphorylation or phosphohydrolase reactions, and to determine the nature of the most favourable conformation of the magnesium-nucleotide complex bound at the active site. The utility of the phosphorothioates stems from two properties. Firstly, unlike non-degraded inhibitors (such as adenosine-β, γ-imido-triphosphate), these compounds are potentially active substrates that can transfer phosphate or thiophosphate groups to acceptor molecules. Secondly, non-terminally substituted phosphorothioates (ATPβS, ATPαS, ADPαS) exist as diastereoisomeric pairs and thus can be used to examine the stereochemical specificity of the enzyme reaction (see Fig. 4). More recently, the terminally substituted phosphorothioates (AMPS, ADPβS, ATPγS) have been synthesised with heavy isotopes of oxygen substituted on the terminal phosphorus, creating diastereoisomeric pairs of these compounds as well (Frey et al. 1982).

The Rp and Sp diastereoisomers of ATPαS and ADPαS are resolved by h.p.l.c., so the possible stereoselectivity of the ecto-triphosphatase and -diphosphatase may be directly tested with unseparated mixtures of the isomers. The isomers of ATPβS are not separable by h.p.l.c., and have to be individually synthesised enzymatically. We tested each pair of isomers as substrates for endothelial ecto-nucleotidases (Cusack et al. 1983), with the following results. In the presence of Mg^{2+}, the catabolism of ADPαS or of ATPβS was virtually stereospecific – the Sp isomer, but not the Rp isomer, of ADPαS was catabolised to form AMPS; and the Rp isomer of ATPβS, but not the Sp isomer, was catabolised to ADPβS. Thus the ecto-triphosphatase and the -diphosphatase each accepts only one isomer of the phosphorothioate nucleotide analogue in which sulphur is bonded to the penultimate

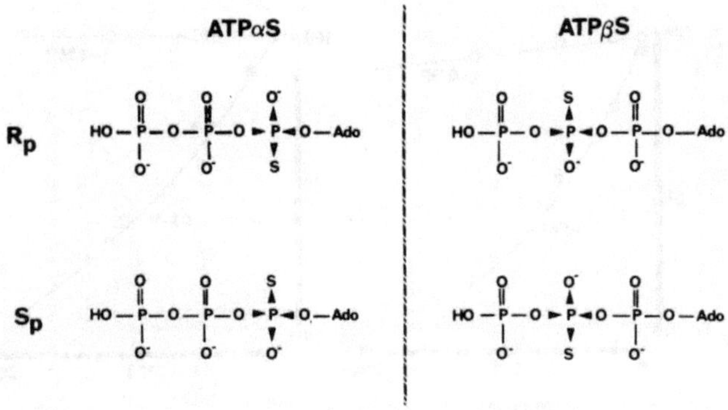

Fig. 4. Two-dimensional representations of the absolute configuration of the Rp and Sp diastereoisomers of ATPαS and ATPβS. The absolute configurations for ADPαS and ADPβS are analogous

phosphorus atom (i.e. the terminal phosphorus atom in the reaction product). It should be noted that the apparently contrasting stereoselectivity shown towards ADPαS and ATPβs is an artefact of the S and R nomenclature; the favoured isomer in each case presents the same configuration to the enzyme (Fig. 4). In contrast to the stereoselectivity exhibited towards ATPβS, the ecto-triphosphatase catabolised both isomers of ATPαS equally well.

The presence of stereoselectivity in the degradation of ATP and ADP phosphorothioate analogues makes it likely that the enzymes recognise a Mg(II)-nucleotide complex at their active sites, because Mg^{2+} preferentially complexes to oxygen rather than sulphur (Fig. 5). This can be demonstrated unequivocally, and the conformation of the Mg(II)-nucleotide complex involved in the enzymatic reaction determined by comparing the observed stereoselectivity in the presence of a series of bivalent metal cations with known preferences for binding to oxygen or sulphur ligands. Reversal of stereoselectivity would be anticipated in the presence of Cd^{2+} rather than Mg^{2+} (Cd^{2+} preferentially complexes to sulphur; Fig. 5) and stereospecificity should be lost in the presence of Co^{2+} (which binds equally well to oxygen or sulphur), if the cation forms part of the enzyme-substrate complex (Eckstein 1979).

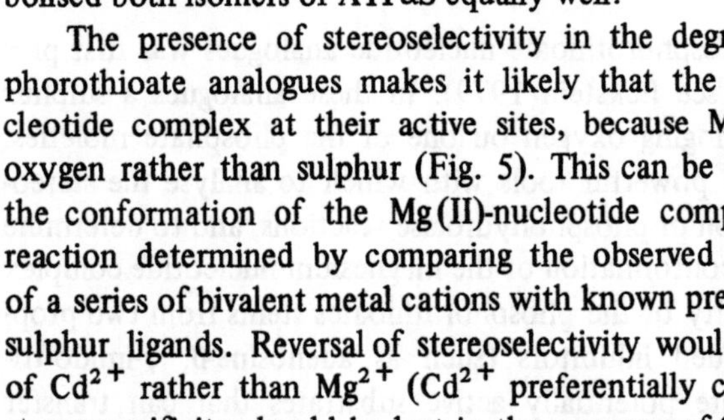

Fig. 5. Two-dimensional representations of the absolute configurations of the Rp and Sp diastereoisomers of ADPαS complexed with either Mg(II) or Cd(II)

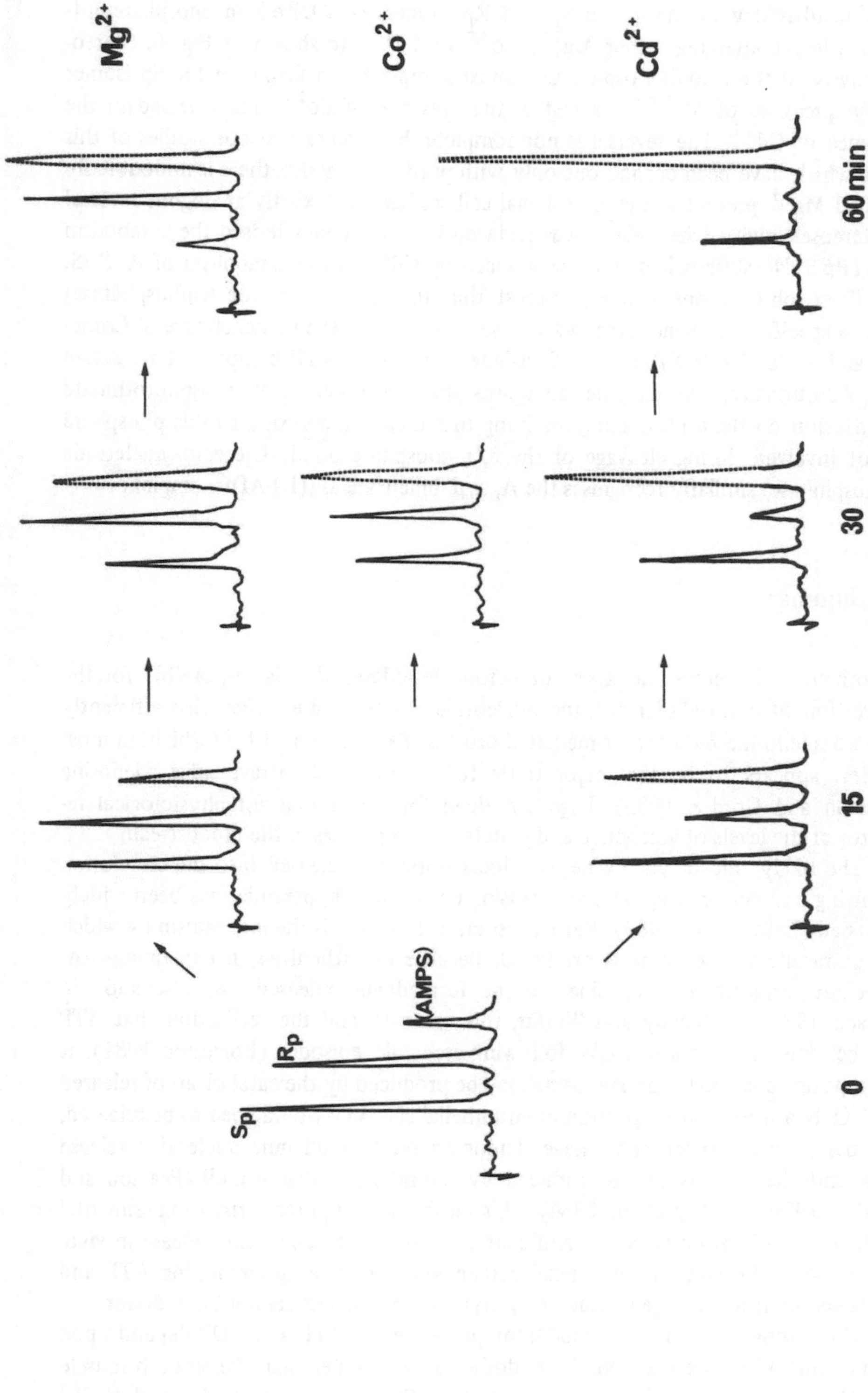

Fig. 6. Catabolism of the Rp and Sp diastereoisomers of ADPαS by endothelial ectoADPase in the presence of Mg^{2+}, Co^{2+}, or Cd^{2+}. Endothelial monolayers were incubated with 100μM ADPαS (Rp) + 100μM ADPαS (Sp), and samples of the supernatant were analysed by h.p.l.c. after the given times

The results from an experiment in which intact endothelial cell monolayers were incubated with the mixed Sp and Rp isomers of ADPαS in phosphate-buffered saline containing either Mg^{2+}, Co^{2+} or Cd^{2+} are shown in Fig. 6. Stereoselectivity of the ecto-diphosphatase, almost completely in favour of the Sp isomer in the presence of Mg^{2+}, was lost in the presence of Co^{2+}, and reversed in the presence of Cd^{2+}. The reversal is not complete, but unlike previous studies of this type, which have been carried out only with purified enzymes, there is undoubtedly residual Mg^{2+} present at the endothelial cell surface. An exactly analogous reversal of stereoselectivity when Mg^{2+} was replaced by Cd^{2+} characterised the catabolism of ATPβS. No stereoselectivity was induced by Cd^{2+} in the catabolism of ATPαS.

These observations strongly suggest that the ecto-(nucleoside triphosphatase) binds a specific conformer (the Λ-screw sense isomer in the nomenclature of Cornelius and Cleland (1978)) of the β, γ-bidentate Mg(II)-ATP complex at its active site. Additionally, the enzyme disregards the conformation of phosphorothioate substitution on the α-phosphate, implying that metal complexing to this phosphate is not involved during cleavage of the β, γ-phosphate bond. The ecto-(nucleoside diphosphatase) similarly recognises the Λ, α, β -bidentate Mg(II)-ADP complex.

7 Summary

Endothelial cells posses an array of ectonucleotidases that is responsible for the conversion of extracellular adenine nucleotides to adenosine. They also efficiently take up adenosine by a carrier-mediated process (Pearson et al. 1978) which, in most species, appears to be the major route for clearance of intravascular adenosine (Pearson and Gordon 1985). Together, these form an important physiological regulator of the levels of vasoactive and platelet-active purines in the bloodstream.

The likely role of adenosine as a local hormone, released into the circulation following ischemia or hypoxia, and causing consequent hyperemia, has been widely discussed (Berne et al. 1983). Rather less clear, however, is the mechanism by which the extracellular adenosine is produced. Because of difficulties in explaining how sufficient intracellular adenosine can be formed and released (see discussion in Pearson 1985 and Newby and Worku, this volume), and the realisation that ATP can be detected extracellularly following ischemic episodes (Forrester 1981), it is tempting to suggest that adenosine may be produced by the catabolism of released ATP. Only a very small proportion of endothelial cell ATP would need to be released, and our in vitro experiments have demonstrated that adenine nucleotide release from endothelial cells can be induced by several non-lethal stimuli (Pearson and Gordon 1979; Le Roy et al. 1984). This mechanism requires further experimental work to be substantiated as a significant contribution to adenosine release in vivo. Nonetheless, the role of endothelial ectoenzymes in clearing circulating ATP and ADP released from damaged tissue or by aggregating platelets cannot be in doubt.

The discoveries that the vasodilator properties of ATP and ADP depend upon interactions with receptors on the endothelial cell, rather than the smooth muscle cell (Furchgott 1983; Gordon and Martin 1983), and that similar endothelial purinoceptors mediate prostacyclin secretion in response to ATP or ADP (Pearson et al. 1983; Hellewell and Pearson 1984), emphasise the importance of the endothelial cell in regulating a wide range of biological responses to adenine nucleotides.

From experiments of the type described in this chapter, we have already begun to distinguish the characteristic features of the ectonucleotidases from those of purinoceptors; e.g. non-hydrolysable ATP analogues can induce endothelial prostacyclin release, and the endothelium-dependent vasodilation produced by phosphorothioate nucleotide analogues is not stereoselective (Pearson et al. 1983; Martin et al. 1985). Further studies will lead to the production of reagents that selectively interact (as substrates or inhibitors) with each ectonucleotidase, and others that distinguish between the characteristics of the ectonucleotidases and purinoceptors. Such agents would allow the nucleotide-dependent pathways of vasoregulation and platelet thrombus formation to be selectively manipulated.

References

Berne RM, Rall TW, Rubio R (eds; 1983) Regulatory function of adenosine. Martinus Nijhoff, Boston

Binet L, Burstein M (1950) Poumon et action vasculaire de l'adenosine triphosphate. Presse Med 58:1201–1203

Burnstock G (ed; 1981) Purinergic receptors. Receptors and recognition series B, vol 12. Chapman and Hall, London

Chan TK, Chan V (1981) Antithrombin III, the major modulator of intravascular coagulation, is synthesised by human endothelial cells. Thromb Haemostasis 46:504–506

Cornuelius RD, Cleland WW (1978) Substrate activity of (adenosine triphosphate) tetraamine-cobalt (III) with yeast hexokinase and separation of diastereomers using the enzyme. Biochemistry 17:3279–3285

Cusack NJ, Pearson JD, Gordon JL (1983) Stereoselectivity of ectonucleotidases on vascular endothelial cells. Biochem J 214:975–981

Eckstein F (1979) Phosphorothioate analogues of nucleotides. Acc Chem Res 12:204–210

Forrester T (1981) Adenosine or adenosine triphosphate. In: Vanhoutte PM, Lensen I (eds) Vasodilation. Raven New York, pp 205–229

Furchgott RF (1983) Role of endothelium in responses of vascular smooth muscle. Circ Res 53:557–573

Frey PA, Richard JP, Ho H-T, Brody RS, Sammons RD, Shen K-F (1982) Stereochemistry of selected phosphotransferases and nucleotidyltransferases. Meth Enzymol 87:213–235

Gillis CN, Roth JA (1976) Pulmonary disposition of circulating vasoactive hormones. Biochem Pharmacol 25:2547–2553

Gordon JL, Martin W (1983) Endothelium-dependent relaxation of the pig aorta: relationship to stimulation of ^{86}Rb efflux from isolated endothelial cells. Br J Pharmacol 70:531–541

Green HN, Stoner HB (1950) Biological actions of the adenine nucleotides. Lewis, London

Hellewell PG, Pearson JD (1984) Purinoceptor-mediated stimulation of prostacyclin release in the piglet pulmonary vasculature. Br J Pharmacol 83:454–462

Jaffe EA (ed; 1984) Biology of endothelial cells. Martinus Nijhoff, Boston

Le Roy EC, Ager A, Gordon JL (1984) Effects of neutrophil elastase and other proteases on porcine aortic endothelial prostaglandin I_2 production, adenine nucleotide release and responses to vasoactive agents. J Clin Invest 74:1003–1010

Lollar P, Owen WG (1980) Clearance of thrombin from circulation in rabbits by high-affinity binding sites on endothelium. J Clin Invest 66:1222–1230

Martin W, Cusack J, Carleton JL, Gordon JL (1985) Specifity of P_2-purinoceptor mediating endothelium-dependent relaxation of the pig aorta. Eur J Pharmacol 108:295–299

Nossel HL, Vogel HJ (eds; 1983) Pathobiology of the endothelial cell. Academic, New York
Pearson JD (1984) Ectonucleotidases: measurement of activities and use of inhibitors. In:Paton DM (ed) Methods in adenosine research. Plenum, New York, pp 83–107
Pearson JD, Gordon JL (1979) Vascular endothelial and smooth muscle cells in culture selectively release adenosine nucleotides. Nature (Lond) 281:384–386
Pearson JD, Gordon JL (1985) Nucleotide metabolism by endothelium. Annu Rev Physiol 47:617–627
Pearson JD, Carleton JS, Hutchings A, Gordon JL (1978) Uptake and metabolism of adenosine by pig aortic endothelial and smooth muscle cells in culture. Biochem J 170:265–271
Pearson JD, Carleton JS, Gordon JL (1980) Metabolism of adenine nucleotides by ectoenzymes of vascular endothelial and smooth muscle cells. Biochem J 190:421–429
Pearson JD, Slakey LL, Gordon JL (1983) Stimulation of prostaglandin production through purinoceptors on cultured porcine endothelial cells. Biochem J 214:273–276
Ryan JW, Ryan US (1984; in press) Endothelial surface enzymes and the dynamic processing of plasma substrates. Int Rev Exp Pathol 26:1–43

Regulation of the Hydrolysis of Adenine Nucleotides at the Surface of Cultured Vascular Cells

L. L. SLAKEY, J. D. EARLS, D. GUZEK, and E. L. GORDON [1]

1 Introduction

The working hypothesis for the experiments presented here is that the reaction sequence: ATP → ADP → AMP → Adenosine occuring *extracellularly*, is a key regulatory sequence. Adenosine and its nucleotides occur extracellularly during many crisis or signalling events. They are agonists for a wide variety of physiologic responses. The antagonistic and competitive interactions among adenosine and its nucleotides, and the kinetic properties of the ectonucleotidases which carry out the sequential hydrolysis of the phosphate groups, provide for great subtlety in regulating the rate and extent of tissue response to the appearance of nucleotides in the pericellular space.

The appearance and effects of extracellular adenine nucleotides during thrombus formation provide a good example of this complexity. ATP is released into the extracellular space upon thrombin binding to vascular cells (Pearson and Gordon 1979; Lollar and Owen 1981). ADP and ATP are released from activated platelets. The transient extracellular nuleotide concentrations in this situation are probably very high. The combined ATP and ADP concentration in the dense granule of the human platelet is greater than 1 *M* (Ugurbil and Holmsen 1981). If we assume that during thrombus formation, each platelet releases its granule contents into a volume roughly equal to its own total volume, which is poorly stirred because of the forming thrombus, the initial nucleotide concentration would be millimolar. When thrombin binds to vascular cells, they release as much as half their ATP (Pearson and Gordon 1979). Again, if this occurs into a volume equal to the cell's own volume which is poorly exchanged with flowing plasma, the initial concentration would be millimolar (Fig. 1).

ADP promotes thromboxane synthesis, platelet aggregation, and release of more ADP. Thus it triggers a positive feedback loop that leads to autocceleration of platelet activation. ATP and AMP inhibit platelet aggregation, perhaps by acting as competitive inhibitors of ADP binding. Adenosine can block platelet aggregation induced by ADP and all other stimuli, by causing an elevation of platelet cAMP. The effect of adenosine is modified in mixtures of ADP and adenosine, however, because ADP can inhibit platelet adenylate cyclase. Similarly, ADP can block the

[1] Department of Biochemistry, University of Massachusetts, Amherst, Massachusetts, 01003, USA and Section of Vascular Biology, MRC Clinical Research Centre, Harrow, HA1 3UJ, UK

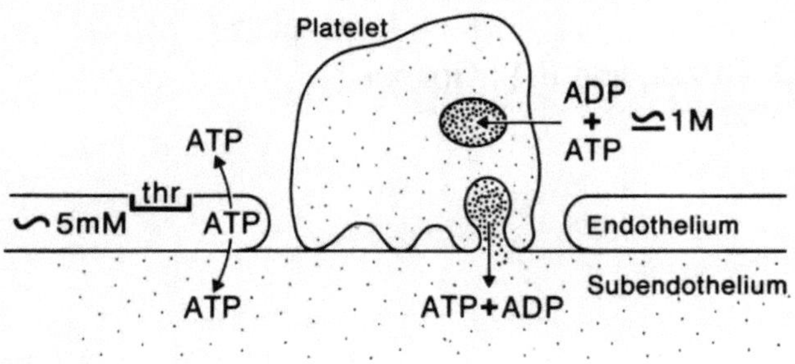

Fig. 1. Release of adenine nucleotides at the vascular surface during thrombus formation

contribution made by prostacyclin toward inhibition of aggregation, since the action of prostacyclin is also mediated by platelet adenylate cyclase. (The effects of ADP and adenosine on platelets have been elegantly reviewed by Haslam and Cusack 1981.) The point is, it is too simple to speak of "the effect" of any component of the adenine nucleotide hydrolysis pathway. Rather one must ask, what are the effects of the relative amounts of all the components at a particular moment and microlocation? Thus it is clear that the *rate of change* in extracellular nucleotide composition can play a key role in determining both the *rate* and *extent* of platelet (and blood vessel wall) response during thrombus formation.

Arterial endothelial and smooth muscle cells have been shown by Pearson et al. (1980) to have ecto-ATPase, -ADPase, and AMPase (5′-nucleotidase) activity. The properties of these enzymes are presented in another paper in this volume. The kinetic properties of homogeneous or partially purified preparations of the membrane-bound 5′-nucleotidase from heart (Naito and Lowenstein 1981), liver (Widnell 1974), lymphocytes (Dornand et al. 1978), and intestinal smooth muscle (Burger and Lowenstein 1970, 1975) have been described in some detail. In addition, there are descriptions of its properties in several whole cell preparations (Stefanovic et al. 1976; Newby et al. 1975). The findings of these studies vary enormously, but some interesting common observations emerge from a comparison of them. Activity is affected by many compounds likely to be present in plasma and interstitial fluid, for example, bicarbonate and free amino acids. It is always found to be inhibited by ATP or ADP; indeed, inhibitability by the ADP analogue APCP has come to be used as a marker to distinguish this enzyme from other phosphatases.

This last property establishes the potential for a particularly interesting feature of the regulation of the time course of overall conversion of ATP or ADP to adenosine. When the extracellular concentration of either rises to the level at which it is an effective inhibitor of 5′-nucleotidase, the rate of appearance of adenosine may actually be diminished. Thus the rate of appearance of adenosine should have a paradoxic dependence upon the size of the initial bolus of ATP or ADP. Our work with cultured vascular endothelial cells shows that this does occur.

The concept that the kinectics of extracellular adenine nucleotide hydrolysis play a central role in the integration of signalling developed as we considered diverse findings and points of view. The work of Lowenstein's group on the kinetics of 5′-nucleotidase (e.g. Burger and Lowenstein 1975), Ryan and Ryan's insight that extracellular nucleotide hydrolysis in lung vasculature should be viewed as *conversion* rather then degradation (Ryan and Smith 1971; Crutchley 1979), Burnstock's description of the properties of the classes of purinergic receptors (Burnstock and Brown 1981) and many conversations with our collaborators, J. Gordon and J. Pearson, have contributed especially to its present form.

2 Procedures

Primary cultures of pig aortic endothelial cells were prepared by brief collagenase digestion of the lumenal surface of the blood vessel. Serial subcultures were grown and maintained in medium supplemented with plasma-derived serum (Dickinson and Slakey 1982). For the experiments described here, cells were transferred to polystyrene beads, as described by Davies (1981). On the day of the experiment, cell-coated beads were transferred to a small column, and perfused without recirculation for 15–30 min with serum-free medium buffered with 20 m*M* HEPES. At time = 0, a reservoir containing substrate in serum-free medium was placed in the line, and this solution was recirculated over the cells for 30–90 min. Aliquots were withdrawn and analyzed for nucleotide composition by thin-layer chromtography (if radiolabelled substrate was used) or by high performance liquid chromatography. Between experiments on the same column it was washed again for 15–30 min with serum-free medium. Such a column can be used for at least five experiments in a day without change in the results between the first experiment and the last. Further, the rates of reaction observed are proportional to the cell number.

3 Kinetics of Nucleotide Hydrolysis by Endothelial Cells

The time course of change in nucleotide composition observed when 700 *μM* ATP is used as the initial substrate is shown in Fig. 2. The concentration of ATP falls very quickly. The pool of ADP builds to a concentration slightly more than half the initial ATP concentration and then falls. The most striking feature of the pattern

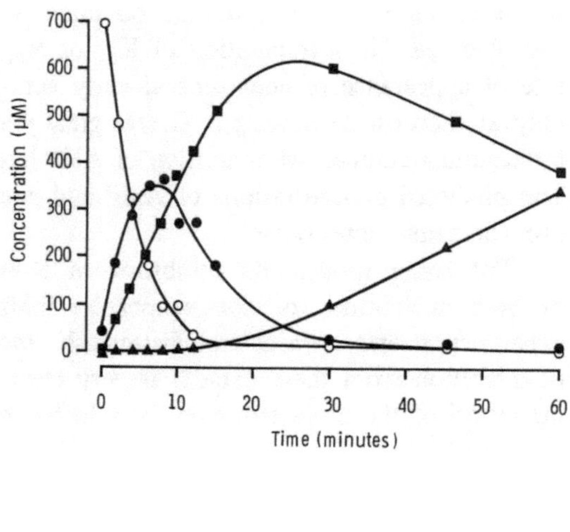

Fig. 2. The hydrolysis of ATP by endothelial cells. Total incubation volume, 0.7 ml; cell number, 2x10^6. ○, ATP; ●, ADP; ■, AMP; ▲, adenosine

is that most of the nucleotide appears in the AMP pool before any significant conversion of AMP to adenosine occurs. It can be seen that 5′-nucleotidase activity (AMP → adenosine) is very low until most of the ADP has disappeared.

Adenosine is further metabolized, presumably following transport into the cells (Pearson et al. 1978), into inosine, hypoxanthine, xanthine, and uric acid. These products are released from the cells and appear gradually in the medium. In the experiments reported here, which focus on nucleotidase kinetics, adenosine and all its metabolites are reported in sum as adenosine. Approximately half the amount shown as adenosine, i.e. the product of 5′-nucleotidase, is actually present as adenosine at later time points.

When Pearson et al. (1980) first examined the pattern of ATP hydrolysis by cultures of pig aortic endothelial cells using 1 *μM* substrate, they found immediate and linear production of adenosine. We reasoned that an initial ATP concentration of 1 *μM* was below the threshold at which inhibition of 5′-nucleotidase would occur, and repeated the experiment using initial concentrations from 100 to 1000 *μM*. Estimation of the kinetic parameters governing a sequence of reactions such as the nucleotidases, operating simultaneously, cannot be done by solving the relevant rate equations except for the first reaction in the sequence. However, the behavior of the system of reactions can be simulated, that is, numerical approximations of the solutions of the rate equations can be computed, and compared with observed behavior. The model we used to simulate ATP hydrolysis has the following properties. The ATPase and the ADPase were assumed to obey simple Michaelis-Menten kinetics over the whole range of substrate and product concentrations observed. The model incorporates a set of choices to describe the behavior of 5′-nucleotidase. It can be set for: (1) no inhibition of 5′-nucleotidase, i.e. simple Michaelis-Menten kinetics; (2) competitive inhibition by ADP or ATP; (3) non-competitive inhibition by ADP or ATP; or (4) competitve inhibition by ADP and ATP.

The simulation was then carried out with test values for K_m and V_{max} for each enzyme, and the results compared with observation until the best fit possible was obtained. The results of such a set of comparisons for the initial ATP concentration of 500 *μM* are shown in Fig. 3. The simulated time course of change in ATP and ADP are well matched to the data except at points at which each of them is nearly depleted and its product concentration is high, suggesting some product inhibition in these reactions. Figure 3A shows the best match of observed and simulated values which could be obtained when no inhibition of 5′-nucleotidase was assumed. No manipulation of K_m or V_{max} improved this fit. The simulated rate of appearance of adenosine at early time points can be matched to the data only at the cost of making it diverge greatly at later time points. The fit obtained by assuming competitive inhibition of AMP hydrolysis by ADP is shown in Fig. 3B. The predicted concentrations of AMP and adenosine are well matched to the data over the whole time course.

The other models for inhibition of 5′-nucleotidase, i.e. inhibition by ATP, or both nucleotides, or non-competitive inhibition, also fit this data set within experimental error. We cannot distinguish among the possible inhibitors and types of inhibition from these data. It is very clear, however, that assumption of some inhibition of the 5′-nucleotidase by a higher nucleotide is required to account for

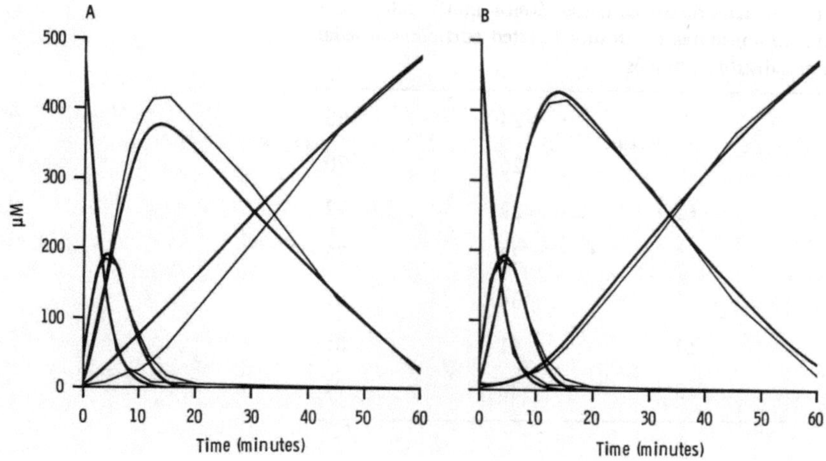

Fig. 3A, B. Hydrolysis of 500 μM ATP by endothelial cells; comparison of real with simulated time course. ***Narrow line*** connects observed values; ***heavy line*** shows simulated time course. **A** Simulation assumes no inhibition of 5′-nucleotidase. **B** Simulation assumes competitive inhibition of 5′-nucleotidase by ADP

the observed patterns of AMP hydrolysis and adenosine appearance. Experiments like those shown in Fig. 2 and 3 have been done at initial ATP concentrations from 100 to 1000 μM. The behavior described above is observed over this entire range, and the lag time before the beginning of rapid conversion of AMP to adenosine increases as the initial ATP increases.

4 Individual Differences in 5 -Nucleotidase

Our interest in cell surface nucleotidase activities began with a serendipitous observation made while using 5′-nucleotidase as a plasma membrane marker, viz., that its total activity was very different in cultured endothelial and smooth muscle cells (Hayes et al. 1979). Further, it appeared to be stable with culture in pig aortic smooth muscle cells [in bovine aortic smooth muscle cells it increases in culture (Fowler et al. 1977)]. In early experiments with pig aortic endothelial cells, activity was found to be high in freshly isolated cells, and low in serial subcultures (Hayes et al. 1979). A consideration of possible roles for this activity led to the hypotheses and experiments described above. We also continued to explore the nature of the change in activity observed when endothelial cells are put in culture.

Total 5′-nucleotidase acitivity was measured on freshly isolated cells from individual aortas. The values observed in eighteen experiments done over an interval of a year are shown in Table 1. Statistical analysis showed that there is a component of variance ($p<0.001$) attached to this grouping. The values observed are not sex-related. They scattered randomly over the time of observation, and did not correlate with batches of reagents used for cell isolation or enzyme analysis.

For seven of the above experiments, the cells were divided into aliquots on the day of isolation, and portions were cultured by our usual procedures. Enzyme

Table 1. 5′-Nucleotidase activities (fmol min^{-1} $cell^{-1}$) observed in homogenates of freshly isolated aortic endothelial cells from individual animals

	7	24	60
	10	26	70
	12	29	72
	13	29	78
	17	34	81
	17	41	85
Mean	13	30	74

activity was measured after 1,4 or 5 days in culture. The data from these early experiments suggest that if activity is in the high or middle range initially, it diminishes under culture conditions. If it is low initially, it is stable in culture.

We have been slow to see the most interesting implications of these data, and tended rather to view both the scatter in values, and the loss of activity in culture, as experimental nuisances. Since we now know that some individual animals have levels of 5′-nucleotidase in freshly isolated cells that are within the range observed for most cultured endothelial cells, we can use the cultured endothelial cells with confidence that they are an appropriate quantitative model for at least a fraction of the animal population. The more speculative and more exciting interpretations are as follows: (1) If 5′-nucleotidase has a major role in regulating the rate of adenosine appearance during thrombus formation, then individual differences in amount of this activity would significantly alter individual capacity to respond to initiating events so as to prevent formation of an occlusive thrombus, or occurrence of sustained vasospasm. (2) Suppose changes in culture model changes that occur in endothelial cell populations which are subjected in vivo to chronic insult. Again, lowered 5′-nucleotidase activity would result in diminished resistance to catastrophic response to further insult.

The modeling program was used to simulate the effect upon adenosine generation of altering 5′-nucleotidase activity over the range of values we have observed. For this simulation, the activities and kinetic parameters of the other enzymes were assumed to be the same, and it was assumed that 5′-nucleotidase was inhibited by ADP.

Two cases were examined: (1) The K_m of 5′-nucleotidase was assumed to be constant, (Fig. 4A) and (2) the K_m of the 5′-nucleotidase was varied proportionally to the change in V_{max}. (Fig. 4B) (Preliminary experiments suggest that the enzyme on freshly isolated cells with high activity has a high K_m.) The model predicts that when V_{max} alone is increased, the lag in adenosine appearance is decreased, although not proportionately. When V_{max} and K_m are varied proportionately, the lag in adenosine appearance is little affected; only the later rate of adenosine appearance rises with increasing enzyme activity.

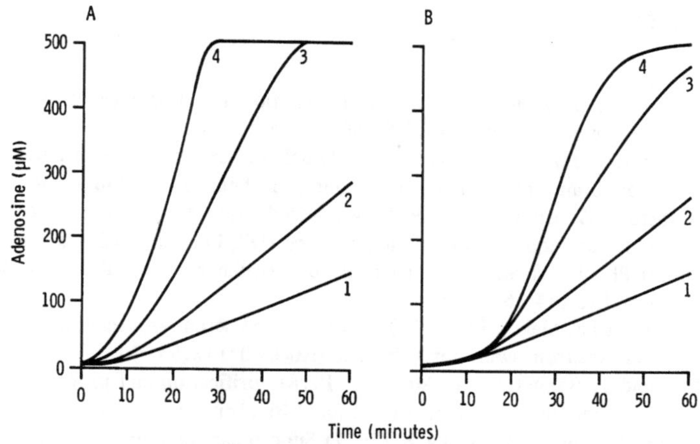

Fig. 4A, B. Simulation of time course of appearance of adenosine from 500 μM ATP, showing dependence on activity and apparent K_m of 5′-nucleotidase. Parameters for simulation: incubation volume, 0.7 ml; cell number, 0.5×10^6; 5′-nucleotidase inhibited by ADP. **A** Activity only changing; *1*, V_{max} = 6 fmol min^{-1} $cell^{-1}$; *2*, V_{max} = 12; *3*, V_{max} = 30; *4*, V_{max} = 74. **B** Velocities as in **A**, and K_m increased proportionally to V_{max}, from *1*, K_m = 20 μM, to 4, K_m = 247 μm

5. Summary

The hydrolysis of extracellular adenine nucleotides all the way to adenosine has been demonstrated in many systems. Since the extent of phosphorylation can profoundly modify the physiologic effect of extracellular adenine nucleotides, one might predict that the regulation of the time course of hydrolysis would be complex and finely tuned. The data presented here show that on cultured pig aortic endothelial cells, the conversion of AMP to adenosine is inhibited in experiments done with initial ATP concentrations likely to occur in a forming thrombus. In a computer simulation of the whole reaction sequence simulated time courses could be well matched to observed ones by assuming that the ATPase and ADPase obeyed simple Michaelis-Menten kinetics. Matching the data for AMP hydrolysis required the introduction of inhibition by at least ADP into the model. Computer simulation of the effect of changing the activity of 5′-nucleotidase over the range we have observed showed that the lag time in adenosine appearance was very sensitive to changes in V_{max} alone, and relatively little affected when V_{max} and K_m were changed proportionately.

All the experiments reported here were done with vascular cells, and we have stated the importance of the kinetics of the ectonucleotidases in the integration of signalling especially in the context of thrombus formation. Adenine nucleotides are released during neurotransmission as well. Several other papers in this volume have shown the presence of ectonucleotidases in the nervous system, and called attention especially to the complexity of their distribution. One may conjecture that in the nervous system spatial separation among the locations of release and hydrolysis of, and receptors for adenine nucleotides adds an additional element of subtlety to the kinetic regulation of the nucleotidases, in determining the net effect of released nucleotides.

References

Burger RM, Lowenstein JM (1970) Preparation and properties of 5′-nucleotidase from smooth muscle of small intestine. J Biol Chem 254:6274–6280

Burger RM, Lowenstein JM (1975) 5′-nucleotidase from smooth muscle of small intestine and from brain: inhibition by nucleotides. Biochemistry 14:2362–2366

Burnstock G, Brown CM (1981) An introduction to purinergic receptors. In: Burnstock G (ed) Purinergic receptors. Chapman and Hall, London, pp 26–35

Davies PF (1981) Microcarrier culture of vascular endothelial cells on solid plastic beads. Exp Cell Res 134:367–376

Dickinson ES, Slakey LL (1982) Plasma-derived serum as a selective agent to obtain endothelial cultures from swine aorta. In vitro (Rockville) 18:63–70

Dornand J, Bonnafous JC, Mani JC (1978) Purification and properties of 5′-nucleotidase from lymphocyte plasma membranes. Eur J Biochem 87:459–465

Fowler S, Shio H, Wolinsky H (1977) Subcellular fractionation and morphology of calf aortic smooth muscle cells. J Cell Biol 75:166–184

Haslam RJ, Cusack NJ (1981) Blood platelet receptors for ADP and for adenosine. In: Burnstock G (ed) Purinergic receptors. Chapman and Hall, London, pp 221–286

Hayes LW, Goguen CA, Stevens AL, Magargal WW, Slakey LL (1979) Enzyme activities in endothelial cells and smooth muscle cells from swine aorta. Proc Natl Acad Sci USA 76: 2532–2535

Lollar P, Owen WG (1981) Active-site-dependent, thrombin-induced release of adenine nucleotides from cultured human endothelial cells. Ann NY Acad Sci 370:51–56

Naito Y, Lowenstein JM (1981) 5′-nucleotidase from rat heart. Biochemistry 20:5188–5194

Newby AC, Luzio JP, Hales CN (1975) Properties and extracellular location of 5′-nucleotidase of rat fat cell plasma membrane. Biochem J 146:625–633

Pearson JD, Carleton JS, Hutchings A, Gordon JL (1978) Uptake and metabolism of adenosine by pig aortic endothelial and smooth muscle cells in culture. Biochem J 170:265–271

Pearson JD, Gordon JL (1979) Vascular endothelial and smooth muscle cells in culture selectively release adenine nucleotides. Nature (Lond) 281:384–386

Pearson JD, Carleton JS, Gordon JL (1980) Metabolism of adenine nucleotides by ectoenzymes of vascular endothelial and smooth muscle cells in culture. Biochem J 190:421–429

Ryan JW, Smith U (1971) Metabolism of adenosine 5′-monophosphate during circulation through the lungs. Trans Assoc Am Physiol 84:297–306

Stefanovic V, Mandel P, Rosenberg A (1976) Ecto-5′-nucleotidase of intact cultured C6 rat glioma cells. J Biol Chem 251:3900–3905

Ugurbil K, Holmsen H (1981) Nucleotide compartmentalization: radioisotopic and nuclear magnetic resonance studies. In: Gordon JG (ed) Platelets in biology and pathology. Elsevier, Amsterdam, pp 146–147

Widnell CC (1974) Purification of rat liver 5′-nucleotidase as a complex with sphingomyelin. Methods Enzymol 32:368–375

Hydrolysis of ATP and Formation of Adenosine at the Surface of Cholinergic Nerve Endings

H. ZIMMERMANN, E. J. M. GRONDAL, and F. KELLER [1]

1 Introduction

Enzymes which hydrolyze extracellularly applied nucleoside triphosphates have been described at the surface of a variety of cell types. These include mammary gland tumor cells (Carraway et al. 1980), blood platelets (Chambers et al. 1967), polymorphonuclear leucocytes (De Pierre and Karnovsky 1974), granulocytes (Smolen and Weissmann 1978), lymphoid cells (see Dornand, Gutensohn, this volume), Ehrlich ascites tumor cells (Ronquist and Ågren 1975), spermatozoa (Majumder 1981), isolated pancreas cells (Hamlyn and Senior 1983), vascular endothelial and smooth muscle cells in culture (see Pearson, Slakey, this volume) as well as glial and neuroblastoma cells in culture (Stefanovic et al. 1976). More recently ectonucleotidase activity has also been described in fractions of intact nerve endings (synaptosomes) isolated from either mammalian (Sorensen and Mahler 1982) or chicken brain (Nagy et al. 1983, see Nagy, this volume) or from the electric organ of the electric ray *Torpedo marmorata* (Keller and Zimmermann 1983).

In a number of cases it has been demonstrated that not only enzyme activity for hydrolysis of ATP, but also of ADP and AMP is present at the cell surface, suggesting that the final hydrolysis product is the nucleoside adenosine. The exact nature of the possible enzyme chain causing formation of adenosine has not yet been established in any of the examples mentioned. Furthermore, the ectonucleotidases appear to have a rather low substrate specificity towards purine and pyrimidine bases. A general characteristic of ectotriphosphatase activities is their activation to a similar extent by either Ca^{2+} or Mg^{2+}.

From a cell biological point of view the existence of these ectonucleotidases raises two main questions. Where does the extracellular substrate come from? What is the function of extracellular hydrolysis of the nucleotide?

Regarding the variety of cell types containing ectonucleotidase activity one may wonder whether ectonucleotidase activity is not a property of all cells (Trams et al. 1980). Since ATP appears to be released from cells by a number of different stimuli including trauma and hypoxia there may be needed for a general mechanism of degradation and salvage of extracellular nucleotide. However, in a number of cases a more specific function of extracellular hydrolysis of nucleotides can be envisaged. This relates to the function of ecto-ATPase activity of blood platelets in platelet

[1] AK Neurochemie, Zoologisches Institut der Universität Frankfurt, Fachbereich Biologie, Siesmayerstr. 70, 6000 Frankfurt/M. 11, FRG

aggregation (Chambers et al. 1967), the hydrolysis of vascular ATP by endothelial cells (Pearson et al. 1980; see also Pearson, Slakey, this volume), and the nervous system (this communication and Nagy, this volume).

For both the central and peripheral nervous system a pharmacological action of ATP and/or adenosine has been described (for review see Stone 1981). It has been suggested that ATP may act as a transmitter in its own right and become released from purinergic neurones (Burnstock 1975). In this case an enzymic mechanism for inactivation of ATP inside the synaptic cleft may be expected similar to the extracellular hydrolysis of the neurotransmitter acetylcholine (ACh) by acetylcholinesterase. In addition it has been shown that ATP is stored together with classical neurotransmitter substances like ACh and noradrenaline inside the same cellular organelles (synaptic vesicles). ATP is also stored together with adrenaline in the chromaffin cells of the adrenal medulla and with serotonin in blood platelets. In all cases ATP is released into the interstitial medium (Zimmermann 1982). For these systems the question of the source of extracellular nucleotide can readily be answered.

The present report will discuss the metabolism and function of ATP at the cholinergic synapse in the electric organ of *Torpedo*. This includes aspects of its synthesis, storage, and release as well as extracellular hydrolysis by ectonucleotidases and the functional role of extracellular nucleotide hydrolysis.

2 The System Used

The electric organ of the electric ray *Torpedo marmorata* carries a very rich innervation of only cholinergic neurons and lends itself favorably to subcellular fractionation (for review see Whittaker and Zimmermann 1976). The system is phylogenetically derived from neuromuscular junctions and the fine structure of the synapses reveal the typical features of an ACh transmitting motor synapse (Fig. 1).

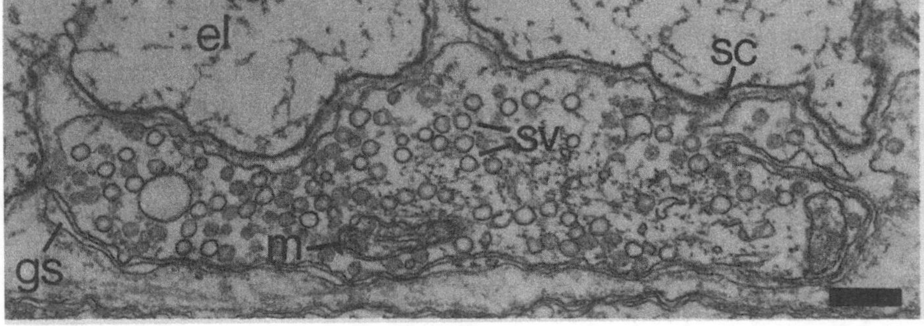

Fig. 1. Finestructure of an electromotor synapse in the *Torpedo* electric organ. The presynaptic nerve ending is separated from the postsynaptic electroplaque cell (*el*) by a small synaptic cleft (*sc*). The nerve terminal contains numerous synaptic vesicles (*sv*) and occasional mitochondria (*m*); *gs*, glial sheath. *Bar* represents 0.5 μM

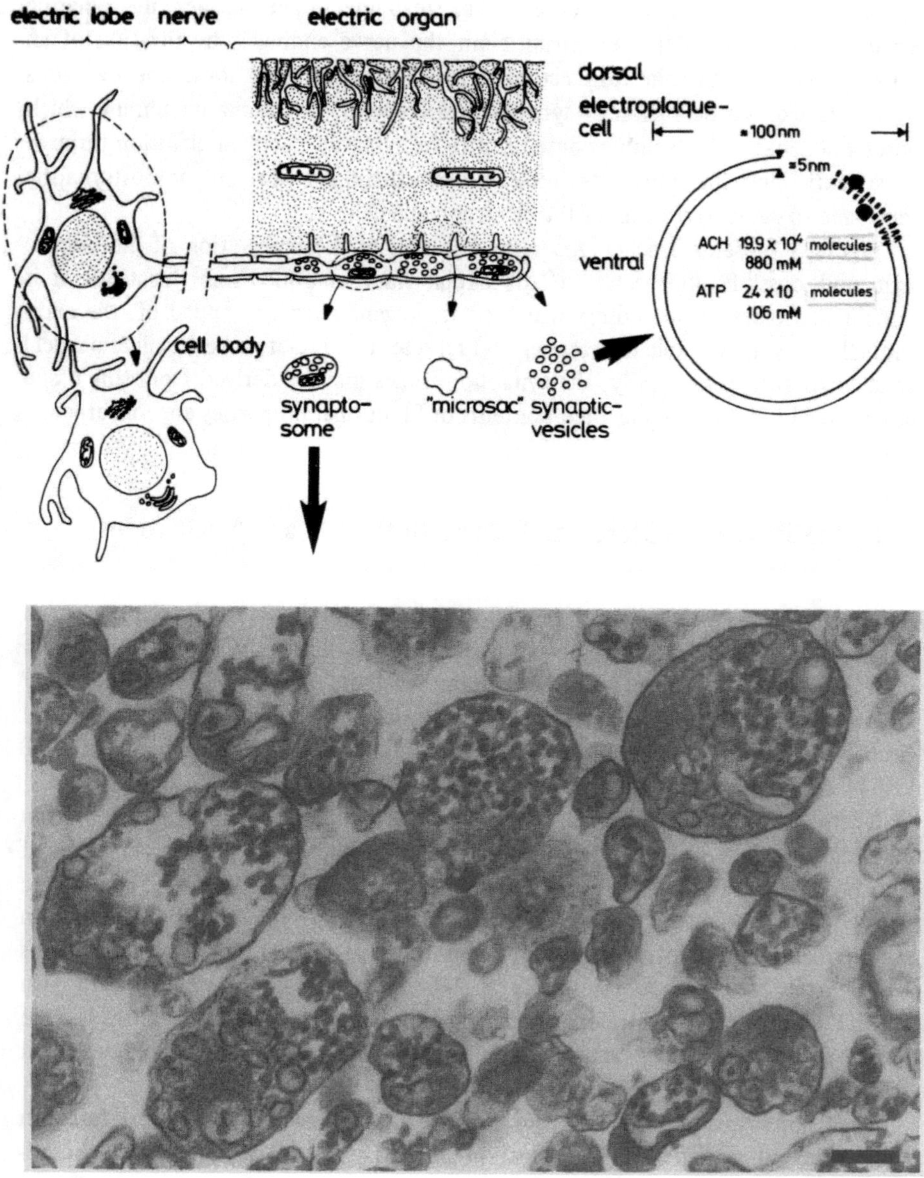

Fig. 2. Subfractionation of the cholinergic electromotor system with the cholinergic neuron innervating the electroplaque cell. Cholinergic cell bodies may be isolated from brain and synaptosomes, synaptic vesicles as well as microsacs containing the postsynaptic acetylcholine receptor from the electric organ. *Right:* A synaptic vesicle with its contents in acetylcholine and ATP. *Bottom:* Electromicrograph of synaptosomes isolated from the electric organ. *Bar* represents 200 nm

The neurotransmitter ACh is stored within small membrane vesicles, the synaptic vesicles. Release of neurotransmitter from the nerve ending is by invasion of an action potential which in turn causes a local influx of Ca^{2+}ions. The Ca^{2+}ions trigger the exocytotic fusion of synaptic vesicles with the plasma membrane which causes the release of its soluble contents into the synaptic cleft. On diffusion through the synaptic cleft the transmitter molecules reach the receptors in the postsynaptic membrane to cause excitation of the electroplaque cell.

Figure 2 presents a schematic representation of the innervation of the electric organ and possible approaches to subcellular fractionation. From this tissue it is possible to isolate nerve endings which on homogenization are ripped off the axon terminal. They are capable of resealing and represent an intact nerve terminal particle for in vitro studies. Similarly, the synaptic vesicles may be derived from this tissue and analyzed in vitro as far as their contents or functional properties are concerned.

3 Synaptic Vesicles Store and Release Both ACh and Nucleotides

As depicted in Fig. 2 synaptic vesicles isolated from the *Torpedo* electric organ contain about 200.000 molecules of the neurotransmitter ACh and in addition about 24.000 molecules of ATP. Furthermore, there are small amounts of ADP (10% of ATP content) and traces of AMP. Vesicles isolated from the closely related *Narcine* electric organ have been shown to contain considerable amounts of GTP (17% of ATP content) (Zimmermann 1982).

There is both indirect and direct evidence for co-release of ACh and ATP from cholinergic nerve terminals. Repetitive stimulation causes correlated depletion of ACh and ATP from synaptic vesicles. Similarly, there is correlated refilling of synaptic vesicles with ACh and ATP during a recovery period following repetitive stimulation (Zimmermann and Whittaker 1974; Giompres et al. 1981). Using synaptosomes isolated from the *Torpedo* electric organ, stimulation-induced release of both ACh and ATP could be demonstrated. The molar ratio of the released substances is close to that of storage inside synaptic vesicles (Morel and Meunier 1981; Schweitzer and Kelly 1982).

These observations raise the question of the extracellular fate and function of the released nucleotide. The question of extracellular metabolism will be addressed first.

4 Fractions of Isolated Synaptosomes Contain Ecto-Nucleoside-Triphosphatase Activity

It is unlikely that ATP released from the nerve terminals into the interstitial medium (synaptic cleft) can be removed by cellular uptake. Therefore, the possibility was investigated that ectonucleotidases situated at the surface of the nerve cell might become instrumental in hydrolysis of extracellular ATP. Figure 3 shows the analysis

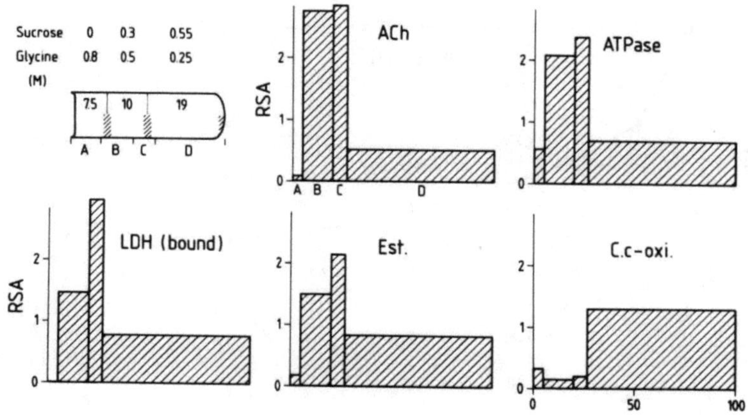

Fig. 3. Analysis of density gradient fractions derived from electric organ for various marker substances (de-Duve-plot). *ACh* acetylcholine; *LDH* lactate dehydrogenase; *Est* cholinesterase; *C.c-oxi* cytochrome-c-oxidase. ATPase activity was determined under isotonic conditions in the presence of Ca^{2+} (4 m*M*), Mg^{2+} (4 m*M*), and ouabain (1 m*M*) (data from Keller and Zimmermann 1983)

of density gradient fractions obtained during isolation of synaptosomes from electric organ. It is obvious that the ATPase activity of intact particles (measured in the presence of ouabain and both Ca^{2+} and Mg^{2+}) is distributed in parallel to the neurotransmitter ACh and also to acetylcholinesterase and bound lactate dehydrogenase (LDH). All the marker substances are enriched most in gradient fraction C (synaptosome fraction). In fraction C 96% of LDH activity was in a bound form, i.e., existed in particles sealed by a plasma membrane. This activity could be released from particles by addition of Triton-X100. The mitochondrial marker cytochrome-c-oxidase is distributed differently and mainly contained in the bottom fraction of the gradient.

Analysis of the topography of synaptosomal acetylcholinesterase of *Torpedo* synaptosomes by two groups has revealed that the enzyme is present as an ecto-enzyme at the surface of the particle (Morel and Dreyfus 1982; Li and Bon 1983; see also Bon, this volume).

From our investigations we infer an identical localization of the ATPase activity described. If an unspecific protease (pronase) is added to a suspension of intact synaptosomes under isotonic conditions there is a time dependent loss of ATPase activity (Ca^{2+}, Mg^{2+}activation), whereas activity of bound LDH even after 3 h of incubation is unaffected (Fig. 4). In control experiments without protease ATPase activity was constant over this period of time. Thus, it may be concluded that the added pronase could not have penetrated the particles to affect intracellular proteins. Furthermore, the exclusion of the substrates for determination of bound LDH activity (pyruvate, NADH) for 3 h suggestes that also low molecular weight substances were unable to penetrate the synaptosomes. A comparable result is obtained when synaptosomes are incubated in the presence of the diazonium salt of sulfanilic acid (DSA) which causes covalent inhibition of enzyme activity. There is a concentration-

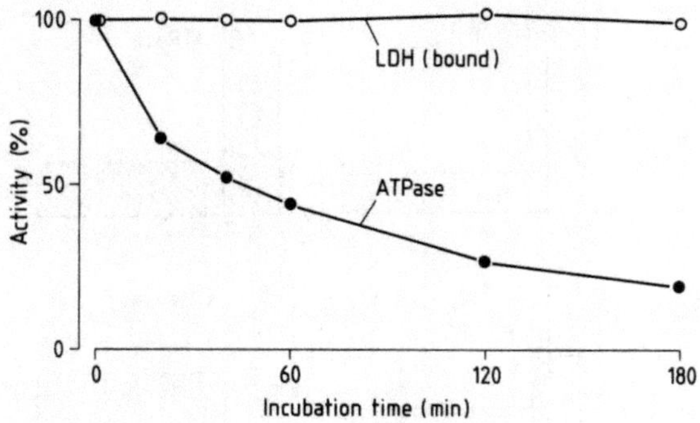

dependent fall in enzyme activity of both Ca^{2+}-ATPase and Mg^{2+}-ATPase activity, whereas activity of bound (intracellular) LDH stays constant. For instance, at a concentration of 2 m*M* DSA and 5 min of incubation, only 45% of Ca^{2+}-ATPase and 42% of Mg-ATPase activity are left (means of two experiments), whereas activity of bound LDH ist unchanged. If for comparison synaptosomes are lysed with 1% Triton-X-100, LDH activity is reduced to the same extent (to 45%). It was also verified that the ATP applied during the assay (1 m*M*) did not increase synaptosomal permeability on its own.

We conclude from this that the catalytic site of the ATPase is localized extracelluary. The possibility that a significant contamination of ATPase containing membrane fragments was responsible for the activity observed is excluded on the grounds of electron microscopical evidence (Fig. 2) which shows that synaptosomal particles are intact.

In order to obtain information on the functional aspect of the ecto-ATPase activity under in situ and in vivo conditions, enzyme activity on sealed synaptosomes was further characterized. Hydrolysis of ATP is activated by either Ca^{2+} or Mg^{2+} to a similar extent (Fig. 5). Kinetic parameters are characterized in Table 1. The low apparent K_m-value for ATP in the presence of either Ca^{2+} or Mg^{2+} (4 m*M*) is in

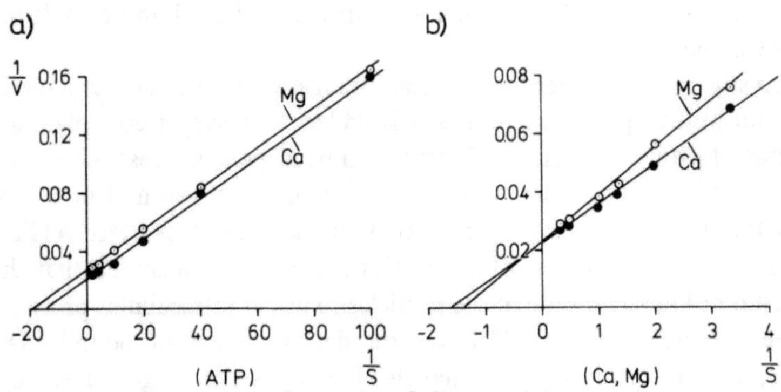

Fig. 5a, b. Activation of ATPase activity in suspensions of intact synaptosomes by increasing concentrations of ATP (Ca^{2+}, Mg^{2+} = 4 m*M*) (**a**) or Ca^{2+} or Mg^{2+} (ATP = 1 m*M*) (**b**) (data from Keller and Zimmermann 1983)

Table 1. Kinetic constants of Ca^{2+}-ATPase and Mg^{2+}-ATPase in intact cholinergic synaptosomes (results from Keller and Zimmermann 1983)

	Km_{app} (m*M*)	V_{max} (nmol Pi x min^{-1} x mg $protein^{-1}$)
ATP_{Ca}	0.079 ± 0.003	45.36 ± 2.06
ATP_{Mg}	0.053 ± 0.004	36.51 ± 2.18
Ca^{2+}	0.61 ± 0.07	36.57 ± 2.31
Mg^{2+}	0.71 ± 0.08	33.61 ± 2.00

the same range (50 *μM*) of that observed for other ecto-ATPases, including those of brain synaptosomes (ref. above, see also Nagy, this volume). The high K_m-values for activitation by Ca^{2+} or Mg^{2+} (ATP 1 m*M*) could reflect the concentrations of the two ions in the extracellular fluid which is in the millimolar range. Maximal initial veleocity is always in favor of Ca^{2+}activation.

Further analysis revealed that intact synaptosomes can hydrolyze a variety of nucleotides. These include tri-, di- and monophosphates (Table 2). Since GTP is stored in synaptic vesicles together with ATP it could also become hydrolyzed extracelluarly. As the fraction of intact synaptosomes also contained enzyme activity for hydrolysis of ADP and AMP and also GDP and GMP, it is likely that the triphosphates may be hydrolyzed to the nucleoside. The exact nature of the enzyme chain is not yet characterized. Rather importantly there is no significant activity towards for unspecific phosphatases, like β-glycophosphate (β-Gly.-P), glucose-6-phosphate (G-6-P), para-nitro phenylphosphate (pNPP), or phosphoenol pyruvate (PEP).

These results suggest that the cholinergic nerve endings in the electric ray contain at their surface ectoenzyme activity for hydrolysis of a variety of nuleotides from triphosphate to nucleoside.

Table 2. Hydrolysis of phosphates by intact cholinergic synaptosomes (nmol Pi x min^{-1} x mg $protein^{-1}$)[a]

ATP	49.1 ± 2.2 (3)	GTP	45.3 ± 1.9 (3)	IMP	14.9 (1)	G-6-P	0.26 (1)
ADP	34.9 (1)	GDP	31.3 (1)	pNPP	0.17 (1)	β-Glyc.-P	0.6 ± 0.25 (2)
AMP	28.9 ± 7.6 (3)	GMP	12.4 (1)	PEP	0.17 (1)		

[a] Values are means ± standard deviation or range. Number of independent experiments in parentheses. Determination was by analysis of total phosphate released

5 Production and Fate of Extracellular Adenosine

When intact synaptosomes are incubated under isotonic conditions with [^{3}H]-ATP there is a time dependent formation of radiolabeled derivatives in the extracellular medium (Fig. 6). Besides ADP and AMP, adenosine is the main hydrolysis product. Only after 30 min of incubation can a significant amount of inosine be detected which has presumably leaked out from the synaptosomes.

This raises the question whether the nucleoside can be salvaged by the nerve terminal. Figure 7 compares the time dependent uptake of [^{3}H]-adenosine and [^{3}H]-adenine into intact synaptosomes in suspension using identical amounts of tracer. Uptake of adenosine exceeds that of adenine by a factor of 10. Concentration-dependent uptake of adenosine and its competitive inhibition by deoxy-adenosine

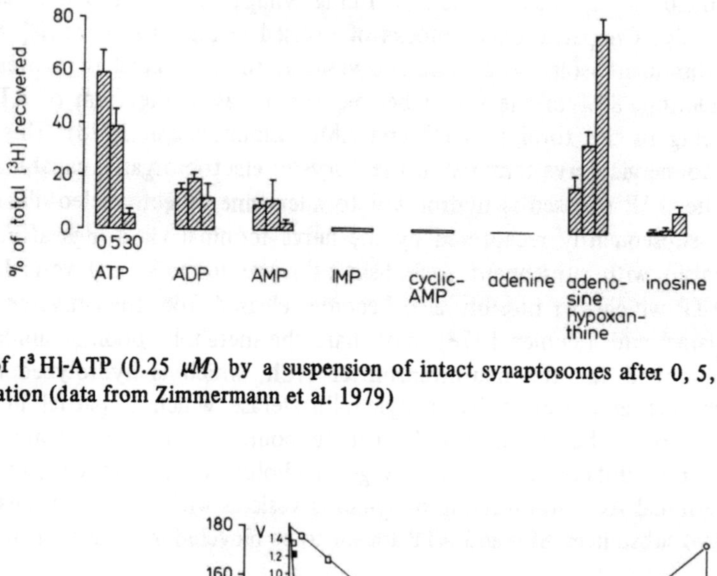

Fig. 6. Hydrolysis of [^{3}H]-ATP (0.25 μM) by a suspension of intact synaptosomes after 0, 5, and 30 min of incubation (data from Zimmermann et al. 1979)

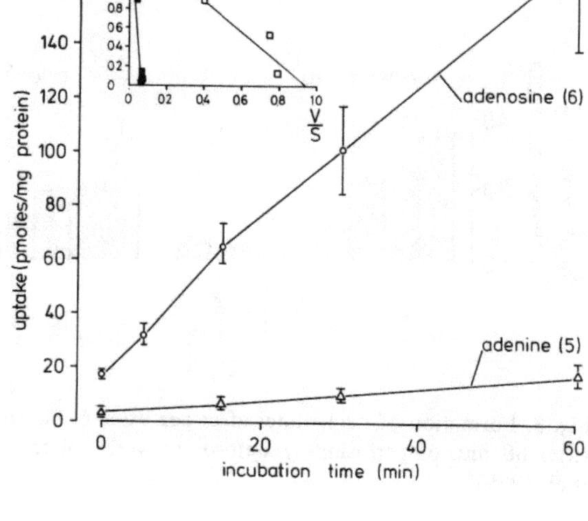

Fig. 7. Time-dependent uptake of [^{3}H]-adenosine (0.31 μM) and [^{3}H]-adenine (0.36 μM) into synaptosomes. *Vertical bars* represent S.E.M. Number of experiments in parentheses. *Inset:* Eadie-Hofstee plot of concentration dependent uptake of [^{3}H]-adenosine (□) into synaptosomes and its competitive inhibition by 2′-deoxyadenosine (1 mM) (■) (data from Zimmermann et al. 1979)

is exemplified in the inset to Fig. 7. The apparent K_m-value for adenosine uptake of 2 μM equals that of the high affinity uptake system for the hydrolysis product of ACh, choline, which has previously been described in a number of cholinergic tissues. The maximal uptake capacity (V_{max} = 30 pmol min^{-1} mg $protein^{-1}$) is by a factor of 1000 lower than the maximal velocity of ATP hydrolysis by the ectoenzymes present in the preparation (compare Table 1).

The adenosine taken up from the extracellular medium is metabolized rapidly inside the nerve endings. This may be shown by either application of [^{3}H]-adenosine to a suspension of intact synaptosomes with subsequent analysis of metabolites in the separated particles. Another possibility (Fig. 8) ist the perfusion of excised blocks of electric tissue with [^{3}H]-adenosine with subsequent isolation and analysis of labeled synaptosomes. The figure shows that adenosine is converted into a number of nucleotides with ATP, ADP, and AMP as major metabolites. A similar result is obtained if [^{3}H]-adenine is applied.

The last question whether nerve terminal ATP derived from extracellular adenosine can also be used for filling synaptic vesicles has also been answered positively. On perfusion of blocks of excised electric tissue with [^{3}H]-adenosine and subsequent isolation of synaptic vesicles, the radiolabel is recovered in the vesicle fraction. 85% of the [^{3}H]-label recovered was in the form of ATP, the remainder being in the form of ADP and AMP (Zimmermann 1978). This suggets that the cholinergic nerve terminal in the *Torpedo* electric organ is capable of purine salvage. The ATP released is hydrolyzed to adenosine by ectonucleotidases. The adenosine is subsequently recaptured by the nerve terminal via a high affinity uptake mechanism with subsequent anabolism of nerve terminal and vesicular ATP (Fig. 9). ATP which can possibly also become released from the activated postsynaptic cell (Israël and Meunier 1978) may share the metabolic pool. A similar cycle has been described for the neurotransmitter ACh, which is hydrolyzed extracellularly to choline and acetate by acetylcholinesterase which is partial in an ecto-position (see Bon, this volume) and partially bound to the basal lamina. There is a high affinity uptake system for salvage of choline with subsequent resynthesis of nerve terminal ACh and refilling of synaptic vesicles with the neurotransmitter. Thus, the two substances ACh and ATP appear to be recycled in parallel at the nerve terminal.

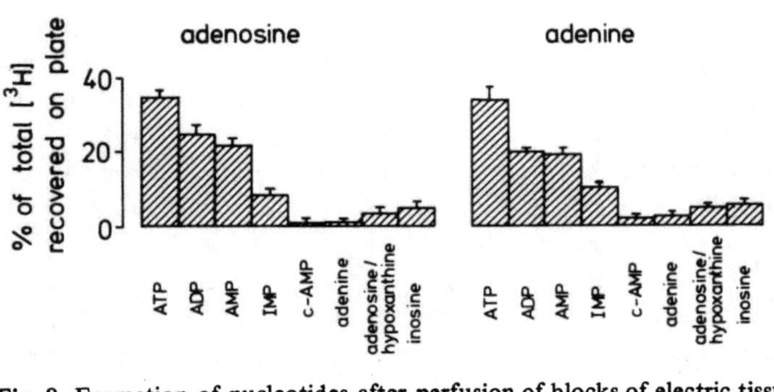

Fig. 8. Formation of nucleotides after perfusion of blocks of electric tissue with [^{3}H]-adenosine. After 60 min of perfusion synaptosomes were isolated and analyzed (data from Zimmermann et al. 1979)

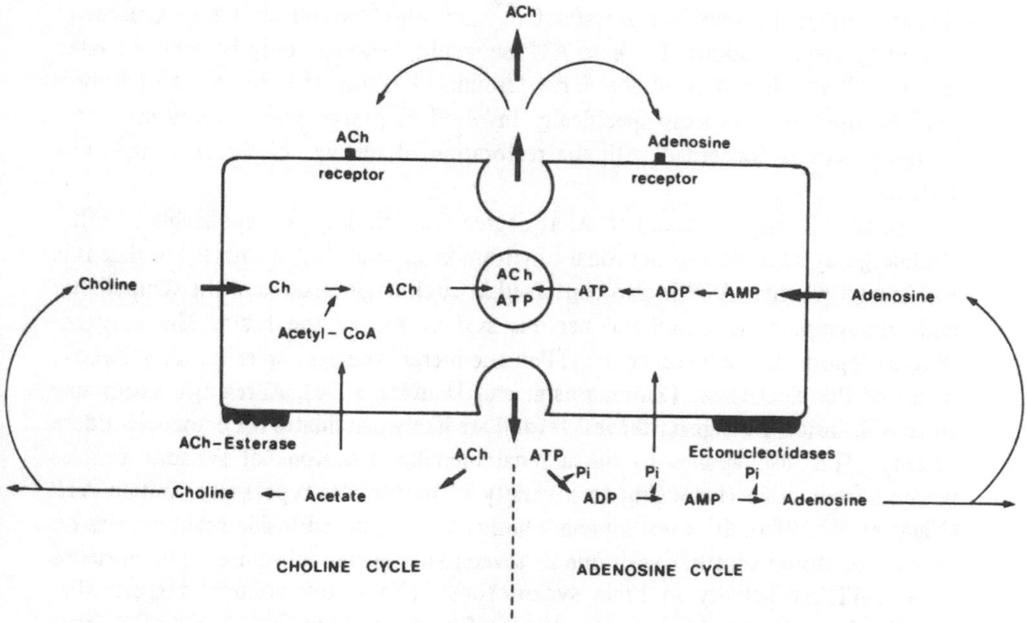

Fig. 9. Model for recycling of acetylcholine and ATP at the cholinergic electromotor nerve terminal: co-storage, co-release, co-hydrolysis by ectoenzymes, co-uptake of choline and adenosine via respective high affinity uptake systems, final synthesis, and vesicular reuptake of original compounds. Both ACh and ATP may cause depression of transmitter release via presynaptic autoreceptors

6 Functional Significance of Ectonucleotidase Activity

Although in this particular example the source of the extracellular substrate ATP can be identified it is still rather difficult to explain the functional role of the released ATP and that of ecto-ATPases in the process of synaptic transmission.

So far a physiological role for co-storage of ATP in the transmitter storing organelle has not been established. Several possibilities may be suggested for an extracellular function of ATP after its release from the cholinergic nerve ending. Thus, ATP may act directly at specific ATP receptors [e.g., frog sympathetic ganglia (Silinsky and Ginsborg 1983)]. After hydrolysis to adenosine it may activate specific adenosine receptors [e.g., autoinhibition via autoreceptors (Silinsky 1984)]. ATP may influence specific cation channels [e.g., in myoblasts (Kolb and Wakelam 1983)]. Finally, ATP may serve as a co-substrate for extracellular protein-phosphorylation which also may occur in nervous tissue (see Kinzel, Kübler, this volume).

Extracellular hydrolysis of released ATP may either terminate specific functions of ATP or produce adenosine from ATP as the physiologically active principle and it will, of course, serve purine salvage and thus recycling of ATP at the synapse. However, the possibility should not be excluded that the energy of hydrolysis of extracellular ATP may be used by the synapse as it is used in a variety of intracellular processes (force generation, ion transport, formation of a pH gradient). In this particular system, the two co-substrates Ca^{2+} and Mg^{2+} would always be present at saturating concentrations. The ecto-ATPase would, however, only become activated when ATP is released from the nerve terminal. Possibly the energy of hydrolysis may be used in a process specifically involved in presynaptic transmitter release or the processes concerned with the restoration of resting conditions at the axon terminal.

Finally, it may be asked to what degree these findings are applicable to other cholinergic systems and to neuronal function, in general. Unfortunately to date it is not known whether ATP is also contained in cholinergic vesicles of the sympathetic and parasympathetic peripheral nervous system and of the brain. The only additional report for the presence of ATP in cholinergic vesicles comes from the electric organ of the electric eel (Zimmermann and Denston 1976). Adrenergic axons also store ATP inside the transmitter vesicles and are likely candidates for ectonucleotidase activity. This also applies to the adrenal medulla. Fractions of synaptic vesicles isolated from brain (belonging to a variety of transmitter types) also contain ATP (Nagy et al. 1976). It is not known whether ATP is stored inside brain vesicles on its own or stored together with one or several transmitter substances. The presence of ecto-ATPase activity in brain synaptosomes (Nagy, this volume) suggests that ATP is also released from certain types of nerve endings in the central nervous system.

7 Summary

Cholinergic synaptosomes isolated from the elctric organ of the electric ray, *Torpedo marmorata,* contain ectonucleotidase activity. A variety of substrates like ATP, GTP, ADP, and AMP are hydrolyzed to a similar extent, whereas unspecific phosphatase activity is absent. ATPase activity is stimulated equally well by either Ca^{2+} or Mg^{2+}ions. ATP is hydrolyzed to adenosine which in turn can be salvaged by the nerve endings by way of a high affinity uptake system. Since ATP is co-stored with the neurotransmitter acetylcholine inside synaptic vesicles and co-released on activation of the axon, the substrate will become available to the ectoenzyme only on synaptic transmission. Ecto-ATPase appears to be involved in purine salvage, but at the same time, the energy of hydrolysis of ATP may be used in a process concerned with the restoration of resting conditions at the axon terminal.

References

Burnstock G (1975) Purinergic transmission. In: Iversen LL, Iversen SD, Synder SH (eds) Handbook of psychopharmacology, vol 5. Plenum, New York, pp 131–194

Carraway CAC, Corrado FJ, Fogle DD, Carraway KL (1980) Ecto-enzymes of mammary gland and its tumors. Ca^{2+}- or Mg^{2+}-stimulated adenosine triphosphatase and its perturbation by concanavalin A. Biochem J 191:45–51

Chambers DA, Salzman EW, Neri LL (1967) Characterization of "ecto-ATPases" of human blood platelets. Arch Biochem Biophys 119:173–178

DePierre JW, Karnovsky ML (1974) Ecto-enzymes of the guinea pig polymorphonuclear leucocyte II. Properties and stability as markers for the plasma membrane. J Biol Chem 249: 7121–7129

Giompres PE, Zimmermann H, Whittaker VP (1981) Changes in the biochemical and biophysical parameters of cholinergic synaptic vesicles on transmitter release and during a subsequent period of rest. Neuroscience 6:775–785

Hamlyn JM, Senior AR (1983) Evidence that Mg^{2+}- or Ca^{2+}-activated triphosphatase in rat pancreas is a plasma-membrane ecto-enzyme. Biochem J 214:59–68

Israël M, Meunier FM (1978) The release of ATP triggered by transmitter action and its possible physiological significance: retrograde transmission. J Physiol (Paris) 74:485–490

Keller F, Zimmermann H (1983) Ecto-adenosine triphosphatase activity at the cholinergic nerve endings of the *Torpedo* electric organ. Life Sci 33:2635–2641

Kolb KH, Wakelam MJO (1983) Transmitter-like action of ATP on patched membranes of cultured myoblasts and myotubes. Nature (Lond) 303:621–623

Li ZY, Bon C (1983) Presence of membrane bound acetylcholinesterase from a preparation of nerve endings from *Torpedo marmorata* electric organ. J Neurochem 40:338–349

Majumder GC (1981) Enzymic characteristics of ecto-adenosine triphosphatase in rat epididymal spermatozoa. Biochem J 195:103–110

Morel N, Meunier FM (1981) Simultaneous release of actetylcholine and ATP from stimulated cholinergic synaptosomes. J Neurochem 36:1766–1773

Morel N, Dreyfus P (1982) Association of acetylcholinesterase with the external surface of the pres-synaptic plasma membrane in *Torpedo* electric organ. Neurochem Int 4:283–288

Nagy A, Baker RR, Morris SJ, Whittaker VP (1976) The preparation and characterization of synaptic vesicles of high purity. Brain Res 109:285–309

Nagy A, Shuster TA, Rosenberg MD (1983) Adenosine triphosphatase activity at the external surface of chicken brain synaptosomes. J Neurochem 40:226–234

Pearson JD, Carleton JS, Gordon JL (1980) Metabolism of adenine nucleotides by ectoenzymes of vascular endothelial and smooth-muscle cells in culture. Biochem J 190:421–429

Ronquist G, Ågren GK (1975) A Mg^{2+} and Ca^{2+} stimulated adenosine triphosphatase at the outer surface of Ehrlich ascites tumor cells. Cancer Res 35:1402–1406

Schweitzer ES, Keller RB (1982) ATP release from cholinergic synapses. Soc Neurosci Abstr 8:493

Silinsky EM (1984) On the mcheanisms by which adenosine receptor activation inhibits the release of acetylcholine from motor nerve endings. J Physiol (Lond) 346:243–256

Silinsky EM, Ginsborg BL (1983) Inhibition of acetylcholine release from preganglionic frog nerves by ATP but not adenosine. Nature (Lond) 305:327–328

Smolen JE, Wiessmann G (1978) Mg^{2+}-ATPase as a membrane ecto-enzyme of human granulocytes. Inhibitors, activators and response to phagocytosis. Biochim Biophys Acta 512: 525–538

Sorensen RG, Mahler HR (1982) Localization of endogenous ATPases at the nerve terminal. J Bioenerg Biomembr 14:527–547

Stefanovic V, Ledig M, Mandel P (1976) Divalent cation-activated ecto-nucleoside triphosphatase activity of nervous system cells in tissue culture. J Neurochem 27:799–805

Trams EG, Kaufmann H, Burnstock G (1980) A proposal for the role of ecto-enzymes and adenylates in traumatic shock. J Theor Biol 87:609–621

Whittaker VP, Zimmermann H (1976) The innervation of the electric organ of Torpedinidae: a model cholnergic system. In: Malins DC, Sargent JR (eds) Biochemical and biophysical perspectives in marine biology, vol 3. Academic, London, pp 67–116

Zimmermann H (1978) Turnover of adenine nucleotides in cholinergic synaptic vesicles of the *Torpedo* electric organ. Neuroscience 3:827–836

Zimmermann H (1982) Coexistence of adenosine 5′-triphosphate and acetylcholine in the electromotor synapse. In: Cuello AC (ed) Co-Transmission. MacMillan, London, pp 243–259

Zimmermann H, Whittaker VP (1974) Effect of electrical stimulation on the yield and composition of synaptic vesicles from the cholinergic synapses of the eletric organ of *Torpedo:* a combined biochemical, electrophysiological and morphological study. J Neurochem 22:435–450

Zimmermann H, Denston CR (1976) Adenosine triphosphate in cholinergic vesicles isolated from the electric organ of *Electrophorus electricus.* Brain Res 111:365–376

Zimmermann H, Dowdall MJ, Lane DA (1979) Purine salvage at the cholinergic nerve endings of the *Torpedo* electric organ: the central role of adenosine. Neuroscience 4:979–993

Enzymatic Characteristics and Possible Role of Synaptosomal Ecto-Adenosine Triphosphatase from Mammalian Brain

AGNES NAGY[1]

1 Introduction

In 1957 Engelhardt described a very active ATPase enzyme on the outer surface of avian (nucleated) red blood cells. He stated that the action of this enzyme is strictly oriented in space towards the surrounding medium: it splits rapidly any ATP that reaches the outside of the cell and leaves the ATP on the inside of the cell unaffected. He clearly distinguished this specific cell surface "ecto-enzyme" from "exo-enzymes" which are secreted into the extracellular space and also from the normal "endo-enzymes" which act within the cell. Prior to the observations of Engelhardt, a fairly active cell surface ATPase had already been reported by Acs et al. (1954) in ascites tumour cells. A detailed analysis of the Ehrlich ascites tumour cell surface ecto-ATPase was later carried out by Ronquist and Agren (1975).

In 1974 Trams and Lauter predicted that most eukaryotic cells would display a non-specific, Mg^{2+}-stimulated ecto-ATPase activity. Since then, the presence of active ecto-ATPase has been observed on the surface of many other cellular membranes: guinea pig polymorphonuclear leukocyte (De Pierre and Karnovsky 1974a), cultured neuroblastoma cells (Stefanovic et al. 1974a; Trams 1974; Rosenblatt et al. 1976), intact human glia and cultured glioma cells (Agren et al. 1971; Trams and Lauter 1974; Stefanovic et al. 1976a), human platelets (Wang et al. 1977), cultured hamster cardiac cells (Coetzee and Gevers 1977), mouse myeloid leukemic cells, normal granulocytes and macrophages (Weiss and Sachs 1977), mouse glia cells (Trams and Lauter 1978), human and rat leukocytes (Smolen and Weissman 1978; Medzihradsky et al. 1980; Wilson et al. 1981), intact spermatozoa from rat caudal epididymis (Majumder and Biswas 1979), mammary-adenocarcinoma ascites cells (Carraway et al. 1980), electrophorus electroplax microsomes (Amende et al. 1983), rat pancreatic cells (Hamlyn and Senior 1983).

The facts that ATP is the principal active substance released by purinergic nerve cells (Burnstock 1972, 1975) and that ATP is also known to be packed and released as a co-transmitter with noradrenaline (Geffen and Livett 1971; Lagerkrantz 1971), 5-hydroxytryptamine (Da Prada and Pletscher 1968) and acetylcholine (Whittaker et al. 1972; Nagy et al. 1976), lead to the conclusion that an ectoenzyme might also exist on the external surface of the nerve endings. It was

[1] Department of Neurology, Reed Neurological Research Center UCLA Center for Health Sciences, Los Angeles, California 90024, USA

suggested, by analogy with the inactivation of acetylcholine, that the ATP released from the nerve endings is rapidly broken down to adenosine by an ectoenzyme-chain consisting of ATPase, ADPase and 5′-nucleotidase (Rosenblatt et al. 1976; Zimmermann et al. 1979; Schubert et al. 1981). In suspensions of hypothalamic synaptosomes, White (1978) reported the very rapid hydrolysis of ATP released by high K^+. ATP has also been shown to be released from the postsynaptic side of cholinergic nerve terminals, and it was suggested that the ATP might act as a retrograde inhibitory messenger (Israel and Meunier 1978; Israel et al. 1980).

Sorensen and Mahler (1982) investigated the localization of ATPase activities associated with purified rat brain synaptic plasma membranes and concluded that a Mg^{2+}-ATPase and an ATPase active at high concentrations of Ca^{2+} in the absence of Mg^{2+} (Ca_H-ATPase) are oriented with their ATP hydrolysis sites on the junctional side of the postsynaptic membrane and are therefore classified as ecto-ATPases. Synaptosomal ecto-ATPase enzyme has been described in intact chicken nerve ending preparations (Nagy and Rosenberg 1981; Nagy et al. 1983), where postsynaptic elements were practically absent. An ectoenzyme activity with similar properties has been reported by Keller and Zimmermann (1983) at the cholinergic nerve endings of the *Torpedo Marmorata* electric organ.

2 Mammalian Brain Tissues

The evidence for an ectoenzyme is difficult to obtain because one must demonstrate that the synaptosomal ATPase activities are not due to (1) contamination by other subcellular particles or (2) leakage or breakage of the isolated nerve endings. The cellular heterogeneity of tissues from the mammalian central nervous system makes it particularly difficult to obtain pure subfractions of brain cells. Furthermore, due to the limitations of methods for measuring cross-contamination, identification and quantification of the contaminating particles in synaptosomal preparations is often difficult. Both mitochondria and endoplasmic reticulum (microsomes) possess abundant ATPase activity and consequently, they are the contaminations of primary concern in the synaptosomal preparations. By using widely accepted marker enzymes for various subcellular particles, we have found that approximately 4% microsomal, 0.5–2% mitochondrial, and 0.5% myelin and glial contaminations are present in our average synaptosomal preparations (Nagy et al. 1983, 1984).

2.1 ATPase Activity of Intact Synaptosomal Preparations

The small amounts of mitochondrial F_1-ATPase and plasma membrane Na^+K^+-ATPase present in our synaptosomal fractions were specifically inhibited (sodium azide or oligomycin and ouabain, respectively). Contaminating microsomal ATPase activity cannot be eliminated from synaptosomal preparations due to the lack of specific inhibitors. Nevertheless, when we consider the total Mg^{2+}-ATPase and Ca^{2+}-ATPase activities contributed by the approximately 4% microsomal contamination in a mammalian brain synaptosomal preparation (specific activities of

Mg^{2+}-, and Ca^{2+}-ATPases of microsomal reference fractions are approximately 200 nmol $P_i min^{-1} mg^{-1}$ of protein), a substantial amount of ATPase activity (approximately 92%) is apparently attributable only to the external surface of the intact synaptosomes. The ecto-ATPase activities measured in intact synaptosomal preparations are in the range of 50–120 nmol $P_i min^{-1} mg^{-1}$ of protein (Table 1).

Table 1. ATPase activities of intact and disrupted synaptosomal fractions from various mammalian brain tissues[a]

Brain tissue	No. of expts.	Mg^{2+}-ATPase in Intact synaptosomes	Mg^{2+}-ATPase in Disrupted[b] synaptosomes	Ca^{2+}-ATPase in Intact synaptosomes	Ca^{2+}-ATPase in Disrupted[b] synaptosomes
Human temporal cortex	7	44.6	128.9	53.4	75.1
Mongolian gerbil (STR/UC) cerebral cortex	8	35.8	113.0	31.1	64.5
Mice (C57/BL6) cerebral cortex	2	48.9	67.4	48.6	44.4
Rat (Sprague-Dawley) cerebral cortex	8	120.0	252.6	113.0	216.1

[a] Units are nmol $P_i min^{-1} mg^{-1}$ of protein. Data represent the mean of the number of experiments shown. SEM are within 15% of the mean values

[b] Synaptosomes were disrupted by freeze-thawing three times

The highest specific ectoenzyme activity is found in rat synaptosomes (120 nmol $P_i min^{-1} mg^{-1}$), while mice, gerbil, and human brain synaptosomes possess 30–40% less activity than that found in the rat. Ecto-ATPase activity represents 30–50% of the total ouabain-insensitive Mg^{2+}-ATPase activities and 75–100% of the total Ca^{2+}-dependent ATPase activity measured in osmotically disrupted synaptosomes. The more than tenfold increase in the Na^+K^+-ATPase activity which resulted from disruption of the synaptosomes indicates that the synaptosomes are intact when expressing the divalent cation-dependent ATPase activities and also that plasma membrane fragments (both pre-, and postsynaptic) represent only a small amount of contamination present. The intact nature of the synaptosomes was regularly checked by monitoring the synaptoplasma marker enzyme, LDH activity.

The kinetics of ATP hydrolysis rules out the possibility that ATP enters the synaptosomes before it is hydrolysed. A lag period in the time curve of ATPase activity, and indication of substrate diffusion into synaptosomes, was clearly lacking in all cases.

All efforts to remove the ecto-ATPase activity from the surface of synaptosomes by repeated washings in isoosmotic media have so far been unsuccessful. This indicates that synaptosomal ecto-ATPase is an integral membrane protein.

2.2 Ion Specificity

Synaptosomal ecto-ATPase requires the presence of divalent cations for activity, and the addition of EDTA produces a reversible 100% loss of activity. Mg^{2+} and Ca^{2+} seem to activate this enzyme equally well and the apparent activation constants for both Mg^{2+} and Ca^{2+} are $1\text{-}4 \times 10^{-4} M$, at pH 7.4 and 0.2 m$M$ ATP concentration. Although the ecto-ATPase of neuroblastoma cells was preferentially stimulated by Ca^{2+} (Stefanovic et al. 1974b), it seems to be a general characteristic of ecto-ATPases that they can be stimulated by either Ca^{2+} or Mg^{2+} (Trams and Lauter 1974; Agren et al. 1976; Carrawy et al. 1980; Hamlyn and Senior 1983; Nagy et al. 1983; Keller and Zimmermann 1983). There is no experimental data revealing whether intact synaptosomes have two independent ATP hydrolysing enzymes with different divalent cation requirements, or rather one enzyme which can utilize equally well both Mg^{2+} and Ca^{2+}-ATP as a substrate.

2.3 pH Dependence

The apparent pH optimum measured in a Tris-Maleate buffer system was in the range of 7.4–7.8 in the presence of either divalent cation in all brain tissues (rat, mice, gerbil, human) studied. This pH optimum does not differ from pH optima found for other ATPases.

2.4 Substrate Specificity

At nucleotide concentrations of 0.5 m*M*, synaptosomal ecto-ATPase shows little substrate specificity. Nucleotide triphosphates (GTP, UTP, ITP) were hydrolysed to a similar extent as ATP, and ADP was hydrolysed at approximately 30% of the rate for ATP, also in the presence of either Mg^{2+} or Ca^{2+}. A similar observation was noted with neuroblastoma cells (Stefanovic et al. 1974a), and with hepatocyte and HeLa cells (Trams and Lauter 1974). The low substrate specificity of the synaptosomal ecto-ATPase hydrolysing activity suggests that the term "ATPase" may be a misnomer and that "ecto-nucleotide-triphosphatase" would be a more suitable name for this particular enzyme activity. Nevertheless, the use of the old nomenclature is convenient for the sake of continuity and probably more relevant to physiological function. To date, it is not known whether intact cells and synaptosomes have separate enzyme mechanisms for the hydrolysis of each nucleotide triphosphate or have only one enzyme with a rather broad range of substrates. The possibility that the ATP hydrolysing activity of intact synaptosomes was due to a non-specific phosphatase activity was ruled out, since p-nitrophenylphosphate was not metabolized at all by synaptosomal ecto-ATPases.

2.5 Enzyme Kinetic Constants

Ecto-ATPases of synaptosomal fractions from various mammalian brain tissues had K_ms for ATP of $2\text{-}5 \times 10^{-5} M$ which fall in the middle of the reported K_m range for other ecto-ATPases (Chambers et al. 1967; De Pierre et al. 1974b; Ronquist and Agren 1975; Majumder 1981). The K_m values of ecto-ATPases are about one order of magnitude lower than those characteristic for the intrasynaptosomal ATPases ($4 \times 10^{-4} M$ for the Na^+K^+-ATPase and $2.5 \times 10^{-4} M$ for the intrasynaptosomal Mg^{2+}-ATPase; Desaiah and Ho 1977). The low K_m value indicates a relatively high affinity of synaptosomal ecto-ATPase for the ATP substrate.

2.6 Inhibitors

Among the tested components, a non-penetrable agent, p-chloromercuribenzoate inhibited the ecto-ATPase activity most effectively (60% inhibition at 0.5 m*M* PCMB concentration), indicating that free thiol groups are essential for the functioning of this enzyme. The non-penetrable nature of this inhibitor serves as further proof for the external location of the ATPase active sites in intact synaptosomes. The insensitivity of synaptosomal ecto-ATPase to inhibitors of ion-transporting ATPases, such as ouabain, sodium vanadate, N-ethylmaleimide (Na^+K^+-ATPase); F^-_1, N-ethylmaleimide (H^+- + K^+-ATPase); N-ethylmaleimide, sodium vanadate ($Ca^{2+}Mg^{2+}$-ATPase); N^-_3, oligomycin (mitochondrial H^+-translocating F_1-ATPase) demonstrates that ATPase activities of intact synaptosomes are different from any known ion-transport ATPases.

3 Possible Physiological Function of Synaptosomal Ecto-ATPases

The ecto-ATPase on nerve endings from mammalian brain tissues is not the only example of a surface-bound enzyme. Observations of a similar kind suggest the localization of invertase, glucosidases and glycosyltransferases on the surface of yeast cells and some bacteria. These enzymes are involved in the vegetative metabolism by changing "food" molecules which are expected to reach the cell from outside. Surface protein kinases and phosphatases most probably have important roles in the phosphorylation-dephosphorylation of extracellularly located plasma membrane components (see Kübler, Kinzel, this volume).

ATPases are usually regarded as enzymes closely concerned with the energetics of the cell. Since the extracellular nucleotide triphosphate levels were known to be very low, it was difficult for some time to understand biologically the sense of localizing this important enzyme at the place where it has little chance to meet its substrate. Presently, we believe that the control of extracellular nucleotide triphosphate levels might be one of the possible functions of ecto-ATPases.

The following functions might also be attributed to synaptosomal ecto-ATPases. Both ATP and adenosine have been postulated as neurotransmitters (Burnstock 1972, 1975). At present, ATP is known to be released into the synaptic cleft either presynaptically or postsynaptically, and the synaptosomal ecto-ATPase may be the first component of an "enzyme chain" responsible for the complete hydrolysis of ATP to adenosine. 5′-Nucleotidase, the last enzyme of this metabolic chain has been demonstrated on the external membrane surface of various plasma membranes (De Pierre et al. 1974a; Trams and Lauter 1974; Stefanovic et al. 1976b; Carraway et al. 1976) including cholinergic plasma membranes (Dowdall 1977). There is not much information available about the adenylate kinase enzyme which would transform ADP to AMP at the outer surface of synaptic plasma membranes. A muscle type of adenylate kinase has been isolated from brain tissue (Khoo and Russel 1972; Pradhan and Criss 1976), but the exact subcellular localization of this enzyme requires further investigation. The fact that adenosine, but not ATP, ADP or AMP is taken up by the nerve endings as the final step of the purinergic effect, strongly suggests the existence of the above degradation pathway.

By modulating the extracellular ATP concentration, we can speculate how synaptosomal ecto-ATPase might participate in the regulation of neurotransmitter release and other cell functions by a variety of mechanisms (Fig. 1):

1) Extracellular ATP increases the passive permeability of the plasma membrane to Na^+ and K^+ which would consequently lower the excitability threshold of the cells (Tower 1960). Ecto-ATPase can terminate this specific effect of external ATP since neither of the metabolic products of ATP degradation causes permeability changes.

2) Adenosine, which is responsible for the modulation of many neuronal functions (Stone 1981) may be derived from the ATP co-secreted with neurotransmitters and subsequently degraded by ectoenzymes (Kuroda 1978; De Mey et al. 1979). The most important action of adenosine may be its ability to depress the evoked release of transmitters, whether from cholinergic or adrenergic neurons (Phyllis

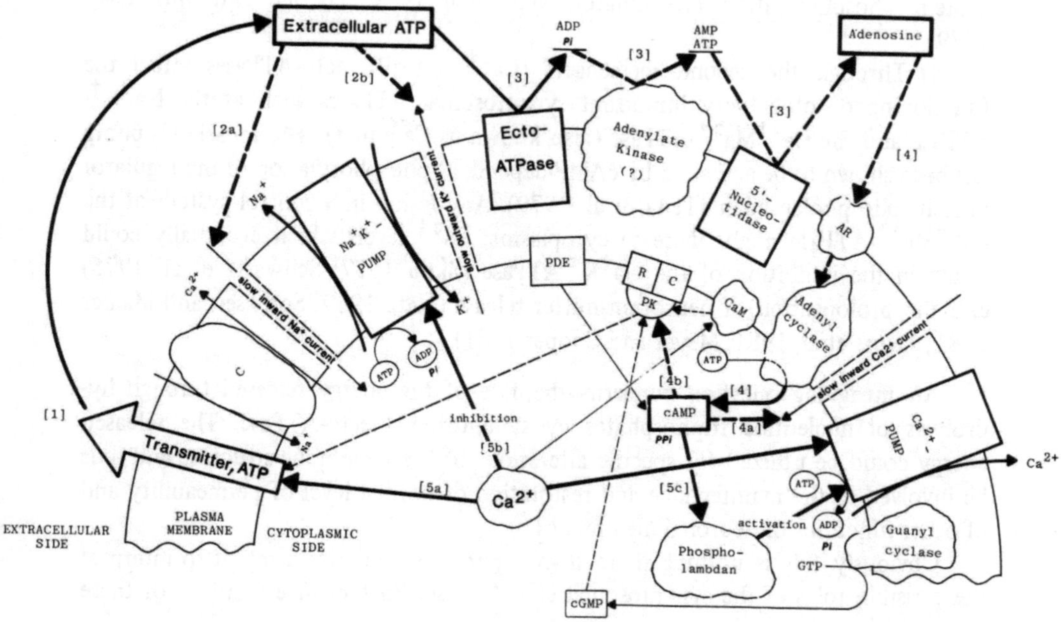

Fig. 1. The possible mechanisms by which ecto-ATPase could modulate the biochemical activities of presynaptic plasma membranes relevant to passive and active ion fluxes and transmitter release. *Route (1)* ATP released as co-transmitter into the extracellullar space; *Route (2)* extracellular ATP effects the passive permeability of the plasma membrane towards: (*a*) Na^+ (*b*) K^+; *Route (3)* extracellular ATP is hydrolysed by ecto-ATPase and eventually results in adenosine as end product; *Route (4)* adenosine interacts with extracellular receptors and modulates the intracellular cyclic AMP level before being taken up by the nerve endings; cAMP can alter the ion movement across the cell membrane: (*a*) directly (Ca^{2+}); (*b*) through protein kinase activity-protein phosphorylation. *Route (5)* Intracellular Ca^{2+} concentration controls neurotransmitter release: (*a*) directly; (*b*) through the inhibition of Na^+K^+-pump. Intracellular Ca^{2+} level is controlled by the Ca^{2+}-pump which is activated by the cAMP-dependent phosphorylation of a regulating protein, phospholambdan

et al. 1974; Hedquist and Fredholm 1976; Stone and Taylor 1977; Schubert and Mitzdorf 1979; Dunwiddie 1980). Therefore, an enzyme such as the ecto-ATPase, which is responsible for the immediate hydrolysis of the released ATP and thus produces adenosine, may be an excellent candidate for participating in the feedback regulation of neurotransmitter release.

3) At central synapses, the action of extracellular ATP becomes more complex when cyclic nucleotides are also considered. By participating in the regulation of extracellular adenosine, ecto-ATPase might exert indirect effects on intracellular events, such as cAMP production, Ca^{2+} movements and protein phosphorylation. Evidence has been presented from many cell types for adenosine action mediated through cyclic AMP (Kuroda 1978; Berne et al. 1983). Adenosine binds to cell surface receptors and modifies adenylate cyclase activity (Schwabe et al. 1973). The altered cAMP level may then modulate a number of events at the synaptic junction either directly (McAfee and Greengard 1972), or through alteration of

protein phosphorylation (Reddington and Mehl 1979; Therien and Mushinsky 1979).

4) Through the second messengers (Ca^{2+}, cAMP), ecto-ATPases affect the functioning of other very important synaptosomal ATPases such as the $Na^{+}K^{+}$-ATPase and the $Ca^{2+}Mg^{2+}$-ATPase (also known as Ca-pump). The muscle Ca-pump has been shown to be activated by cAMP-dependent phosphorylation of the regulator protein phospholambdan (Tada et al. 1979). Any lesion in a control system of the $Ca^{2+}Mg^{2+}$-ATPase might increase cytoplasmic Ca^{2+} levels which eventually could result in the inhibition of the $Na^{+}K^{+}$-ATPase (Skou 1957; Schwartz et al. 1975) and the prolongation of neurotransmitter release (Katz 1969; Sorensen and Mahler 1981; Javors et al. 1981; Meyer and Cooper 1981).

An intriguing question concerns the fate of the energy released through hydrolysis of nucleotide triphosphates by synaptosomal ecto-ATPase. The released energy could be utilized for specific alterations of the membrane structure and thus be involved in the maintenance and restoration of normal level of permeability and of other functions of neuronal membranes.

Obviously this is nothing more than a purely speculative attempt to interpret the possible roles of the synaptosomal ecto-ATPase. Further investigations of these outlined interactions will be necessary.

4 Summary

In summary, we found that similarly to our earlier observations made on chicken synaptosomes, nerve endings from various mammalian brain tissues also possess a nucleotide triphosphate hydrolysing enzyme activity (EC 3.6.1.15) which has its active site facing the external medium. The detected enzyme activity is not due to leakage of intracellular enzyme activities or to contaminating subcellular particles.

The enzymatic properties of the synaptosomal ecto-ATPase are very similar to those which were described for various other cell surface nucleotide triphosphatases. Synaptosomal ecto-ATPase has a high affinity for ATP and it is apparently stimulated equally well either by Mg^{2+} or Ca^{2+} in the absence of other cations. The activation constants for both divalent cations are approximately $4 \times 10^{-4} M$ in mammalian brain tissues studied. Although synaptosomal ecto-ATPase shows a rather broad substrate specificity, the involvement of a non-specific phosphatase in the hydrolysis of externally added ATP is unlikely. The apparent pH optimum is in the range of 7.4–7.8 for both divalent cation-dependent ATPase activities.

The rapid degradation of ATP co-released with transmitters from nerve endings is a possible function of this enzyme. Metabolites of extrasynaptosomal ATP hydrolysis can be involved in the regulation of neurotransmission in various ways. The energy released during the enzymatic reaction could be used for the maintenance and restoration of the normal membrane permeability of nerve endings.

Abbreviations used: AR, adenosine receptor; *C* presumptive Na^{+}-Ca^{2+}-exchange carrier; *CaM*, calmodulin or calmodulin-like protein; *PDE*, phosphodiesterase; *PK*, protein kinase; (*C*), Catalytic subunit and (*R*) Regulatory subunit of PK; P_i, inorganic phosphate

References

Acs G, Ostrowski W, Straub FB (1954) Über die Adenylpyrophosphatase-aktivität an der Oberfläche der Aszites-Krebszellen. Acta Physiol Acad Sci Hung 6:260–263

Agren G, Ponten J, Ronquist G, Westermark B (1971) Demonstration of an ATPase at the cell surface of intact normal and neoplastic tumour cells in culture. J Cell Physiol 78:171–176

Agren G, Ponten J, Ronquist G, Westermark B (1976) Comparison between Ca^{2+} and Mg^{2+} on surface located ATPase of intact normal and neoplastic human cells in culture. Acta Physiol Scand 98:263–265

Amende LM, Chock SP, Albers RW (1983) Characterization of the Ca^{2+}- and Mg^{2+}-dependent ATPases in Electrophorus electroplax microsomes. J Neurochem 40:1040–1047

Berne RM, Rall TW, Rubio R (eds; 1983) Regulatory function of adenosine. Nijhoff, The Hague Boston, London

Burnstock G (1972) Purinergic nerves. Pharmacol Rev 24:509–581

Burnstock G (1975) Purinergic transmission. In: Iversen LL, Iversen SD, Snyder SH (eds), vol 5. Plenum, New York, pp 131–194

Carraway CAC, Corrado FJ IV, Fogle DD, Carraway KL (1980) Ecto-enzymes of mammary gland and its tumours. Biochem J 191:45–51

Carraway KL, Fogle DD, Chesnut RW, Huggins JW, Carraway CAC (1976) Ecto-enzymes of mammary gland and its tumours. J Biol Chem 251:6173–6178

Chambers DA, Salzman EW, Neri LL (1967) Characterization of "ecto-ATPase" of human blood platelets. Arch Biochem Biophys 119:173–178

Coetzee GA, Gevers W (1977) 5-bromo-2′-deoxyuridine-stimulated calcium ion- or magnesium ion-dependent ecto-(adenosine triphosphatase) activity of cultured hamster cardiac cells. Biochem J 1964:645–652

Da Prada M, Pletscher A (1968) Isolated 5′-hydroxytryptamine organelles of rabbit blood platelets: physiological properties and drug-induced changes. Br J Pharmacol 34:591–597

De Mey J, Burnstock G, Vanhoutte PM (1979) Modulation of the evoked release of noradrenaline in canine saphenous vein via presynaptic receptors for adenine but not ATP. Eur J Pharmacol 55:401–405

De Pierre JW, Karnowsky ML (1974a) Ecto-enzymes of the guinea pig polymorphonuclear leukocyte I. Evidence for an ecto-adenosine monophosphatase, -adenosine triphosphatase, and -p-nitrophenyl phosphatase. J Biol Chem 249:7111–7120

De Pierre JW, Karnowsky ML (1974b) Ecto-enzymes of the guinea pig polymorphonuclear leukocyte II. Properties and suitability as markers for the plasma membrane. J Biol Chem 249:7121–7129

Desaiah D, Ho IK (1977) Kinetics of catecholamine sensitive $Na^{+}K^{+}$-ATPase activity in mouse brain synaptosomes. Biochem Parmacol 26:2029–2035

Dowdall MJ (1977) Adenine nucleotides in cholinergic transmission: presynaptic aspects. In: Nucleotides and neurotransmission. Conf Neurobiol de Gif, Dec 1977, pp 7–8

Dunwiddie TV (1980) Endogenously released adenosine regulates excitability in the in vitro hippocampus. Epilepsia 21:541–548

Engelhardt WA (1957) Enzymes as structural elements of physiological mechanisms. Proc Int Symp Enzym Chem (Tokyo) 2:163–166

Geffen LB, Livett BG (1971) Synaptic vesicles in sympathetic neurons. Physiol Rev 51:98–157

Hamlyn JM, Senior AE (1983) Evidence that Mg^{2+}- or Ca^{2+}-activated adenosine triphosphatase in rat pancreas is a plasma-membrane ecto-enzyme. Biochem J 214:59–68

Hedquist P, Fredholm BB (1976) Effects of adenosine on adrenergic neurotransmission. Prejunctional inhibition and postjunctional enhancement. Naunyn-Schmiedebergs Arch Pharmacol 293:217–223

Israel M, Meunier FM (1978) The release of ATP triggered by transmitter action and its possible physiological significance: retrograde transmission. J Physiol (Paris) 74:485–490

Israel M, Lesbats B, Manaranche R, Meurier FM, Franchon P (1980) Retrograde inhibition of transmitter release by ATP. J Neurochem 34:923–932

Javors MA, Bowden CL, Ross DH (1981) Kinetic characterization of Ca^{2+} transport in synaptic membranes. J Neurochem 37:381–387

Katz B (1969) The release of neural transmitter substances. Liverpool University Press, Liverpool

Keller F, Zimmermann H (1983) Ecto-adenosine triphosphatase activity at the cholinergic nerve endings of the Torpedo electric organ. Life Sci 33:2635–2641

Khoo JC, Russel PJ (1982) Isoenzymes of adenylate kinase in human tissue. Biochem Biophys Acta 268:98–101

Kuroda Y (1978) Physiological roles of adenosine derivatives which are released during neurotransmission in mammalian brain. J Physiol (Paris) 74:463–470

Lagerkrantz H (1971) Isolation and characterization of sympathetic nerve trunk vesicles. Acta Physiol Scand Suppl 366:1–44

Majumder GC (1981) Enzymic characters of ecto-adenosine triphosphatase in rat epididymal intact spermatozoa. Biochem J 195:110–118

Majumder GC, Biswas R (1979) Evidence for the occurence of an ecto-(adenosine triphosphatase) in rat epididymal spermatozoa. Biochem J 183:737–743

McAfee DA, Greengard P (1972) Adenosine 3′-5′-monophosphate electrophysiological evidence for a role in synaptic transmission. Science (Wash DC) 178:310–312

Medzihradsky F, Cullen EI, Lin HL, Bole GG (1980) Drug-sensitive ecto-ATPase in human leukocytes. Biochem Pharmacol 29:2285–2290

Meyer EM, Cooper JR (1981) Correlation between $Na^{+}K^{+}$-ATPase activity and acetylcholine release in rat cortical synaptosomes. J Neurochem 36:467–475

Nagy A, Rosenberg MD (1981) Adenosine triphosphatase activity at the external surface of synaptosomes. 8th Meet Int Soc Neurochem, Nottingham, p 114

Nagy A, Baker RR, Morris SJ, Whittaker VP (1976) The preparation and characterization of synaptic vesicles of high purity. Brain Res 109:285–309

Nagy A, Escueta AV (1984) Rapid preparation of synaptosomes from mammalian brain using non-toxic isoosmotic gradient material (Percoll). J Neurochem 43:1114–1123

Nagy A, Shuster TA, Rosenberg MD (1983) Adenosine triphosphatase activity at the external surface of chicken brain synaptosomes. J Neurochem 40:226–232

Phyllis JW, Kostopoulos GK, Limacher JJ (1974) Depression of corticospinal cells by various purines and pyrimidines. Can J Physiol Pharmacol 52:1226–1229

Pradhan TK, Criss WE (1976) The major forms of adenylate kinase from adult and fetal rat tissues. Enzyme (Basel) 21:327–331

Reddington M, Mehl E (1979) Synaptic membrane proteins as substrates for cyclic AMP-stimulated protein phosphorylation in various regions of rat brain. Biochim Biophys Acta 555:230–238

Ronquist G, Agren GK (1975) A Mg^{2+}- and Ca^{2+}-stimulated adenosine triphosphatase at the outer surface of Erlich ascites tumour cells. Cancer Res 35:1402–1406

Rosenblatt DE, Lauter CJ, Trams EG (1976) Deficiency of a Ca^{2+}-ATPase in brains of seizure prone mice. J Neurochem 27:1299–1304

Schubert P, Mitzdorf U (1979) Analysis and quantitative evaluation of the depressive effect of adenosine on evoked potentials in the hippocampal slice. Brain Res 172:186–190

Schubert P, Lee K, Kreutzberg GW (1981) Formation and function of adenosine in the CNS I. Release and modulatory action. 8th Meet Int. Soc Neurochem, Nottingham, p 111

Schwabe U, Ebert R, Erbler HC (1973) Adenosine release from isolated fat cells and its significance for the effect of hormones on 3′,5′-AMP levels and lipolysis. Naunyn-Schmiedebergs Arch Pharmacol 276:133–148

Schwartz A, Lindenmayer GE, Allen JC (1975) The sodium-potassium adenosine triphosphatase: pharmacological physiological and biochemical aspects. Pharmacol Rev 27:3–134

Skou JC (1957) The influence of some cations on an adenosine triphosphatase from peripheral nerves. Biochim Biophys Acta 23:394–401

Smolen JE, Weissmann G (1978) Mg^{2+}-ATPase as a membrane ecto-enzyme of human granulocytes. Inhibitors, activators and response to phagocytosis. Biochim Biophys Acta 512: 525–538

Sorensen RG, Mahler HR (1981) Calcium stimulated adenosine triphosphatase in synaptic membranes. J Neurochem 37:1407–1418

Sorensen RG, Mahler HR (1982) Localization of endogenous ATPase at the nerve terminal. J Bioenerg Biomembr 14:527–547

Stefanovic V, Ciesielski-Treska J, Ebel A, Mandel P (1974a) Nucleoside triphosphatase activity at the external surface of neuroblastoma cells. Brain Res 81:427–441

Stefanovic V, Ciesielski-Treska J, Ebel A, Mandel P (1974b) Ca^{2+}-activated ATPase at the external surface of neuroblastoma cells in culture. FEBS Lett 49:43–46

Stefanovic V, Ledig M, Mandel P (1976a) Divalent cation activated ecto-nucleoside triphosphatase activity of nervous system cells in tissue culture. J Neurochem 27:799–805

Stefanovic V, Mandel P, Rosenberg A (1976b) Ecto-5′-nucleotidase of intact cultured Cc rat glioma cells. J Biol Chem 251:3900–3905

Stone TW (1981) Physiological roles for adenosine and adenosine 5′-triphosphatase in the neuron system. Neuroscience 6:523–555

Stone TW, Taylor DA (1977) Microiontophoretic studies of the effects of cyclic nucleotides on excitability of neurons in rat cerebral cortex. J Physiol (Lond) 266:523–543

Tada M, Yamada M, Ohmori F, Kuzuya T, Abe H (1979) In: Mukohata Y, Parker L (eds) Cation flux across biomembranes. Academic, London, pp 179–190

Therien HM, Mushinski WE (1979) Distribution and properties of protein kinase and protein phosphatase activities in synaptosomal plasma membranes and synaptic junctions. Biochim Biophys Acta 585:188–200

Tower DB (1960) In: Thomas J (ed) Neurochemistry of epilepsy, Thomas CC, Springfield, USA

Trams EG (1974) Evidence for ATP action on the cell surface. Nature (Lond) 252:480–482

Trams EG, Lauter CJ (1974) On the sidedness of plasma membrane enzymes. Biochim Biophys Acta 345:180–197

Trams EG, Lauter CJ (1978) Ecto-ATPase deficiency in glia of seizure-prone mice. Nature (Lond) 271:270–271

Wang TY, Hussey CB, Sasse EA, Hause LL (1977) Platelet aggregation and the ouabain-insensitive ATPase. Ecto-ATPase, reflection of membrane integrity. Am J Clin Pathol 67:528–532

Weiss B, Sachs L (1977) Differences in surface membrane ecto-ATPase and ecto-AMPase in normal and malignant cells I. Decrease in ecto-ATPase in myeloid leukemic cells and the independent regulation of ecto-ATPase and ecto-AMPase. J Cell Physiol 93:183–188

White TD (1978) Release of ATP from a synaptosomal preparation by elevated extracellular potassium and by veratridine. J Neurochem 30:329–336

Whittaker VP, Essmann WE, Dowe GHC (1972) The isolation of pure cholinergic synaptic vesicles from the electric organs of elasmobranch fish of the family Torpedinidae. Biochem J 128:833–846

Wilson PD, Rustin GJ, Smith GP, Peters TJ (1981) Electron microscopic cytochemical localization of nucleoside phosphatases in normal and chronic granulocytic leukemic human neutrophyls. Histochem 13:73–84

Zimmermann H, Dowdall MJ, Lane DA (1979) Purine salvage at the cholinergic nerve endings of the Torpedo electric organ: the central role of adenosine. Neuroscience 4:979–993

Properties and Expression of an ATP-Degrading Ectoenzyme Cascade on Human Lymphoid Cells

W. GUTENSOHN and J. RIEGER [1]

1 Introduction

Human lymphoid cells seem to be interesting models for the study of the expression of nucleotide-degrading ectoenzymes, since extreme differences in enzyme activities are observed between subpopulations of normal peripheral blood lymphocytes (Silber et al. 1975; Thompson et al. 1983), different human T- or B-lymphoblastoid cell lines (Gutensohn et al. 1980), or malignant cells from patients with various forms of leukemias and lymphomas (Gutensohn and Thiel 1981; Gutensohn et al. 1983a). However, most of these investigations have sofar only included ecto-5'-nucleotidase. An interesting correlation of the expression of this enzyme with the presence of the so-called common ALL antigen (a 100 kD membrane glycoprotein) was observed in certain froms of acute leukemias (Gutensohn and Thiel 1981; Gutensohn et al. 1983a). However, this is obviously not reflected in a close association of these two markers on the plasma membrane level (Gutensohn et al. 1983b). In search for other enzymes which might be in closer contact with 5'-nucleotidase ATPase, ADPase and a nucleosidediphosphate kinase were studied and characterized on the surface of human lymphoid cells.

2 Enzymes of Purine Nucleotide Metabolism on the Surface of Lymphoid Cells

The following experiments for the characterization of the ectoenzymes were performed with the human lymphoblastoid B-cell line BHG-83-1. To distinguish ATPase and ADPase on these cells from unspecific phosphatases competition experiments with typical phosphatase substrates like β-glycerophosphate (2.5 m*M*) or ρ-nitrophenylphosphate (2.5 m*M*) were carried out. Inhibition of ATPase and ADPase by these additions is marginal. These results were confirmed for both enzymes in Lineweaver-Burk plots using 5 m*M* concentrations of β-glycerophosphate and ρ-nitrophenylphosphate. Thus, the contribution of unspecific ectophosphatases to the breakdown of nucleotides is negligible. Likewise ATPase and ADPase are not in-

[1] Institut für Anthropologie und Humangenetik der Universität München, Arbeitsgruppe Biochemische Humangenetik. Goethestr. 31, 8000 München 2, FRG

Fig. 1. Dependence of nucleosidediphosphate kinase on ATP concentration. BHG-83-1,5x10[3] cells per assay, were tested under isotonic conditions

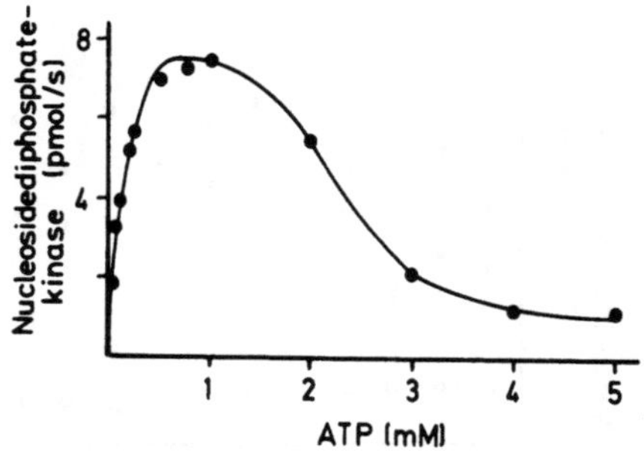

hibited by the ADP-analog α, β-methylene-adenosine diphosphate which is a strong and specific inhibitor of ecto-5′-nucleotidase. Insensitivity to ouabain in concentrations from 0.01–10 m*M* clearly distinguishes this ATPase from Na/K-ATPase and shows that this enzyme represents the major membrane-bound ATP-degrading activity in these cells. Preincubation of the cells with inhibiting titers of specific polyclonal (Gutensohn et al. 1980) or monoclonal (Kummer et al. 1984) anti-human-5′-nucleotidase sera also not block ATPase or ADPase activity.

The nucleosidediphosphate kinase activity was detected during studies on inhibition of ADPase by ATP (see blow). Dependent on the concentration of unlabeled ATP (Fig. 1), [^{3}H]-ATP is formed from [^{3}H]-ADP. The identity of this product was confirmed (1) by its R_f-value in the PEI-cellulose system (see Methology); (2) by using this labeled ATP recovered from the thin-layer sheet as substrate in a subsequent ATPase reaction; (3) by high pressure liquid chromatography on a Polygosil 60–10 NH_2 column[2]. The same reaction and the same dependence on ATP as in Fig. 1 was also demonstrated in a microsomal fraction from BHG-83-1 cells showing that this nucleosidediphosphate kinase is a membrane-bound enzyme. On this cell line this kinase is the most active of the three ectoenzymes as seen in Table 1. On the other hand, an adenylate kinase (myokinase-like) activity could not be demonstrated on the surface of these cells. Starting from [^{3}H]-AMP no [^{3}H]-ADP is formed in the presence of unlabeled ATP and starting from [^{3}H]-ADP no [^{3}H]-ATP is found in the absence of unlabeled ATP.

Table 1. Ectoenzyme activities on human peripheral blood lymphocytes, granulocytes, and a lymphoblastoid B-cell line

Enzyme	PBL	Granulocytes (pmol/s^{-1} ∘ 10^6 cells)	BHG-83-1 cells
ATPase	39	30	500
ADPase	107	n.d.	510
5′-Nucleotidase	1.9	0	10.6
Nucleosidediphosphate kinase	270	5	1250

[2] Kindly performed by Dr. Becker, Institut für Physiologie, München

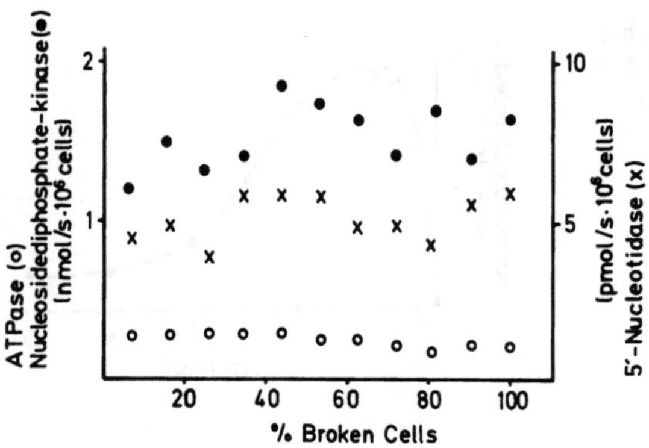

Fig. 2. Contribution of intracellular enzymes to ectoenzyme-activities. The viability of freshly harvested BHG-83-1 cells was determined. Part of the cell suspension was sonicated to achieve complete lysis. Artificial mixtures of viable and dead cells were prepared and enzyme activities tested under isotonic conditions

3 Ectoenzyme Nature

The fact that ATPase, ADPase, and nucleosidediphosphate kinase are true ecto-enzymes is shown by the following experiments: (1) All enzyme activities are measured on intact cells under isotonic conditions with the impermeable nucleotide substrates given from the outside. Viability of the cell samples is always 95% or higher and a possible contribution of intracellular enzymes from 5% dead cells is excluded by the next experiment. (2) Enzyme activities were measured in varying mixtures of viable and deliberately broken cells under isotonic conditions (Fig. 2). Although the data show some scatter there is definitely no slope which would indicate a sizeable contribution of intracellular enzymes with an increasing percentage of broken cells in the mixture. (3) For the ATPase covalent modification by the impermeable reagent diazotized sulfanilic acid (according to DePierre and Karnovsky 1974) was achieved on intact cells, whereas an intracellular enzyme (LDH) of the same cells remained unchanged.

4 Characterization of the Enzymes

All three enzymes have a requirement for a divalent metal ion as shown by a marked inhibition of ATPase (85%), ADPase (37%), and nucleosidediphosphate kinase (50%) in the presence of 5 m*M* EDTA. The influence of ions on enzyme activity is summarized in Table 2. There is considerable activity even in the absence of any additions pointing towards firmly bound endogeneous metal ions. Even after dialysis of microsomal fractions against EDTA, activity is immediately restored upon addition of Mg^{++}. The stimulating effect of Mg^{++} and Ca^{++} reached saturation at about 2 m*M* (data not shown here).

As shown in Fig. 3 ATPase and nucleosidediphosphate kinase exhibit a rather broad range of substrate specificities. In these experiments other nucleotides were introduced either to compete with the standard radioactive substrates (A and B) or to replace one substrate (C). The data in Fig. 3 C are similar to the curve for ATP in Fig. 1, however, UTP does not show the marked inhibition of nucleosidedi-

Table 2. Metal ion dependence of ATPase and ADPase on BHG-83-1 cells[a]

Ion in assay (2 m*M*)	ATPase (pmol • s^{-1})	ADPase
–	333	686
$MgCl_2$	880	978
$CaCl_2$	592	955
$MnCl_2$	511	733
$Pb(NO_3)_2$	211	186
$ZnSO_4$	189	180
$CuCl_2$	133	89

[a] Cells (1.7×10^6 per assay) were tested under isotonic conditions

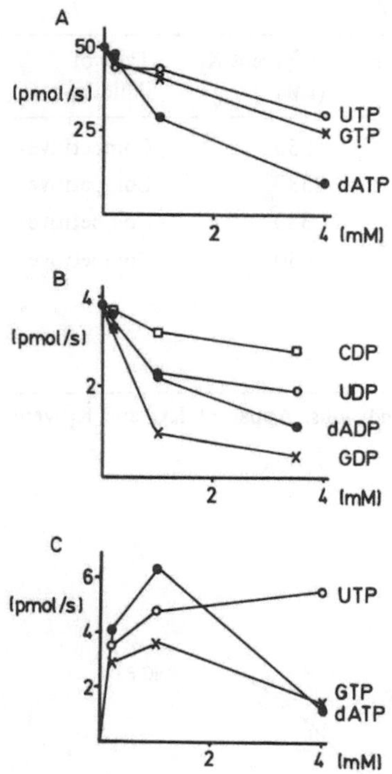

Fig. 3 A–C. Influence of nucleoside-triphosphates and -diphosphates on ectoenzymes on BHG-83-1 cells. **A** ATPase (1×10^4 cells/assay); **B** and **C** nucleosidediphosphate kinase (2.5×10^3 and 5×10^3 cells/assay, resp.). Nucleoside-phosphates were added to the standard assay mixture (**A** and **B**). In **C** nucleoside-triphosphates replace ATP in the assay

phosphate kinase at higher concentrations. Inhibition of ADPase by other nucleosidediphosphates is not as marked as that of ATPase by triphosphates.

The enzymes can be solubilized from the membrane in a detergent extract. In general, noninonic detergents like Triton X 100, Nonidet P 40, or Tween 20 have a stimulating effect, whereas ionic detergents inhibit enzyme activity. Bromelain was reported to liberate membrane-bound alkaline phosphatase from HeLa cells in an active form by proteolytic cleavage (Hanford and Fishman 1983). We have confirmed this for human placental microsomes. On the other hand, bromelain treatment of intact BHG-83-1 cells has no demonstrable effect on the ectoenzymes described here. There is no degradation of enzyme activity on the cell surface, but also no liberation into the extracellular fluid.

Kinetic data are given in Table 3. The apparent constants given in this table should be viewed with some caution, since they were determined on intact cells where competing reactions cannot be excluded or suppressed. This is especially true whenever ATP and ADP are given together, because of the high activity of nucleosidediphosphate kinase. The possibility that the activity of this latter enzyme is just a reverse reaction of another enzyme (e.g., ATPase) was excluded by comparing two different B-cell lines (Fig. 4).

Table 3. Kinetic parameters of ectoenzymes on BHG-83-1 cells[a]

Enzyme	Substrate	Apparent K_M (μM)	Inhibitor	Apparent K_i (μM)	Type of inhibition
ATPase	ATP	20	ADP	250	Competitive
			AMP	2330	Competitive
ADPase	ADP	50	ATP	330	Competitive
			AMP	550	Competitive
Nucleosidediphosphate kinase	ATP	80			
	ADP	150			

[a] Enzymes were assayed on intact cells under isotonic conditions. Apparent K_M and K_i were derived from Lineweaver Burk plots

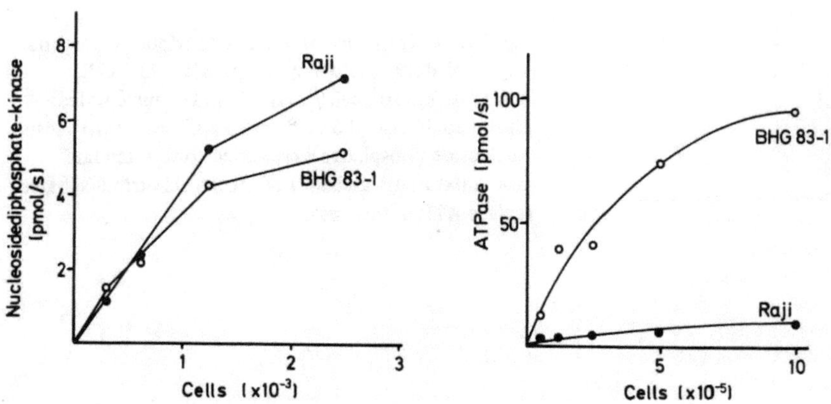

Fig. 4. Comparison of ATPase and nucleosidediphosphate kinase in two different B-cell-lines

5 Subcellular Localization and Plasma-Membrane Subfractionation

Both ATPase and 5′-nucleotidase can be enriched in plasmamembrane preparations obtained from BHG-83-1 cells. However, if microsomes are analyzed in detail on a five-step sucrose gradient according to Bretz et al. (1980), the profiles of distribution of ATPase, 5′-nucleotidase, and the typical Golgi enzyme galactosyltransferase are very similar. Obviously, the method of homogenization and microsome preparation produces aggregates between plasma membrane and denser membrane fractions, which subsequently can no longer be resolved by the sucrose gradient centrifugation.

As an alternative method, induced shedding of membrane components from intact cells according to Ferber et al. (1980) was applied. The detergent ET-12-H is an analog of lysophosphatidyl-choline. Figure 5 shows the release of a number of membrane-bound enzymes dependent on the detergent concentration. In the same experiment an intracellular soluble (LDH) and an intracellular membrane-bound enzyme (galactosyltransferase) are only liberated at higher ET-12-H concentrations proving the intactness of the cells up to about 0.3 m*M* ET-12-H. ATPase, ADPase, and 5′-nucleotidase are released to about the same degree at each detergent concentration tested. The ATPase liberated in these experiments is identical with the enzyme on intact cells in its sensitivity to EDTA and its insensitivity to phosphatase substrates and ouabain. If, as was suggested (Ferber et al. 1980), this detergent in the sublytic range selects special membrane areas preexisting in the undisturbed cell for shedding, the results given in Fig. 5 would suggest a close neighborhood of ATPase, ADPase, and 5′-nucleotidase on the cell surface. The experiment with 0.3 m*M* ET-12-H was conducted in a preparative scale and the sedimentable portion of the shed membrane material as well as a microsomal fraction of the remaining cells were obtained. When enrichment of the three enzymes was followed in these fractions, it was always higher than in a control experiment performed in the absence of detergent. However, although the figures for ATPase and ADPase run parallel enrichment of 5′-nucleotidase is always two- to three- fold higher, Thus, 5′-nucleotidase does not seem to be very closely coupled to the other enzymes.

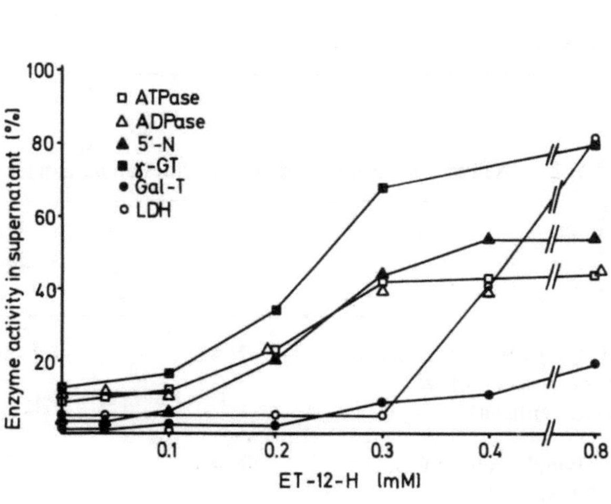

Fig. 5. Release of enzymes from BHG-83-1 cells induced by detergent. Cells were treated with the given concentrations of ET-12-H as described under Methology. Enzymes released into the supernatant are expressed as percent of total cellular content

6 Expression of the Ectoenzymes in Various Lymphoid Cell Types

Since in earlier studies wide variations in the expression of 5′-nucleotidase in different lymphoid cells were observed, it was interesting to see whether a coordinated pattern of expression of ATPase and 5′-nucleotidase could be found in various lymphoid cell types. These results are summarized in Tables 1, 4, and 5. High activities of both enzymes are exclusively found in the B-cell series and here generally a good correlation between the two enzymes is observed (r = 0.8). It should be noted that in this series the Burkitt lymphoma-derived lines are all characterized by low enzyme values. ATPase in all other cell lines, including the leukemic cells, is low or rather in the range found for normal peripheral blood lymphocytes (cf. Table 1). In all these cases, ATPase expression is independent of 5′-nucleotidase, which is well documented by cells with extremely high or absent activity of the latter enzyme (see Tables 4 and 5).

Table 4. Ecto-ATPase and -5′-nucleotidase activities on permanent human lymphoblastoid cell lines

Cell line	ATPase	5′-Nucleotidase	Cell line	ATPase	5′-Nucleotidase
	(pmol/s · 10^6 cells)			(pmol/s · 10^6 cells)	
B-cell lines			Pre-B-cell line		
Raji	22	0	Nalm 1	244	1.2
Daudi	20	0	Null-cell line		
BJAB	6	0	Reh	13	4.3
Just B2	600	2.0			
MD 52-1	822	2.1	T-cell lines		
MD 52-2	939	2.3	CCRF-CEM	69	8.8
W.P.	694	1.1	Molt-4	36	0.4
S.C.	456	2.1	RPMI-8402	20	0.1
B.J.	456	1.5	JM	11	0.2
Wi	11	0	Erythroblastoid cell line		
IDU-78	539	2.6	K 562	24	8.5
BHG-83-1	500	10.6			

Table 5. Ecto-ATPase and -5′-nucleotidase activities on human leukemic cells

Cell type		Enzyme activities (range) ATPase	5′-Nucleotidase
		(pmol/s · 10^6 cells)	
Acute lympohycytic leukemia (common type, positive for cALL-antigen)	N = 6	7.2–39.4	8.3–44.2
Acute lymphocytic leukemia (T-cell type)	N = 6	7.5–33.6	0

7 Methodology

Enzyme assays: ATPase, ADPase, 5′-nucleotidase, and nucleosidediphosphate kinase were all tested in a final volume auf 100 μl. Final concentrations for the standard assay were: Tris-HCl, pH 7.4, 50 m*M*; $MgCl_2$ 5 m*M*; and (2-[^{3}H])ATP (2.5 mCi mmol^{-1}) 1 m*M*; (2-[^{3}H])ADP (2.5 mCi • mmol^{-1}) 1 m*M*; (2-[^{3}H])AMP (6.25 mCi mmol^{-1}) 0.4 m*M* for ATPase, ADPase, and 5′-nucleotidase, respectively. For nucleosidediphosphate kinase concentrations were (2-[^{3}H])ADP (125 mCi • mmol^{-1}) 1m*M* and unlabeled ATP 1 m*M*. When tests under isotonic conditions were performed all ingredients including the cell sample were in 0.9% NaCl. Standard incubation times at 37°C were 15 min for ATPase and ADPase, 30 min for 5′-nucleotidase, and 10 min for nucleosidediphosphate kinase. Reactions were stopped by addition of 5μl of 4.2 *M* perchloric acid and neutralization by 5 μl of 4.45 N KOH in nonisotonic assays and of 5 μl of 0.1 *M* EDTA in isotonic assays. Samples were centrifuged and a 20 μl aliquot of the supernatant was applied to PEI-cellulose thin-layer sheets together with unlabeled carrier substances. Plates were developed first in 0.1 *M* acetic acid, pH 3.5, then dried and redeveloped in 3 *M* ammonium acetate, pH 6.7. The following compounds could be separated: ATP (R_f 0.15), ADP (0.31), AMP (0.45), adenosine (0.65), inosine (0.81). Spots were localized under UV-light, cut out, immersed in scintillation fluid and counted. Alternatively, a standard assay for 5′-nucleotidase was performed as previously described (Gutensohn et al. 1980). Galactosyltransferase (Gutensohn et al. 1983b) and lactate dehydrogenase (Bergmayer 1974) were tested by published procedures. γ-Glutamyltransferase was assayed following the method given by the supplier of a test kit (Boehringer). Protein was determined by the Lowry procedure.

Lymphoblastoid cell lines: T-cell lines: CCRF-CEM[3], JM, Molt 4, RPMI-8402. Pre-B-cell line: Nalm 1. B-cell lines (derived from Burkitt lymphoma): BJAB, Daudi Raji. B-cell lines (Epstein-Barrvirus-infected PBL): Just B2, MD-52-1, MD-52-2, W.P., S.C., B.J., Wi, IDU-78, BHG-83-1 (the origin of the latter two lines is unknown; they are, however, characterized as human B-cell lines by their surface expression of HLA, Ia-antigens, and immunoglobulins). Null-cell line: Reh. Erythroblastoid cell line: K 562. Cell lines were either kept and grown in our own laboratory or obtained from other sources (Dr. U. Kummer, Institut für Hämatologie der GSF, Dr. J. Johnson, Institut für Immunologie, Munich). Cells were grown in RPMI 1640 medium containing 10% fetal calf serum in the absence of antibiotics either in stationary (plastic flasks) or rolling (glass bottles) suspension culture. Peripheral blood lympocytes (PBL) of normal donors were separated from whole blood over Ficoll-Paque. Granulocytes were obtained from the same source using a Percoll step gradient as described by Hjorth et al. (1981). Leukemic cell samples were obtained from the laboratory of Dr. E. Thiel (Institut für Hämatologie der GSF) rethawed from stocks frozen in liquid nitrogen. Viability of all lymphoid cells was assessed by a trypan-blue exclusion test.

[3] Recent marker analyses have shed some doubt as to the true T-cell nature of this particular cell line grown in our laboratory

Cellfractionation: Microsomal fractions from homogenized cell suspensions were prepared by standard procedures. Further subfractionation of these membranes was achieved either on a two-step density gradient according to Touster et al. (1970) or on a five-step gradient following Bretz et al. (1980). Extraction of membrane proteins was achieved by incubation for 30 min at 37°C in 5 m*M* Tris-HCl, pH 7.4/0.5% Triton X 100 followed by centrifugation for 30 min at 180,000 x g_{max}. The supernatant was used as extract. Treatment of cells with sublytic concentrations of ET-12-H (1-Dodecyl-propanediol-3-phosphocholine) as adapted from Ferber et al. (1980) was performed at a cell concentration of 2.5×10^7 ml^{-1} in PBS with the detergent concentrations given in the experiments. The mixture was incubated for 10 min at 37°C with mild shaking, followed immediately by centrifugation at 400 x g for 10 min and removal of the supernatant. The sedimented cells were resuspended in PBS. After removal of aliquots the cells were used for a microsome preparation and the supernatant of the ET-12-H treatment was centrifuged for 90 min at 322,000 x g_{max} to sediment membrane particles.

8 Discussion

Permanent human lymphoblastoid cell lines can only serve as model systems. However, they represent defined stages of differentiation within certain lymphocyte subpopulations and for biochemical studies they are a material of practically unlimited availability and stability in the expression of various surface markers and enzyme activities.

Using one Be-cell line (BHG-83-1) the same set of ectoenzymes (ATPase, ADPase, 5′-nucleotidase, and nucleosidediphosphate kinase) could be characterized, which was previously shown on cultured vascular endothelial and smooth muscle cells (Pearson et al. 1980). The lymphoblast enzymes are very similar in their true ectoenzyme nature, their metal ion requirements, their insensitivity to certain inhibitors, and their rather broad substrate specificities.

In contrast to other tissues, where we expect rather uniform distributions of these enzymes, lymphoid cells show marked differences between subpopulations or cell lines. Since the physiological role of these ectoenzymes is still largely obscure, lymphoid cells may be a good model for their study. A question addressed here was whether the enzymes of such a set could act in a concerted fashion for the complete breakdown of extracellular ATP or ADP to adenosine, which would then be rescued by transport into the cell or transferred to closely associated adenosine receptors (Dornand et al. 1980, Bruns 1980). The experiments using sublytic concentrations of ET-12-H to induce membrane shedding (Fig. 5) indicate a possible close neighborhood of ATPase, ADPase, and 5′-nucleotidase within special membrane areas on the cell surface. However, some doubts as to the specificity of this shedding procedure remain: with lymphoblastoid cell lines the detergent concentration required to achieve measurable release to membrane-bound enzymes is much higher than that used originally by Ferber et al. (1980) for thymocytes. During our own studies (Gutensohn et al. 1983b, and this paper) the only clear-cut example of a plasma membrane protein not released under these conditions is the common ALL antigen. As mentioned, enrichment of 5′-nucleotidase in the shed membrane material is always higher than that of ATPase and ADPase. Moreover, when the shed membrane material is analyzed by SDS polyacrylamide gel electrophoresis, it does not show a selection of only a few membrane proteins, but has a composition more or less similar to that of whole plasmamembranes (Gutensohn et al. 1983b). This was also reported for thymocytes at lower ET-12-H concentrations (Ferber et al. 1980) where differences in protein pattern between shed membranes and plasma membranes are mostly quantitative.

Our data on ATPase activities in the lymphoblastoid cell lines are at variance in some details with a previous report (Fox et al. 1981), although the general picture of high ATPase in EBV-transformed B-cell lines and low enzyme levels in all other cells is confirmed. However, since these lower ATPase levels are also found in normal lymphocytes they should not be classified as a "deficiency" as has been proposed (Fox et al. 1981).

Another requirement for a concerted action of such an enzyme cascade would be a coordinated pattern of expression. Such a pattern could at best be postulated for the lymphoblastoid B-cell lines, but not for the other cells tested. In addition, looking at absolute activities per cell, ATPase and ADPase are always much more active than 5′-nucleotidase even in those cells where ATPase levels are low or normal compared to control PBL. So, most probably these enzymes have independent functions in addition. Whether the ATPase and ADPase are two activities of one enzyme or two separate enzymes cannot be decided on the basis for our data. The kinetic parameters (Table 3) would favor the latter possibility, however, as mentioned these figures may be misleading because of the presence of competing enzymes on the intact cells.

It is even more difficult to suggest a plausible role for the nucleosidediphosphate kinase which was described here and in other systems (Pearson et al. 1980). Whatever the importance of these ectoenzymes may be, their activity and widespread occurrence in various tissues suggests a major regulatory function. This function should be looked for in conjunction with processes, which under physiological conditions liberate nucleosidetriphosphates and -diphosphates into the extracellular space, for example, synaptic transmission or platelet aggregation.

9 Summary

Three plasma membrane-bound enzymes, ATPase, ADPase, and a nucleosidediphosphate kinase were studied on the surface of a permanent human lymphoblastoid B-cell line. They are characterized by various criteria as ectoenzymes. All three enzymes are dependent on Mg^{++}ions and show a rather broad substrate specificity. On the other hand, they can be clearly distinguished from other enzymes like unspecific phosphatases or Na/K-ATPase. Treatment of intact cells with sublytic concentrations of the detergent ET-12-H induces shedding of membrane components and liberates ATPase, ADPase, and ecto-5′-nucleotidase in a parallel fashion. This points towards a possible association of these enzymes in special membrane areas. ATPase and 5′-nucleotidase activites were determined in 19 different lymphoid cell lines, in human leukemic cells, and in normal lymphocytes and granulocytes. A good correlation in the expression of the two enzymes is found in the group of the B-cell lines, but this does not hold for T-cell lines or leukemic cells.

Acknowledgments. We thank Mrs. H. Jahn and D. Neuser for skillful technical assistance and Ms. A. Brandhofer for her diligent work with the cell lines. Thanks are also due to Dr. E. Thiel, Dr. U. Kummer, and Dr. J. Johnson for providing cell samples and to the Deutsche Forschungsgemeinschaft for financial support (Gu 123/13–6).

References

Bergmayer HU (1974) Methoden der enzymatischen Analyse, Vol 91. Verlag Chemie, Weinheim, pp 607–611

Bretz R, Bretz H, Palade GE (1980) Distribution of terminal glycosyltransferases in hepatic Golgi fractions. J Cell Biol 84:87–101

Bruns RF (1980) Adenosine receptor activation by adenine nucleotides requires conversion of the nucleotides to adenosine. Arch Pharm 315:5–13

DePierre JW, Karnovsky ML (1974) Ectoenzymes of the guinea pig polymorphonuclear leukocyte. Evidence for an ectoadenosine monophosphatase, -adenosine triphosphatase and -p-nitrophenyl phosphatase. J Biol Chem 249:7111–7120

Dornand J, Bonnafous JC, Mani JC (1980) 5′-Nucleotidase adenylate cyclase relationships in mouse thymocytes. A reevaluation of the effects of concanavalin A on cyclic AMP levels. FEBS Lett 110:30–34

Ferber E, Schmidt B, Weltzien HU (1980) Spontaneous and detergent-induced vesiculation of thymocyte plasma membranes. Biochim Biophys Acta 595:244–256

Fox RM, Piddington SK, Tripps EH, Tattersall MHN (1981) Ecto-adenosine triphosphatase deficiency in cultured human T and null leukemic lymphocytes. A biochemical basis for thymidine sensitivity. J Clin Invest 68:544–552

Gutensohn W, Thiel E (1981) High levels of 5′-nucleotidase activity in blastic chronic myelogenous leukemia with common ALL-antigen. Leu Res 5:505–510

Gutensohn W, Gürtler LG, Siegert W, Eichler E, Ernst S (1980) Ectoenzymes on the surface of cells from human lymphoblastoid lines. 5′-Nucleotidase and phosphatase. Blut 41: 411–420

Gutensohn W, Thiel E, Emmerich B (1983a) Evaluation of 5′-nucleotidase as biochemical marker in leukemias and lymphomas. Klin Wochenschr 61:57–62

Gutensohn W, Wünsch B, Rodt H (1983b) Independent expression of the surface markers 5′-nucleotidase and cALLA on leukemic cells. Blut 46:199–207

Hanford WC, Fishman WH (1983) Measurement of biosynthetic rates and intracellular transit times for a cell-surface membrane glycoprotein, alkaline phosphatase in HeLa cells. Anal Biochem 129:176–183

Hjorth R, Jonsson AK, Vretblad P (1981) A rapid method for purification of human granulocytes using Percoll. A comparison with dextran sedimentation. J Immunol Methods 43:95–101

Kummer U, Mysliwietz J, Gutensohn W, Buschette S, Jahn H, Neuser D, Munker R (1984) Development and properties of a monoclonal antibody specific for human ecto-5′-nucleotidase. Immunobiology 166:203–211

Pearson JD, Carleton JS, Gordon JL (1980) Metabolism of adenine nucleotides by ectoenzymes of vascular endothelial and smooth muscle cells in culture. Biochem J 190:421–429

Silber R, Conklyn M, Grusky G, Zucker-Franklin D (1975) Human lymphohcytes: 5′-Nucleotidase-positive and -negative subpopulations. J Clin Invest 56:1324–1327

Thompson LF, Saxon A, O'Connor RD, Fox RI (1983) Ecto-5'-nucleotidase activity in human T-cell subsets. Decreased numbers of ecto-5'-nucleotidase positive cells from both OKT 4^+ and OKT 8^+ cells in patients with hypogammaglobulinemia. J Clin Invest 71:892–899

Touster O, Aronson NN, Dulaney JT, Hendrickson H (1970) Isolation of rat liver plasma membranes. Use of pyrophsphatase and phosphodiesterase I as marker enzymes. J Cell Biol 47:604–618

Study of Ecto-5′-Nucleotidase, Ecto-ATPase, and Adenosine Deaminase Activities in Relation to Lymphoid Cell Differentiation

J. DORNAND, J. C. BONNAFOUS, A. GARTNER, J. FAVERO, and J. C. MANI [1]

1 Introduction

Normal lymphocytes have very low levels of de novo purine biosynthesis and are highly dependent on the purine salvage pathway. The discovery that several immunodeficiencies are linked to deficiencies in some of the enzymes of this metabolic pathway (Giblett et al. 1972) have focused attention on the previously unknown relationship between nucleotide metabolism and lymphocyte differentiation and function.

The enzyme 5′nucleotidase (E.C. 3.1.3.5.) which catalyses the dephosphorylation of nucleoside 5′monophosphates has been demonstrated both biochemically and cytochemically to be associated with the plasma membrane of many mammalian cell types. De Pierre and Karnovsky (1974) have defined 5′nucleotidase (5′N) in guinea pig leukocytes as an ectoenzyme, associated with the external face of the plasma membrane and capable of reacting with the substrate outside the cell. Such "ectoenzyme" characteristics have been demonstrated for this enzyme in human (Quagliata et al. 1974), pig (Dornand et al. 1978a), rabbit (Gregory and Kern 1979), rat (Barton and Goldschneider 1978) and mouse (Uusitalo and Karnovsky 1977) lymphocytes.

Interest in lymphocyte 5′nucleotidase was enhanced with the finding of a decreased activity in lymphocytes from patients with chronic lymphocytic leukemia (Quagliata et al. 1974) and in a great number of immune deficiency diseases (Johnson et al. 1977; Edwards et al. 1978). Since the first report of Quagliata, striking variations in the levels of 5′nucleotidase in different leukemic cells have been reported: low levels of 5′nucleotidase are found in cells from patients suffering from acute (BALL) or chronic (BCLL) lymphocytic leukemia, of hairy cell leukemia, and of the more primitive T acute lymphocytic leukemia (ThyALL), while high levels of 5′nucleotidase are expressed in common acute lymphocytic leukemia (CALL) (Reaman et al. 1981; Gutensohn and Thiel et al. 1981; Webster et al. 1984). In blast crisis of chronic granulocytic leukemia, myeloblasts show low 5′nucleotidase activity in contrast to the relatively high 5′nucleotidase activity expressed in lymphoblasts (Koya et al. 1981). It was concluded that 5′nucleotidase could not be considered as a specific marker for a leukemia or lymphoma subtype, but rather as a normal marker for certain stages of differentiation during lymphopoiesis. An arrest

[1] Laboratoire de Biochimie des Membranes, E.R.CNRS 228, E.N.S.C.M., 8, Rue de l'Ecole Normale, 34075 Montpellier Cedex, France

of B- and T-lymphocytes at an early level of differentiation is also the prevailing view proposed to explain the low 5′nucleotidase measured in patients with immune deficiency diseases like congenital X-linked agammaglobulinemia, adult variable hypogammaglobulinemia and severe immunodeficiency (Thompson et al. 1979; Webster et al. 1984). Recently, it was found that the 5′nucleotidase activity of T- and B-lymphocytes decreases with advancing age, suggesting that this enzymatic activity might be a biochemical marker of immune system function (Boss et al. 1980). All these hypothesis correlate with the low 5′nucleotidase activity of immature cord blood lymphocytes (Rowe et al. 1980) and of the circulating lymphocytes of infants before these cells acquire the functional capacity for IgG synthesis (Bastian et al. 1984).

The relative distribution of 5′nucleotidase among lymphocyte populations and subpopulations could provide information on the stages of lymphocyte differentiation which are affected by a deficiency in this enzyme and on the biochemical mechanism by which a genetically determined deficiency of 5′nucleotidase affects lymphocyte development and function. Nonuniform distribution of 5′-nucleotidase has been described among lymphoid cells (Gregory and Kern 1979; Barton and Goldschneider 1978; Uusitalo and Karnovsky 1977; Rowe et al. 1979), but these results are sometimes contradictory. Moreover, completely different distributions are reported for different animal species. We recently showed that some of the discrepancies apparent among these data could arise from the method of enzyme determination (Dornand et al. 1982a). The use of 5′AMP radiolabeled on the adenosine moiety is unsuitable for 5′nucleotidase measurement on intact cells since the derived adenosine enters the cells where it is instantaneously metabolized into adenine nucleotides (Dornand et al. 1979). Surprisingly, this substrate is often used, thus leading to greatly underestimated values of 5′nucleotidase activity, specially for lymphoblastoid cells which have low 5′nucleotidase levels but cell surfaces fivefold larger than lymphocytes (Dornand et al. 1982a) and which incorporate all the adenosine derived from 5′AMP. These data were recently confirmed by Chiba (Chiba et al. 1984).

In contrast with other systems (Carraway and Carraway 1976), intact lymphocytes display cooperativity for the concanavalin A inhibition of 5′nucleotidase (Dornand et al. 1977, 1978b) which suggests 2h uncoupling of the 5′-nucleotidase from the cytoskeleton, thus, reflecting an altered cytoskeleton interaction important in maintaining the lymphocyte in a nonproliferating state. Care must therefore be taken when comparing the 5′nucleotidase activities of lymphocyte subsets which have received different treatments during their isolation. The enzyme, which is not firmly anchored, can be easily detached from the cell surface. This phenomenon has been described for leukemic lymphocytes (Petitou et al. 1978). NH_4 Cl treatment, which is often used for erythrocyte lysis, partially releases 5′nucleotidase from lymphocyte plasma membranes (Dornand et al. unpublished results). For this reason, the 5′nucleotidase activities which are frequently reported for human T-lymphocytes isolated by the rosetting technique with sheep erythrocytes (Rowe et al. 1979; Boss et al. 1980) are doubtful.

Another cause of discrepancies concerns the expression of 5′nucleotidase activity. Because of the different sizes of lympoid cell subpopulations, comparisons of 5′

nucleotidase specific activities expressed in nmol of substrate hydrolyzed per hour per number of cells are not reliable; this is particularly true when 5′nucleotidase activities of lymphoblasts are compared with those of lymphocytes. For that reason our results are expressed in units per mg of cell protein (Dornand et al. 1982a).

Low 5′nucleotidase levels are unambiguously reported in thymocytes from various species (Barton and Goldschneider 1978; Gregory and Kern 1979; Uusitalo and Karnovsky 1977). Like other workers we found that thymocyte 5′nucleotidase is six- to ten fold lower than that of splenocytes in mice (Dornand et al. 1981) or than that of circulating lymphocytes in humans (Dornand et al. 1982b). We confirmed the hypothesis that this low 5′nucleotidase may reflect the fact that the cortical cell subpopulation which is immunoincompetent and represents 80–90% of total thymocytes, displays very low 5′nucleotidase activity. This hypothesis conflicted with previously published data (Barton and Goldschneider 1978) which reported that rat thymocyte subpopulations express similar 5′nucleotidase activities.

Cultured lymphoblastoid cell lines are often used as models for lymphocyte subpopulations. Depending on their origin and on the presence of certain surface markers, they can be classified as B, T, or null lymphoblasts. Such lympoblastoid cell lines have been used in some of the investigations on the immunodeficiencies (Wortmann et al. 1979). One must consider, however, that it is difficult to compare established cell lines and normal cells. It would appear better to compare specific markers among the various lymphoblastoid cells. We have measured 5′nucleotidase of such cell lines displaying either mature or immature properties.

Adenosine deaminase is an intracellular enzyme leading to inosine formation; its activity also differs among lymphoid cell populations and varies during T cell differentiation (Barton et al. 1979). High adenosine deaminase activities are reported in patients with B-ALL (Reaman et al. 1981) or T-ALL (Ganeshaguru et al. 1981). The inherited deficiency of adenosine deaminase is associated with a significant proportion of patients with severe combined immunodeficiency, an inherited defect in both T and B cell development (Giblett et al. 1972). We also measured the adenosine deaminase activities of various lymphocyte subpopulations and of different lymphoblastoid cells. An inverse relationship appeared between 5′nucleotidase and adenosine deaminase activities of lymphoid cells and a good correlation was found between the ratio of these enzymatic activities and some characteristics currently used to define the maturation stages of these cells.

Several years ago, we showed the presence on the lymphocyte plasma membranes of an ecto-ATPase activated by either Ca^{++} or Mg^{++} ions (Dornand et al. 1974). Although the physiological role of this enzyme is still poorly understood, it is now admitted that ATPase, like 5′nucleotidase, appears to be involved in a cascade of nucleotide hydrolyzing enzymes leading to adenosine. Comparison of the various cell lines and different lymphocyte subpopulations showed a good correlation of ATPase and 5′nucleotidase expression in the lympoid cells. Since exogenous adenosine appears to be correlated with lymphocyte proliferation (Bonnafous et al. 1979a) and differentiation (Astaldi et al. 1981), the local concentration of these ectoenzymes might be an important factor affecting the biochemical events involved in these processes through adenosine production. We have previously reported that lymphocyte 5′nucleotidase is involved in the uptake of adenosine derived from AMP (Dornand et al. 1979) and in the regulation of the intracellular cAMP level (Dornand et al. 1980b) through external membrane-bound adenosine receptor sites (Bonnafous et al. 1979b).

2 Lymphocyte 5'Nucleotidase and Adenosine Deaminase Activities

Measurements of 5'nucleotidase and adenosine deaminase activities in mouse lymphoid cells reveal that thymocytes have the lowest 5'nucleotidase and the highest adenosine deaminase activities, whereas low adenosine deaminase levels and high 5'nucletidase levels were found in splenocytes. Lymph node lymphocytes and bone marrow cells display intermediary values.

In order to determine whether the differences in 5'nucleotidase and adenosine deaminase levels resulted from differences in the percentage of T and B cells in the different organs, we separated T and B splenocytes by agglutination techniques (Dornand et al. 1982a). Wheat germ agglutinin (WGA)- or soybean agglutinin (SBA)-agglutinated cell fractions, which represent a B cell-enriched population, display higher adenosine deaminae and lower 5'nucleotidase activities than nonagglutinated cells (T cell-enriched population. These differences, however, never exceeded 50% so that the percentages of T and B cells among the lymphocyte populations did not explain the differences of 5'nucleotidase and adenosine deaminase levels (Table 1).

Table 1. Ecto-5'nucleotidase (5'N) adenosine deaminase (ADA) and ecto-ATPase of mouse and human lymphocyte subsets[a]

Cell population	5'N nmol h^{-1} mg^{-1}	ADA nmol mn^{-1} mg^{-1}	5'N/ADA	Ecto-ATPase nmol h^{-} mg^{-1}
Mouse unseparated thymocytes	20– 30	280–400	0.07 ± 0.01	360
Mouse cortical thymocytes	8– 25	350–500	0.05 ± 0.01	180
Mouse medullary thymocytes	130–210	180–250	1.2 ± 0.3	680
Mouse lymph node lymphocytes	80–110	80–120	1.0 ± 0.1	1450
Mouse bone marrow lymphocytes	180–220	25– 35	3.3 ± 0.9	ND
Mouse T cell precursors	40– 52	300–320	0.08 ± 0.01	210
Mouse unseparated splenocytes	250–350	40– 55	6.3 ± 0.1	1350
Mouse T enriched splenocytes	350–420	45– 65	5.0 ± 0.5	1380
Mouse B enriched splenocytes	210–320	40– 50	4.1 ± 0.8	1200
Human unseparated thymocytes	64– 72	ND	ND	680
Human cortical thymocytes	40– 60	450–490	0.08 ± 0.01	ND
Human medullary thymocytes	204–300	200–210	1.4 ± 0.2	ND
Human blood lymphocytes	270–320	53– 82	3.9 ± 0.8	1650
Human tonsil lymphocytes	230–280	80–110	2.5 ± 0.1	1700

[a] 5'N and ecto-ATPase were measured on intact cells, ADA on permeabilized cells as previously mentioned (Dornand et al. 1974, 1978a, 1982a)

The separation of mouse thymocytes into cortical and medullary cell subpopulations was performed by a method based on the availability of binding sites for peanut agglutinin (PNA) (Reisner et al. 1976). The agglutinated PNA^+ cortical thymocytes represent the majority of the thymic cells (80–90%) and seem to be functionally immature: they do not respond to phytohemagglutinin A (PHA) and to allogeneic stimulation. They are characterized by high TL and Thy-l and low H_2D antigens and by a high activity of terminal deoxynucleotidyl transferase (Tdt). In contrast, the nonagglutinated PNA^- cells which represent only 5–10% of the total thymocyte population are functionally mature. They respond to PHA and display membranal and functional properties similar to those of peripheral lymphocytes, i.e., low TL and Thy-1, high H_{2D} antigens, and low Tdt activity. (Reisner et al. 1976; Roelants et al. 1979). 5′Nucleotidase activity was ten fold higher in PNA^- (200 nmol $h^{-1} mg^{-1}$) than in PNA^+ (15 nmol h^{-1} mg^{-1}) cells, while adenosine deaminase activity was two- to three fold higher in PNA^+ than in PNA^- cells. PNA^- thymocytes have one-half as much 5′nucleotidase and twice as much adenosine deaminase activities as T splenocytes (Table 1).

T cell precursors isolated from mouse bone marrow cells display high Tdt activity (Pazmino et al. 1977). They showed low 5′nucleotidase and high adenosine deaminase activities when compared to more mature T or B cells. The human thymocytes can also be separated into mature and immature populations (Dornand et al. 1982b). Very low 5′nucleotidase levels were measured in immature thymocytes which exhibit the highest adenosine deaminase activity. The high level of 5′nucleotidase in medullary cells is similar to that of tonsil lymphocytes and slightly lower than that of circulating lymphocytes (Table 1).

Table 2. Correlation between 5′nucleotidase activity of mouse thymocytes and the relative number o PNA-positive and PNA-negative cells in thymus[a]

	Pregnant mice				Virgin mice	
	Pre-partum day	Days post-partum				
	20	1	20	30	Untreated	Hydrocortisone treated
Number PNA^- cells / Number PNA^+ cells	2 ± 1	6,2 ± 2	0,9 ± 0,5	0,10 ± 0,05	0,10 ± 0,05	8 ± 2
Thymocyte 5′N Activity nmol h^{-1} mg^{-1}	100–135	160–226	70–80	15–30	20–30	180–220

[a] PNA^+ and PNA^- thymocytes were isolated from thymus of pregnant nice (day 20 of gestation day 1 or 20, or 30 of the post-partum period) and from thymus of untreated or hydrocortisone-treate mice (Reisner et al. 1976). Their relative number was determined. 5′Nucleotidase activity of the un separated population was measured in parallel. As previously mentioned (Phuc et al. 1981), the numbe of PNA^+ cells was greatly reduced in the pre- and post-partum period, this involution being reversibl 30 days after delivery

Thymus can be depleted of immature cells by intraperitoneal injection of hydrocortisone acetate 48 h before sacrifice (Dornand et al. 1980a). Similarly, pregnancy-induced thymic atrophy observed in mice during the course of gestation and in the postpartum period was shown to be due to the reversible reduction in the number of cortical cells, while the pool of medullary cells appears unchanged (Phuc et al. 1981). Measurements of 5′nucleotidase activities of such thymocytes which were partially or completely depleted of cortical cells revealed that 5′nucleotidase levels increase with the percentage of mature cells (Table 2). This result rouled out the possibility that the low 5′nucleotidase level of PNA^+ thymocytes could be due to the separation process. Moreover, the two steps of this process (incubation with PNA and incubation with 0.2 *M* galactose) do not affect the 5′nucleotidase activity of the membrane bound or purified enzyme.

3 5′Nucleotidase and Adenosine Deaminase Activities in Human Lymphoid Cell Lines

5′Nucleotidase activity has also been measured in several T, B, or null human lymphoblastoid cell lines. Low, but detectable 5′nucleotidase levels were found in T-ALL cells with the exception of $MOLT_3$ and HSB_2 cells which completely lack this activity (Table 3). These cells, which are considered to be immature cells (high Tdt levels), behave like cortical cells having also high adenosine deaminase activities. Epstein Barr virus (EBV)-induced lymphoblasts displayed higher 5′nucleotidase activity and lower adenosine deaminase activity close to that of normal lymphocytes. These cells are derived from normal mature B lymphocytes transformed by EBV; they display low Tdt activity like mature cells and their 5′nucleotidase and adenosine deaminase activities can be explained by this characteristic rather than by their B cell character. The pre-B cell line NALM-1 of immature origin has low 5′ nucleotidase and high adenosine deaminase activities like T-ALL cells. The two cell lines HSB_2 and CCRF-SB are derived from the same patient. The immature cell line HSB_2 completely lacks 5′nucleotidase activity, while CCRF-SB cells of mature origin have a relatively high 5′nucleotidase level; an inverse relationship was found between adenosine deaminase and 5′nucleotidase for these cell lines.

The non-T, non-B (null) leukemia cell lines REH, KM_3, $NALM_{16}$ all of which have a high Tdt activity also show low 5′nucleotidase and high adenosine deaminase levels, while K562 cells which express low Tdt activity have a very low 5′nucleotidase/adenosine deaminase ratio. A good correlation appears between a high Tdt activity and low 5′nucleotidase/adenosine deaminase ratio and vice versa. Since Tdt is one of the features of immature cells this might also apply to the 5′nucleotidase/adenosine deaminase ratio. The 5′nucleotidase activity or the 5′nucleotidase/adenosine deaminase ratio does not correlate with any other well-known phenotypic markers. High levels of 5′nucleotidase and common ALL antigen (CALLA) are coexpressed on leukemic blast cells in common acute lymphocytic leukemia (Gutensohn and Thiel 1981). However, an independent expression of these two surface markers was recently reported (Gutensohn et al. 1983). Our data confirm

Table 3. 5′Nucleotidase, adenosine deaminase, and ecto-ATPase, activities of human lymphoblas cell lines

Cell lines	Origin	5′N (nmol h^{-1} mg^{-1})	ADA (nmol min^{-1} mg^{-1})	5′N / ADA	TDT[a] ND	Ecto-ATPase (nmol h^{-1} mg^{-1})
ICHIKAWA	T-ALL[b]	10	200	0.05	ND	300
MOLT-3	T-ALL	0	803	0	H	
MOLT-4	T-ALL	11	740	0.015	H	320
H-SB2	T-ALL	0	360	0	H	220
CEM	T-ALL	22	150	0.15	H	350
1301	T-ALL	27	180	0.15	ND	
RPMI 8402	T-ALL	40	250	0.32	H	
JM	T-ALL	35	240	0.15	H	
JURKAT	T-ALL	50	200	0.25	ND	
REH	Null[c]-ALL	41	210	0.19	H	560
K562	Null-CML[d]	74	39	1.88	L	900
KM_3	Null-ALL	20	180	0.11	H	
NALM 16	Null-ALL	12	ND	ND	H	
RAJI	Burkitt Lymphoma	130	70	1.85	L	
DAUDI	Burkitt Lymphoma	10	25	0.40	L	
DROTINON	EBV-B-Lymphoblast	172	50	3.44	L	1800
TOM	EBV-B-Lymphoblast	121	70	1.73	L	2000
CHAPUIS	EBV-B-Lymphoblast	210	77	2.72	L	2200
REEP	EBV-B-Lymphoblast	157	52	3.0	L	
CCRF-SB	B Cell[e]	110	35	3.1	L	2100
NALM 1	B CML BP[f]	14	300	0.04	H	

[a] Terminal deoxynucleotidyl transferase level: H(high) > 10 U mg^{-1} cell protein or L (low) < 2 U

[b] All acute lymphoblastic leukemia

[c] Null, non-T non–B

[d] CML, chronic myelocytic leukemia

[e] CCRF-SB appears as normal B cells from T-ALL patients

[f] BP blast phasis

(Methods of enzyme determination are given in Dornand et al. 1982a)

these results: the cell lines $NALM_1$, CEM, RPMI 8402, KM_3, REH, $NALM_{16}$ which are CALLA positive express a low 5′nucleotidase activity. It was already mentioned that the cell lines CEM (Gutensohn et al. 1980) and $NALM_1$ (Sun et al. 1982) had no detectable 5′nucleotidase activity.

4 Properties and Immunofluorescent Determination of Cell Surface 5′Nucleotidase

Although variations in 5′nucleotidase activity of lymphoblastoid cell populations have been reported, no studies to date allow determination of whether the differences are due to differences in enzyme amount or to the effect of activity-modifying factors.

A comparison of the substrate preference, effects of inhibitors, pH optimum, and subcellular localization revealed no differences between enzymes from mature and immature mouse lymphocytes or enzymes from human lymphocytes and leukemic cell lines. The K_m values for 5′nucleotidase, measured on intact cells with 5′AMP as substrate ranged between 40–60 *μM* for peripheral blood lymphocytes and tonsil lymphocytes, 40–50 *μM* for medullary thymocytes, 45–65 *μM* for null cells or T-ALL lymphoblasts, 70–100 *μM* for EBV lymphoblasts or pre-B cells. All these values are quite similar to the K_m value of the cortical thymocytes (50–70 *μM*) suggesting that the mature cells have more 5′Nucleotidase activity due to the presence of more enzyme rather than to the development of a more active molecule.

The IgG fraction of a rabbit anti-pig lymphocyte 5′nucleotidase serum cross-reacts with the enzyme of human lymphocytes. After elimination of the nonspecific determinants, it specifically inhibits the 5′nucleotidase and appears to be a good reagent for the immunofluorescent determination of 5′nucleotidase on human cell surfaces (Dornand et al. 1984a). We previously showed that CCRF-SB cells bound the antiserum on their surface, whereas HSB_2 cells did not. Additional results on other lymphoid cells confirm that a good correlation exists between the 5′nucleotidase activity and the binding of anti-5′nucleotidase IgG to the cell surface ($r = 0.72$). A good correlation between the absence of anti-5′nucleotidase activity is consistent with the hypothesis that the low 5′nucleotidase of immature cells is due to the absence of enzyme molecule rather than to an inhibited state of the enzyme, even if it is not a real proof of this absence.

5 The 5′Nucleotidase Proteic Inhibitor in Lymphoblasts: An Artifact

Recently, it was reported that the low 5′nucleotidase level in human leukemic cells resulted from the presence of a proteic 5′nucleotidase inhibitor in these cells (Sun et al. 1983). Since we did not observe such an inhibitor (Dornand et al. 1982a) we have repeated the experiments of these authors and have compared our results with theirs. Our conclusion was that the use of [^{3}H]-5′AMP as substrate and an inadequate analysis of the products formed have led to a misinterpretation. In the experimental procedure, increasing concentrations of a lymphoblastoid cell cytosol added to a 5′nucleotidase preparation, decreased its enzymatic activity. In fact, the [^{3}H]-adenosine derived from [^{3}H]-5′AMP was transformed into [^{3}H]-inosine, by the adenosine deaminase of the leukemic cells and we could show that inosine in contrast to adenosine, coprecipitated with the excess substrate after addition of $BaSO_4$, following Avruch's method. Therefore, it was not taken into account in the 5′nucleotidase determination and an apparent inhibition was noticed. This led these authors to confuse the adenosine deaminase of the leukemic cells with a proteic 5′nucleotidase inhibitor (Dornand et al. 1984b). Such an inhibitor is not observed when a more convenient experimental procedure was used for 5′nucleodidase determination.

6 Induction of Ecto-5′Nuleotidase in Cortical Thymocytes

Previous studies have shown that thymosin (fractions 5) and interleukin 2 (TCGF) were able to induce the maturation of cortical thymocytes. After incubation with TCGF, the changes in surface antigens of PNA^+ thymocytes correlate with their normal maturation (i.e., H_2D increases and TL decreases) (Chen et al. 1983). We simultaneously observed an increase of the 5′nucleotidase activity of these cells depending on the TCGF preparation used in the assay; TCGF by itself had no effect on the 5′nucleotidase of the PNA^- thymocytes (Table 4). These data correlate with those of Cohen et al. (1981) who reported an increase of 5′nucleotidase levels during T cell maturation by factors derived from thymic epithelial cells and postulated that only one subset of the thymic T cells was sensitive to the maturation effects. We can conclude that the elevation of 5′nucleotidase is a marker of cell maturation. The corollary of this observation may be that the subnormal levels of 5′nucleotidase demonstrated in lymphocyte subpopulations, in lymphoblasts, or in pathological lymphocytes may indicate an abnormal or incomplete maturation of these cells.

Table 4. 5′Nucleotidase activity of mouse thymocytes treated with interleukin-2 for 24 h[a]

Thymocytes	TCGF Dilution	5′N nmol h^{-1} mg^{-1}
PNA^+	0	18.6 ± 12
PNA^-	0	190 ± 20
PNA^+	(EL_4) 1/10	90 ± 8
PNA^+	(H) 1/20	80 ± 15
PNA^+	(H) 1/10	46 ± 5
PNA^-	(EL_4) 1/10	200 ± 10

[a] 10^7 cortical (PNA^+) or medullary (PNA^-) mouse thymocytes were incubated for 24 h at 37°C in 2 ml RPMI 1640–5% FCS, with or without different dilutions of partially purified preparations of interleukin (TCGF), gift from Dr. Dupuy D'Angeac, INSERM U236 Montpellier, France. After three successive washings, the 5′nucleotidase activity of the cells was measured.
TCGF (EL_4) was prepared from the murine lymphoma EL_4; TCGF (H) from human circulating lymphocytes

7 Nucleotide-Hydrolyzing Ectoenzymes of Lymphoid Cells

It is now well established that ecto-ATPase and ecto-ADPase, which are involved, with the 5′nucleotidase in a chain of nucleotide hydrolyzing enzymes are not unspecific phosphatases (Dornand et al. 1974; De Pierre and Karnovsky 1974; Smith et al. 1981). A nonuniform distribution of the ecto-ATPase in different lymphoblastoid cells has been already reported (Fox et al. 1981). High levels of this enzyme were found in EBV lymphoblasts, while malignant T and null cell lines were deficient in ecto-ATPase. The T, B, or null characters of these cells were claimed to determine the differences. We found similar results (Tables 1 and 2). However, when we compared several lymphoid cells, we found a good correlation between ecto-ATPase and 5′nucleotidase expressions in mouse lymphocyte subpopulations (Table 1, $r = 0.75$ for medullary and cortical thymocytes, T or B splenocytes, node and bone marrow lymphocytes) and also in human cells [$r = 0.85$ for tonsil and PBL cells, thymocytes subpopulations (Table 1) and lymphoblastoid cell lines (Table 2)]. The ecto-ADPase activity was only measured in mouse thymocytes: cortical cells express an activity (65 nmol h^{-1} mg^{-1}) which is threefold lower than medullary cells (195 nmol h^{-1} mg^{-1}). These results seem to established that the ecto-ATPase, and probably the ecto-ADPase are lower in immature than in mature cells, independently of their T, B, or null character. In fact, through the expression of each ectoenzyme it is the activity of the nucleotide-degrading chain which appears to be lowered in immature cells. Whether or not this chain can be considered as the real marker of cell immaturity requires further investigation.

8 Nucleotide-Hydrolyzing Ectoenzyme Functions in Lymphoid Cells

Lymphocytes are impermeable to nucleotides; adenosine uptake from 5′AMP necessitates a previous cleavage of the nucleotide. By comparing the uptake of adenosine resulting from 5′AMP hydrolysis with the direct incorporation of free adenosine, we established that 5′nucleotidase facilitated the nucleoside uptake, probably because the concentration of adenosine which was produced near the cell surface was higher than the bulk concentration (Dornand et al. 1979). When 5′AMP was added to mouse thymocytes, medullary cells incorporated three times more adenosine than cortical cells, while they hydrolyzed ten times more 5′AMP (Fig. 1A). If the adenosine uptake depended upon the concentration of 5′nucleotidase molecules on the cell surface, the number of adenosine carriers was one limiting step of this uptake (Dornand et al. 1982a). Inosine and hypoxanthine synthesized in the cells are found in cell supernatant with adenosine, showing that adenosine uptake did not represent adenosine transport, but rather the capacity of the cells to incorporate and metabolize adenosine: 95% of the incorporated adenosine was metabolized into nucleotides, the free cytoplasmic adenosine representing less than 2% of the metabolized adenosine. On the other hand, cortical cells incorporated all the adenosine they produced into nucleotides; the presence of inosine, hypoxanthine and adenosine could not be demonstrated in cell supernatant (Fig. 1B). This result

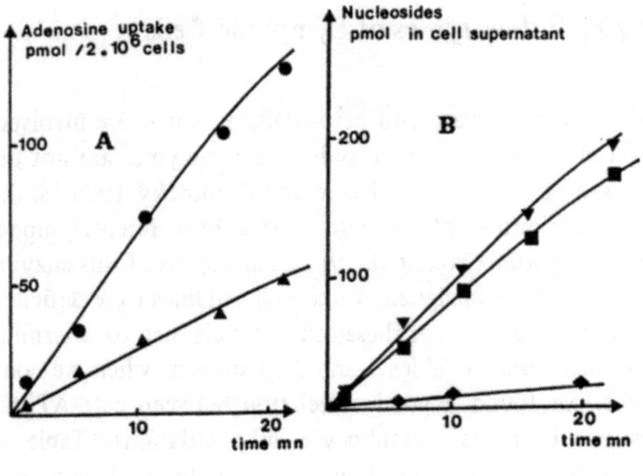

Fig. 1A, B. Time course of adenosine uptake from 5′AMP by mouse lymphocytes. **A** 2.10^6 cortical (▲–▲) or (●–●) medullary thymocytes were incubated at 37° in 200 μl RPMI1640, with 20 *μM* [^{3}H]5′AMP for different periods of time. Incorporated [^{3}H]-adenosine was then measured after cell filtration through an oil cushion, and in some experiments the relative amounts of different metabolites was determined (Dornand et al. 1979). In parallel experiments, (1) 5′AMP hydrolysis was measured with [^{32}P]5′AMP as substrate (medullary cells hydrolyzed 560 pmol substrate in 20 min; cortical cells 60 pmol in the same time.) (data not shown); (2) **B** the presence of nucleosides (adenosine▼—▼, inosine+hypoxanthine ■–■) was analyzed in medullary cell supernatant or in cortical cell supernatant (adenosine ◆–◆, we could not detect other nucleosides)

is explain by the low K_m of adenosine kinase (0.5 *μM*): the amount of adenosine which is produced and incorporated is not sufficient to saturate this enzyme. No difference in adenosine uptake and metabolism were noticed when free adenosine was added to cortical and medullary mouse thymocytes (data not shown), probably because these cells have the same number of adenosine carriers on their surface, and because the adenosine cytoplasmic metabolizing enzymes adenosine kinase (K_m = 0.5 *μM*, V_m = 120 nmol h^{-1} mg^{-1}), and S-adenosyl homocysteine hydrolase (K_m = 12*μM*, V_m = 56 nmol h^{-1} mg^{-1}) were found to be identical in both subsets; although differences were noticed in adenosine deaminase activity, this enzyme displayed the same K_m in both subpopulations (50 *μM*).

The occurrence of external adenosine receptor sites coupled to adenylate cyclase in lymphocytes was first established for thymocytes and splenocytes (Bonnafous et al. 1979b); more recently it was shown that all lymphocyte classes, including mature and immature thymocytes, contain adenosine receptor sites of the R_a type coupled to adenylate cyclase and inducing the stimulation of this enzyme (Bonnafous et al. 1982). Nucleotide-hydrolyzing ectoenzymes which differentialy control adenosine levels could play an important role in vivo in regulating cAMP levels of lymphocyte subpopulations. The cAMP content of medullary cells increased ten fold when these cells were incubated for 15 min with 20 *μM* 5′AMP, while under the same conditions, the cAMP level of the cortical cells was practically unchanged. The cAMP levels of both subsets were increased when the cells were incubated with adenosine (Bonnafous et al. 1983). This result is easily explained by adenosine accumulation, resulting from 5'AMP hydrolysis, in medullary cell supernatants (Fig. 1). It could reflect the situation occurring in vivo in thymus where a great number of cells die and where the nucleotide concentrations are certainly high. Increases in cAMP content have been shown to occur under the influence of hormonal factors during induction of cortical thymocyte maturation (references in: Bonnafous et al. 1983). Low nucleotide-hydrolyzing ectoenzyme activities, coupled with low adenosine deaminase activity might be required to maintain low cAMP levels in cells at an early stage of differentiation.

9 Summary

Ecto-5'nucleotidase levels are linked to the differentiation stage of lymphoid cells. An increasing activity of the enzyme is observed as lymphocytes mature within the thymus. A low activity, probably due to a deficiency in the number of enzyme molecules, is found in immature lymphoblasts.

A good correlation is found between ecto-5'nucleotidase and ecto-ATPase activities, while an inverse relationship is observed between ecto-5'nucleotidase and adenosine deaminase activities of lymphoid cells. These enzymatic activities can be considered as markers of cell maturity. If we consider that PNA receptors, high TdT activity, and immunoincompetence are features of immature cells, it might be the same for low values of the ratio of 5'nucleotidase activity to adenosine deaminase activity.

It is unclear whether the low 5'nucleotidase activity reflects a failure of lymphocyte maturity or is a cause of immaturity. This question is unlikely to be resolved until the physiological function of ecto-5'nucleotidase is understood.

References

Astaldi AA, Leupers CJM, Schellekens PTA (1981) Adenosine induces immunological maturation of thymocytes. Immunobiology. 159:93

Barton RW, Goldschneider I (1978) 5'nucleotidase activity on subpopulations of rat lymphocytes. J Immunol 121:2329–2334

Barton RW, Martiniuk F, Hirschorn R, Goldschneider I (1979) The distribution of adenosine deaminase among lymphocyte populations in the rat. J Immunol 122:216–220

Bastian JF, Ruedi JM, McPherson GA, Colembesky HE, O'Connor RD, Thompson LF (1984) Lymphocyte ecto 5'nucleotidase in infancy: increasing activity in peripheral blood B cells precedes their ability to synthesize IgG in vitro. J Immunol 132:1767–1772

Bonnafous JC, Dornand J, Mani JC (1979a) Adenosine induced cAMP increase in pig lymphocytes is not related to adenylate cyclase stimulation. Biochim Biophys Acta 587:180–191

Bonnafous JC, Dornand J, Mani JC (1979b) Hormone like action of adenosine in mouse thymocytes and splenocytes. Evidence for the existence of membrane adenosine coupled to adenylate cyclase. FEBS Lett 107:95–99

Bonnafous JC, Dornand J, Favero J, Mani JC (1982) Lymphocyte membrane adenosine receptors coupled to adenylate cyclase. Properties and occurence in various lymphocyte subclasses. J Recept Res 2:347–366

Bonnafous JC, Dornand J, Favero J, Mani JC (1983) Role of adenosine in lymphocyte maturation: possible involvement of adenosine receptors coupled to adenylate cyclase. In: Parker JW, O'Brien RL (eds) Intercellular communication in leucocyte function. Wiley, New York, pp 437–440

Boss GB, Thompson LF, Spieglberg HL, Pichler WJ, Seegmiller JE (1980) Age dependency of lymphocyte 5′nucleotidase. J Immunol 125:679–682

Carraway C, Carraway KL (1976) ConcanavalinA perturbation of membrane enzymes of mammary gland. J Supramol Struct 4:121–126

Chen SS, Tung JJ, Gillis S, Good RA, Hadden JW (1983) Changes in surface antigens of immature thymocytes under the influence of T cell growth factor and thymic factors. Proc Natl Acad Sci USA 80:5980–5984

Chiba P, Kraupp M, Rumpold H, Muller MM (1984) Uptake of AMP into K562 and HL–60 leukemic cell lines. In: DeBruyn CHMM, Simmonds HA, Muller MM (eds) Advances in experimental medicine and biology, vol 165B. Plenum, New York, pp 273–278

Cohen A, Dosh HM, Gelfand EW (1981) Induction of ecto 5′nucleotidase activity in human thymocytes. Clin Immunol Immunopathol 18:287–290

DePierre JW, Karnovsky ML (1974) Ectoenzymes of the guinea pig polynuclear leukocyte I. Evidence for an ecto adenosine monophosphatase, adenosine triphosphatase, and p-nitrophenylphosphatase. J Biol Chem 249:7111–7120

Dornand J, Mani JC, Mousseron-Canet M, Pau B (1974) Propriétés d'une ATPase Ca^{+} ou Mg^{2+} dépendante des membranes plasmiques de lymphocytes. Effet de la concanavalineA sur les ATPases membranaires. Biochimiè (Paris) 56:851–857

Dornand J, Reminiac C, Mani JC (1977) Activité 5′AMPase des membranes plasmiques de lymphocytes. Effet de la concanavalineA. Biochimie (Paris) 59:425–432

Dornand J, Bonnafous JC, Mani JC (1978a) Purification and properties of 5′nucleotidase from lymphocyte plasma membranes. Eur J Biochem 87:459–465

Dornand J, Bonnafous JC, Mani JC (1978b) Effects of ConA and other lectins on pure 5′nucletotidase isolated from lymphocyte plasma membranes. Biochem Biophys Res Commun 82:685–692

Dornand J, Bonnafous JC, Gavach C, Mani JC (1979) 5′Nucleotidase-facilitated transport by mouse lymphocytes. Biochimie (Paris) 61:973–978

Dornand J, Bonnafous JC, Mani JC (1980a) 5′Nucleotidase of two populations of mouse thymocytes separated by PNA agglutination. FEBS Lett 118:225–228

Dornand J, Bonnafous JC, Mani JC (1980b) 5′nucleotidase-adenylate cyclase relationships in mouse thymocytes. A reevaluation of the effects of concanavalinA on cyclic AMP levels. FEBS Lett 110:30–34

Dornand J, Bonnafous JC, Favero J, Mani JC (1982a) Ecto-5′nucleotidase and adenosine deaminase activities of lymphoid cells. Biochem Med 28:144–156

Dornand J, Bonnafous JC, Favero J, Mani JC (1982b) Purine production in mature and immature human lymphocytes. FEBS Lett 142:88–92

Dornand J, Bonnafous JC, Favero J, Gartner A, Dauthuille I, Poncelet P, Carayon P, Pau B, Mani JC (1984a) An immunofluorescent technic for the determination of cell surface 5′nucleotidase. In: Rosenfeld C, Serrou B, Vaillet P (eds) Fluorescent technic and membrane markers in cancer and immunology: membrane dynamics, cellular characterization and cell sorting pp 349–354

Dornand J, Bonnafous JC, Favero J, Gartner A, Mani JC (1984b) The proposed 5′nucleotidase inhibitor in human cells is an artefact. Biochim Biophys Acta 804:398–403

Edwards NL, Magilavy DB, Cassidy JT, Fox IH (1978) Lymphocyte ecto-5′nucleotidase deficiency in agammaglobulinemia. Science (Wash DC) 201:628–630

Ganeshagunu K, Lee N, Llewellin P, Prentice HG, Hoffrand AV, Catovsky D, Habeshaw JA, Robinson J, Greaves MF (1981) Adenosine deaminase concentrations in leukemia and lymphoma: relation to cell phenotypes. Leuk Res 5:215–222

Giblett ER, Anderson JE, Cohen F, Pollara B, Meurvissen HJ (1972) Adenosine deaminase deficiency in two patients with severely impaired cellular immunity. Lancet 2:1067–1069

Gregory SH, Kern M (1979) The heterogeneous distribution of 5′nucleotidase among rabbit lymphocytes. J Immunol 123:1978–1982

Gutensohn W, Gurtler LG, Siegert W, Eichler E, Ernst S (1980) Ectoenzymes on the surface of cells from human lymphoblastoid lines: 5'-nucleotidase and phosphatase. Blut 41:411–420

Gutensohn W, Thiel E (1981) High levels of 5'-nucleotidase activity in blastic chronic myelogenous leukemia with common ALL-antigen. Leuk Res 5:505–510

Gutensohn W, Wunsch B, Rhodt H (1983) Independent expression of the surface markers 5' nucleotidase and CALLA on leukemic cells. Blut 46:119–207

Johnson SM, Noerth MS, Asherson GL, Allsop J, Watts RWE, Webster ADB (1977) Lymphocyte purine 5'nucleotidase deficiency in primary hypogammaglobulinemia. Lancet i:168–170

Koya M, Kanoth T, Sawada H, Uchina H, Ueda K (1981) Adenosine deaminase and ecto 5' nucleotidase activities in various leukemias with special reference to blast crisis of chronic myeloid leukemia. Blood 58:1107–1111

Pazmino N, McEwan R, Ihle J (1977) Distribution of terminal deoxynucleotidyl transferase in bovine serum albumin gradient fractionated thymocytes and bone marrow cells of normal and leukemic mice. J Immunol 119:494–499

Petitou M, Tuy F, Rosenfeld C, Mishal Z, Paintrand M, Jasmin C, Mathe G, Inbar M (1978) Decreased microviscosity of membrane lipid in leukemic cells: two possible mechanism. Proc Natl Acad Sci USA 75:2306–2310

Phuc LH, Papiernik M, Berrih S, Duval D (1981) Thymic involution in pregnant mice. Clin Exp Immunol 44:247–252

Quagliata F, Conklyn M, Sibler R (1974) Studies on the lymphocyte 5'nucleotidase in chronic lymphocyte leukemia, infectious mononucleosis, normal subpopulations, and phytohemagglutinin stimulated cells. Cancer Res 34:3197–3202

Reaman GH, Blatt J, Poplack DG (1981) Lymphoblast purine pathway enzymes in B cell acute lymphoblastic leukemia. Blood 58:330–332

Reisner Y, Linker Israeli M, Sharon N (1976) Separation of mouse thymocytes into the two subpopulations by the use of the peanut agglutinin. Cell Immunol 25:129–134

Roelants GE, London J, Major-Withey KS, Serrano B (1979) Peanut agglutinin II- Characterization of the Thy-I, Tla and Ig phenotypes of peanut agglutinin positive cells in adult, embryonic and nude mice using double immunofluorescence. Eur J Immunol 9:139–145

Rowe M, De Gast CG, Platts-Mills TAE, Asherson GL, Webster ADB, Johnson SM (1979) 5' nucleotidase of B and T lymphocytes isolated from human peripheral blood. Clin Exp Immunol 36:97–101

Rowe M, De Gast CG, Platts-Mills TAE, Asherson GL, Webster ADB, Johnson SM (1980) Lymphocyte 5'nucleotidase deficiency in primary hypogammaglobulinemia and cord blood. Clin Exp Immunol 39:337–343

Smith GP, Shah T, Webster ADB, Peters TJ (1981) Studies on the kinetic properties and subcellular localization of adenosine diphosphatase activity in human peripheral blood lymphocytes. Clin Exp Immunol 46:321–326

Sun AS, Holland JF, Ohnuma T, Slankard-Chahinian (1982) 5'nucleotidase activity in permanent human lymphoid cell lines. Biochim Biophys Acta 714:530–535

Sun AS, Holland JF, Lin K, Ohnuma T (1983) Implications of a 5'nucleotidase inhibitor in human leukemic cells for cellular aging and cancer. Biochim Biophys Acata 762:577–584

Thompson LF, Boss GR, Spiegelberg HL, Jansen IV, O'Connor RD, Waldmann TA, Hamburger RN, Seegmiller JE (1979) Ecto-5'nucleotidase activity in T and B lymphocytes from normal subjects and patients with congenital X-linked agammaglobulinemia. J Immunol 123: 2475–2478

Uusitalo RJ, Karnovsky MJ (1977) Surface localization of 5'nucleotidase on the mouse lymphocytes. J Histochem Cytochem 25:87–96

Webster ADB, Shah T, Peters TJ (1984) Lymphocyte ecto-5'nucleotidase in immunodeficiency and leukemia. In: De Bruyn CHMM, Simmonds HA, Muller MM (eds) Advances in experimental medicine and biology. vol 165 A. Plenum, New York, pp 67–72

Wortmann RL, Mitchell BS, Edwards NL, Fox IM (1979) Biochemical basis for differential deoxyadenosine toxicity to T and B lymphoblasts: Role for 5'nucleotidase. Proc Natl Acad Sci USA 76:2434–2437

Ectonucleotidases: 5′Nucleotidase

The Properties, Structure, Function, Intracellular Localisation and Movement of Hepatic 5′-Nucleotidase

J. P. LUZIO[1], E. M. BAILYES[1], M. BARON[1], K. SIDDLE[1], B. M. MULLOCK[2], H. J. GEUZE[3], and K. K. STANLEY[4]

1 Introduction

5′-Nucleotidase (EC 3.1.3.5) was first described 50 years ago (Reis 1934) and was subsequently shown to be widely distributed in animal tissues (Reis 1951) as an intrinsic plasma membrane enzyme (Essner et al. 1958; Segal and Brenner 1960; Song and Bodansky 1967). Its ease of assay and stability made it a convenient and popular marker for the plasma membrane in subcellular fractionation (Evans 1978) and its measurement has also been found useful over many years as an indicator of cholestasis when raised levels of 5′-nucleotidase activity are present in patients' sera (Ellis et al. 1978). The discovery that 5′-nucleotidase was an ectoenzyme in several cell types (De Pierre and Karnovsky 1974a, b; Trams and Lauter 1974; Newby et al. 1975) and the subsequent observation of its internalisation from the cell surface and re-cycling in the absence of protein synthesis (Stanley et al. 1980; Widnell et al. 1982) has led to more active investigation of its properties. This has been fuelled by a growing interest in its function which remains enigmatic (Stanley et al. 1982).

The present article reviews the properties, structure, function, intracellular localisation and movement of liver ecto-5′-nucleotidase and indicates some areas where development of our understanding of the enzyme will occur. In common with most other tissues liver contains two isoenzymes of 5′-nucleotidase, the intrinsic membrane enzyme and a soluble cytosolic form. The latter, which will be largely ignored in the remainder of this article, constitutes only a small proportion of the total cell activity and is a different gene product with different substrate specificity, kinetics and function (Itoh 1981; Newby 1980; Worku and Newby 1983).

[1] Department of Clinical Biochemistry, University of Cambridge, Addenbrooke's Hospital, Hills Road, Cambridge CB2 2QR, UK

[2] Robens Institute, University of Surrey, Guildford, Surrey, GU2 5XH, UK

[3] Centre for Electron Microscopy, Medical Faculty, University of Utrecht, 3511 HG Utrecht, The Netherlands

[4] European Molecular Biology Laboratory, Meyerhofstraße 1, 6900 Heidelberg, FRG

2 Assay of 5′-Nucleotidase

The usual substrate for the assay of 5′-nucleotidase is AMP which is converted to adenosine and inorganic phosphate. The assay is usually carried out within the pH range 7.4–8.5 in the presence of Mg^{2+}, and is monitored by measuring one of the following:

1) release of inorganic phosphate, P_i – a traditional method, particularly for subcellular fractionation studies (Widnell 1975; Evans 1978), though it is difficult to control for non-specific phosphatase activity. Trapping released P_i as lead phosphate has always been widely used as a cytochemical stain for the enzyme. However, problems of enzyme inactivation during fixation and inhibition of enzyme activity by lead salts have been experienced (Widnell 1972; El-Aaser et al. 1973; Uusitalo and Karnovsky 1977; Widnell and Little 1977).

2) release of [^{3}H]adenosine from [^{3}H]AMP – a sensitive assay in which [^{3}H]-adenosine is separated from substrate either by chromatography or more simply precipitating all phosphates by addition of $ZnSO_4$ and Ba $(OH)_2$ (Avruch and Wallach 1971; Newby et al. 1975). The assay is normally performed in the presence of an adenosine trap to prevent subsequent conversion of radioactive adenosine to hypoxanthine and other metabolites. In addition an excess of sodium β glycerophosphate may be added to prevent interference of the assay by other phosphatases (Stanley et al. 1980).

3) release of [^{32}P]Pi from [^{32}P]AMP with subsequent formation of [^{32}P] phosphomolybdate which is extracted into an organic solvent (Widnell 1975; Newby et al. 1975).

4) conversion of adenosine formed to inosine and NH_3 with adenosine deaminase – this type of enzyme-linked assay is particularly useful when measuring column fractions during purification (Bailyes et al. 1982) and for kinetic experiments. It may be followed either by measuring the reduction in A_{265} on forming inosine from adenosine (Ipata 1968) or by assaying the NH_3 formed. The NH_3 may be estimated chemically or by linking to NAD reduction with glutamate dehydrogenase. Enzyme-linked assay of 5′-nucleotidase using adenosine deaminase and glutamate dehydrogenase is popular for clinical samples (Arkestjn 1976).

3 Liver 5′-Nucleotidase as an Ectoenzyme

In many eukaryotic cells plasma membrane 5′-nucleotidase has been formally shown to be an ectoenzyme, i.e. an intrinsic membrane protein whose catalytic site occurs at the external face of the plasma membrane (Stanley et al. 1982). Although the term was devised by Engelhardt (1957) to describe the ATPase activity on the external surface of avian erythrocytes, the existence of ectoenzymes was not generally accepted until sufficient criteria were proposed and demonstrated. 5′-Nucleotidase was shown to be an ectoenzyme in human and murine cultured cell lines (Trams and Lauter 1974), guinea-pig polymorphonuclear leukocytes (De Pierre and Karnovsky 1974a, b) and rat adipocytes (Newby et al. 1975) by demonstrating that:

1) intact cells metabolized externally added, non-penetrating substrates,
2) non-penetrating reagents inhibited the enzyme activity of intact cells.

Localisation of liver 5′-nucleotidase by cytochemistry gave an early indication that the enzyme was an ectoenzyme since the lead phosphate deposit formed was most concentrated on the plasma membrane (at the bile canalicular front) and was observed on the outside of the bilayer (Essner et al. 1958; Farquhar et al. 1974). Similar data can be obtained using collagenase-isolated hepatocytes (Fig. 1). Stronger evidence was provided by the observation that both collagenase-isolated rat hepatocytes (Stanley et al. 1980) and cultured rat liver cell lines (Stanley et al. 1983)

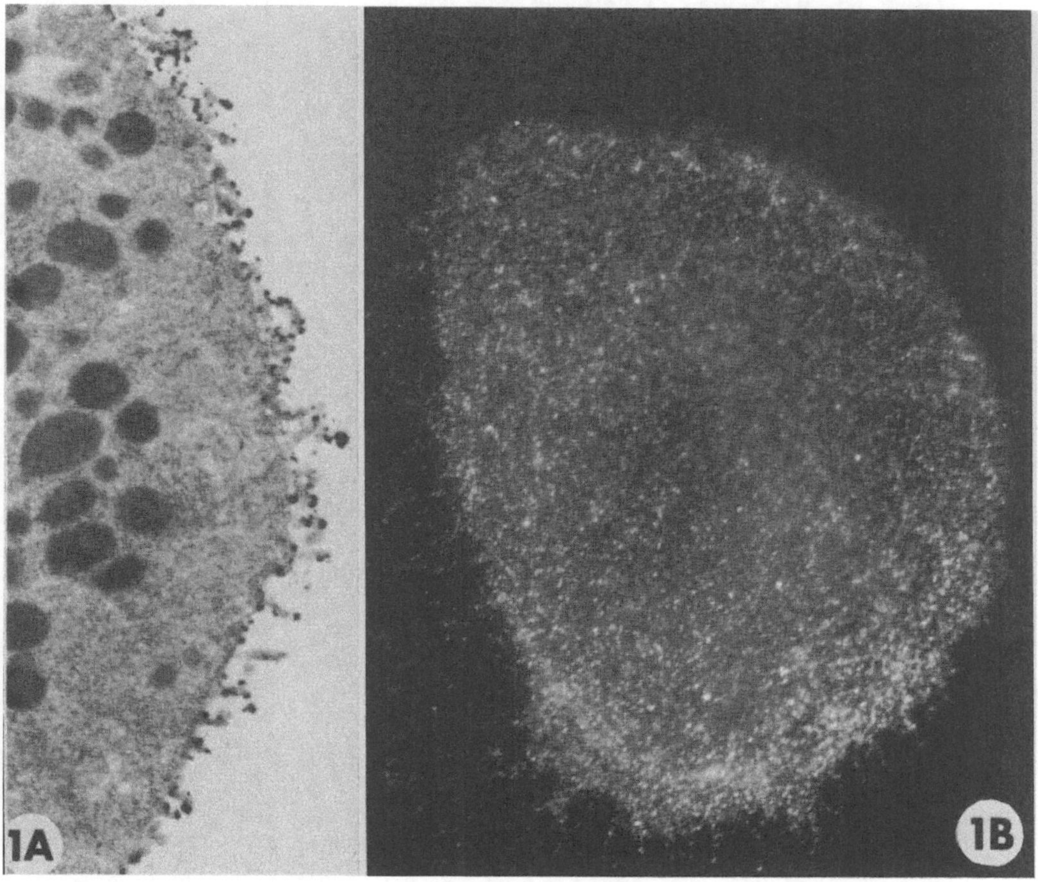

Fig. 1A, B. A Histochemical localisation of 5′-nucleotidase on the surface of collagenase isolated hepatocytes. Cell were fixed with 1% wt/vol gluteraldehyde, pH 6.0, 4°C, washed at pH 7.2 and incubated 15 min 37°C in the presence of 1 m*M* AMP and 1 m*M* Pb $(NO_3)_2$. They were washed and post-fixed in 2% wt/vol glutaraldehyde and 1% O_SO_4 before transmission electron microscopy. Gift from Prof. L. Herman. **B** Indirect immunofluorescence of 5′-nucleotidase on the surface of cultured C3 liver cells. Cells were fixed with formaldehyde treated, with monoclonal anti-5′-nucleotidase antibody 5N4-2 and subsequently with rhodamine-labelled sheep anti-mouse IgG. A punctate distribution of the enzyme is seen

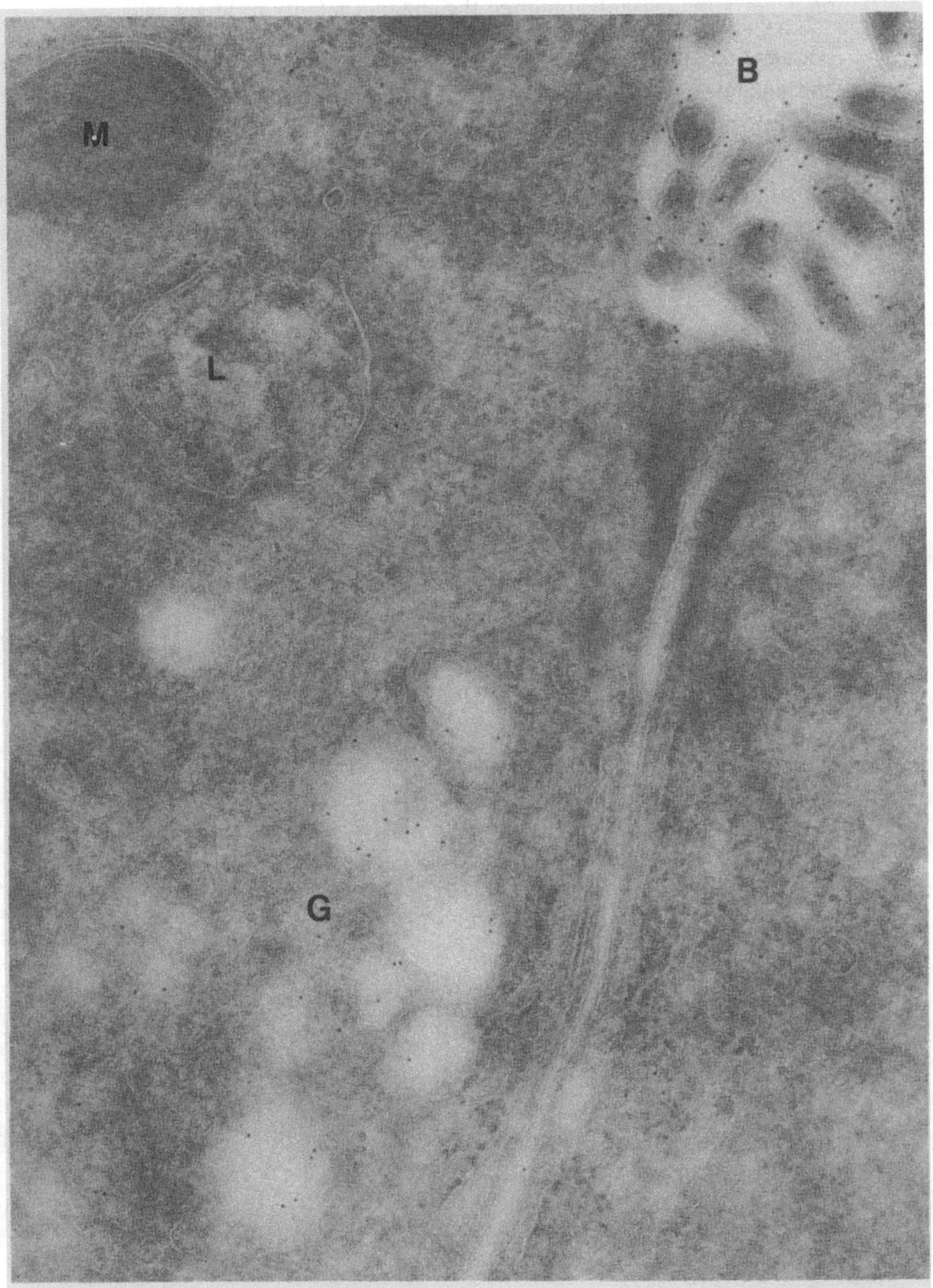

Fig. 2. Immunogold labelling of 5′-nucleotidase at the bile canalicular surface of hepatocytes in situ. Cryosection of rat liver treated with gold-labelled rabbit antibodies to immunoaffinity purified 5′-nucleotidase as described by Geuze et al. (1984a)

could hydrolyse exogenously added substrate and that surface activity was inhibited by the addition of polyclonal anti-5′-nucleotidase antibodies. Indirect immunofluorescence using a monoclonal first antibody on intact cultured rat liver cells (Stanley et al. 1983, Fig. 1) or polyclonal first antibody on liver sections (Geuze et al. 1984a) has confirmed the existence of antigenic enzyme at the cell surface, a result shown even more dramatically using immunogold methods and electron microscopy (Geuze et al. 1984a; Fig. 2).

4 Purification of 5′-Nucleotidase

Purification of minor intrinsic membrane proteins requires optimised detergent solubilisation and an affinity step during the subsequent purification protocol (Luzio et al. 1984). Partial purification of liver 5′-nucleotidase was achieved by several groups (Song and Bodansky 1966; Evans and Gurd 1973; Widnell 1975; Slavik et al. 1977) before successful attempts using affinity techniques. One method resulting in approx. 2800-fold purification from whole rat liver used ammonium sulphate precipitation steps to isolate a sphingomyelin-enzyme complex in reasonable yield (7%; Widnell 1975).

Whilst a variety of detergents were used in early efforts to solubilise the enzyme, a rigorous study of the effects of detergents on rat liver 5′-nucleotidase suggested that the zwitterionic detergent sulphobetaine 14 was best for this purpose (Bailyes et al. 1982). Using the criteria of non-sedimentability at high g forces and formation of a discrete active band on polyacrylamide gel electrophoresis this detergent was found to produce the "smallest and least polydisperse active protein" (Newby et al. 1978). Sulphobetaine 14 has also been successfully used to solubilise 5′-nucleotidase from both bovine liver (Harb et al. 1983) and brain (Meflah et al. 1984a).

Two types of affinity steps have been used in the purification of 5′-nucleotidase, either nucleotide-sepharose columns (Dornand et al. 1978; Fukui and Schichi 1981; Naito and Lowenstein 1981; Harb et al. 1983) or a monoclonal antibody covalently linked to a solid support (Bailyes et al. 1982). In experiments to purify rat liver 5′-nucleotidase from intact liver we have obtained >12000-fold purification with high yields (16%) using a combination of ammonium sulphate precipitation, sulphobetaine 14 solubilisation, binding to a monoclonal antibody immunoadsorbent and elution with 50 m*M* diethylamine at 4°C (Bailyes et al. 1982). The preparation of monoclonal antibodies to human (Kummer et al. 1984) and chicken (Dieckhoff et al., this volume) 5′-Nucleotidase has been reported suggesting that an immunoaffinity approach to purification of the enzyme from other species will be possible. Using AMP-sepharose affinity chromatography Harb et al. (1984) have reported a 16 000-fold purification of bovine liver 5′-nucleotidase.

5 Kinetic Properties of 5′-Nucleotidase

5.1 Substrates

5′-Nucleotidase is a phosphomonoesterase that hydrolyses all ribonucleoside 5′-monophosphates, the activity in human and rat liver towards AMP, UMP and CMP being approximately twice that towards GMP and IMP (Song and Bodansky 1966; Widnell 1975; Nakamura 1976). Little activity towards 5′-deoxyribonucleotides was found in a partially purified enzyme preparation from rat liver (Widnell 1975) and 2′ and 3′ nucleotides were not hydrolysed by the liver enzyme (Song and Bodansky 1966; Widnell 1975; Harb et al. 1983). The K_m for AMP is in the micromolar range with the values obtained from rat liver ranging from 8 to 47 *μM* (Segal and Brenner 1960; Song and Bodansky 1967; Widnell 1975; Nakamura 1976; Bailyes et al. 1982). It is generally assumed that AMP is the normal physiological substrate for the enzyme.

5.2 Metal Ion Requirements

Metal chelators such as EDTA inhibit 5′-nucleotidase and activity can be restored by adding magnesium ions (Widnell 1975; Harb et al. 1983, 1984). Manganese and cobalt ions can substitute for magnesium at low concentrations, but inhibit at higher concentrations (Widnell 1975). The effect of magnesium ions on 5′-nucleotidase is complex. In the absence of magnesium the pH optimum is around 7.5; in its presence, a second pH optimum at around pH 9.5 has been reported in human, rat and mouse liver (Song and Bodansky 1966, 1967; Evans and Gurd 1973).

5.3 Inhibition by Nucleoside Di- and Tri-Phosphates

5′-Nucleotidase is strongly inhibited by micromolar concentrations of nucleoside di- and tri-phosphates. The rat liver enzyme has reported values for the K_i for ATP of 75 to 180 *μM* (Widnell 1975; Nakamura 1976; Bailyes et al. 1982). There may be a tissue difference since the values for rat liver are higher than those found in other rat tissues (2–9 *μM*; Edwards and Maguire 1970; Newby et al. 1975). The inhibition appears to be competitive in the presence of magnesium ions (Newby et al. 1975; Nakamura 1976) though in heart is mixed in its absence (Edwards and Maguire 1970). Part of the inhibition in the absence of magnesium may be due to ATP acting as a metal chelator of enzyme-bound magnesium. The inhibition observed with ADP is in each case greater, with K_i values ranging from 0.04 to 50 *μM*. It thus appears that ecto-5′-nucleotidase could not be fully active within the cell since the intracellular ATP concentration (6 m*M* in rat liver cytosol; Krebs et al. 1974) is far higher than the K_i for ATP. Interestingly the soluble cytosolic isoenzyme of 5′-nucleotidase is activated by ATP and ADP and a useful discriminatory test of the two isoenzymes is the effect of $\alpha\,\beta$ methylene ADP which only inhibits the ectoenzyme (Worku and Newby 1983).

6 Structure of 5 -Nucleotidase

6.1 Protein

SDS-polyacrylamide gel electrophoresis of immunoaffinity-purified rat liver 5′-nucleotidase has revealed a major protein band of apparent M_r 70 000 (α) and a minor band of apparent M_r 38 000 (β) (Bailyes et al. 1982). This pattern was not affected by reduction and alkylation (Bailyes et al. 1984). Using Ferguson plot analysis and comparison with standard proteins, the detergent-solubilised active enzyme was found to have a minimum apparent radius of 3.6 nm when subjected to electrophoresis into polyacrylamide gels under non-denaturing conditions (Bailyes et al. 1982). This size may be converted to an apparent M_r of 155 000 assuming that the solubilised enzyme is a globular protein of partial specific volume 0.74 mlg^{-1}.

The observation that the enzyme may be radiolabelled with ^{125}I without loss of enzyme activity (Siddle et al. 1981) has greatly facilitated the study of its structure. Cross-linking experiments on iodinated immunoaffinity purified enzyme have suggested the presence of a major homodimer ($\alpha\alpha$) of M_r 140 000 accounting for approx. 80% of the enzyme activity and a minor heterodimer ($\alpha\beta$) of M_r 108 000

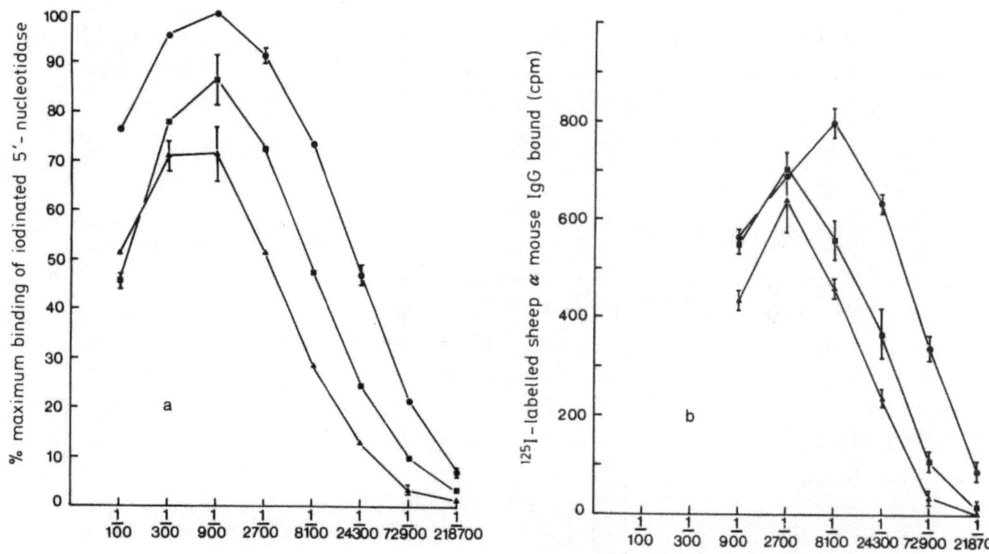

Fig. 3a, b. Monoclonal antibody binding to 5′-nucleotidase. **a** Iodinated 5′-nucleotidase was incubated with tripling dilutions in DMEM 10(Dulbecco's Modified Eagle Medium containing 10% vol/vol foetal calf serum) from a 1% dilution of ascites fluid of 5 N4-2 (●), 5 NE5 (▲) and 5 NH3 (■): Bound enzyme was separated by the addition of sheep anti-mouse IgG immunoadsorbent as described by Siddle et al. (1981). The values are expressed as a % of the maximum binding obtained with 5N4-2. **b** The binding of 5N4-2 (●) 5NE5 (▲) and 5NH3 (■) to the cell surface of Buffalo rat liver cells was detected by the addition of a second antibody, ^{125}I-labelled sheep anti-mouse IgG using the same dilution of ascites fluid used in **a**

accounting for the remaining 20% (Bailyes et al. 1984). Both dimers have been shown to exist in membrane preparations by charge shift analysis of solubilised enzyme subjected to electrophoresis into polyacrylamide gels under non-denaturing conditions (Bailyes et al. 1984). In these experiments ionic detergents were used to increase the electrophoretic mobility of 5′-nucleotidase solubilised in the zwitterionic detergent sulphobetaine 14. Intrinsic membrane proteins are incorporated into detergent micelles and their electrophoresis may be modified by the presence of charged detergent molecules in the micelles. As expected a charge shift was

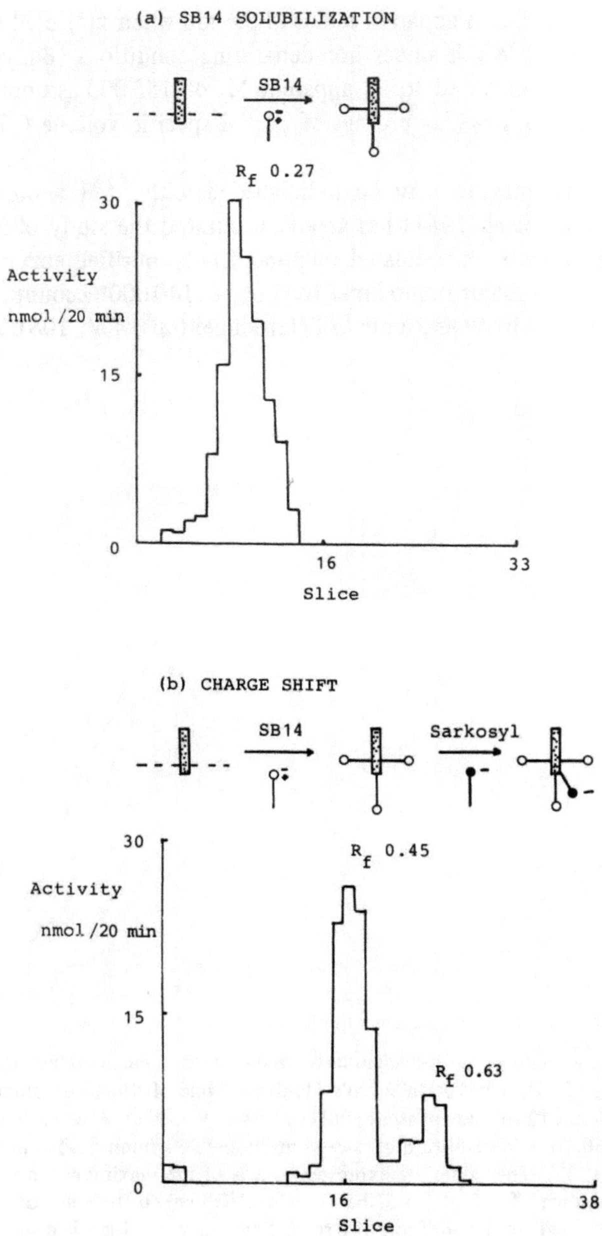

Fig. 4. Polyacrylamide gel electrophoresis of 5′-nucleotidase under non-denaturing conditions. 10% (wt/vol) acrylamide gels in the presence of sulphobetaine 14 ± sarkosyl, the latter detergent providing a charge shift

observed for 5′-nucleotidase but in addition the single peak of enzyme activity was split into two (Fig. 4). The same experiment was conducted with ^{125}I-labelled enzyme and subsequent SDS polyacrylamide gel electrophoresis confirmed that the major peak of activity represented the $\alpha\alpha$ dimer and the minor peak the $\alpha\beta$ dimer. Estimates of the specific enzyme activity in the $\alpha\alpha$ and $\alpha\beta$ polyacrylamide gel peaks showed that the β subunit was inactive (Bailyes et al. 1984).

Peptide mapping of the ^{125}I-labelled enzyme and immunoblotting with polyclonal antibodies showed that the β subunit of rat liver 5′-nucleotidase was homologous with a part of the α subunit. Concanavalin A and three monoclonal antibodies which bind to two cell surface domains of 5′-nucleotidase (Fig. 3) were shown to bind to the α subunit but not the β subunit. Thus it was concluded that the β subunit was a fragment of an α subunit which had lost an extracellular domain. This was consistent with its loss of enzyme activity. A diagrammatic representation of the structure of rat liver 5′-nucleotidase is shown in Fig. 5.

Little is known about the relative amounts of 5′-nucleotidase protein on either side of the membrane bilayer or the nature of its folding. Attempts to isolate active fragments of the enzyme by proteolytic digestion of liver membranes with a variety of microbial and mammalian proteases (Nakamura 1976; Bailyes 1982) have been largely unsuccessful, the proteases mostly have little effect on 5′-nucleotidase activity. This data would be consistent with 5′-nucleotidase having a short-stalked cell surface domain like some endopeptidases (Kenny et al. 1983) though other

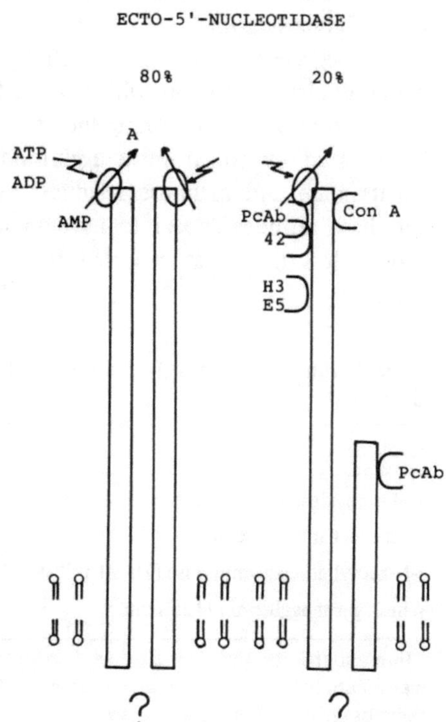

Fig. 5. The structure of rat liver ecto-5′-nucleotidase. Active sites plus inhibitor and antibody binding sites are diagrammatically shown. The drawing is not to scale. PcAb, rabbit polyclonal antiserum to rat liver 5′-nucleotidase; 42, H3 and E5, mouse monoclonal antibodies to rat liver 5′-nucleotidase (Bailyes et al. 1984)

interpretations are possible (Kenny and Maroux 1982) and analysis by negative staining of purified 5′-nucleotidase reconstituted into liposomes is required to provide firm evidence. There is no convincing evidence for, nor information about, the putative cytoplasmic tail of liver 5′-nucleotidase. Immunological evidence for 5′-nucleotidase being a transmembrane protein in lymphoma cell lines has been obtained (Zachowski et al. 1981).

Analyses of affinity purified 5′-nucleotidase from rat heart (Naito and Lowenstein 1981) and bovine liver (Harb et al. 1984) have also shown the enzyme to be a homodimer of approx. M_r 70 000 subunits though evidence for a monomeric enzyme has been presented for pig lymphocytes (Dornand et al. 1978) and bovine retinal rods (Fukui and Schichi 1981).

6.2 Carbohydrate

5′-Nucleotidase is a glycoprotein which can be inhibited by concanavalin A (Riordan and Slavik 1974; Slavik et al. 1977). A recent study of the effect of lectins and glycosidases on the bovine brain enzyme has provided evidence for a complex-type glycosylation with a fucose residue on the internal β-N-acetyl glucosamine of the pentasaccharide core (Meflah et al. 1984b). In contrast the same group have found that the bovine lymphocyte plasma membrane enzyme is of the high-mannose type and not inhibited by wheat germ agglutinin unlike the brain enzyme (Harb et al. 1983). With the bovine liver plasma membrane enzyme they have found lower inhibition by wheat germ agglutinin than for the brain enzyme and that this inhibition disappears on neuraminidase treatment (Harb et al. 1983; Meflah et al. 1984b). They suggest that bovine liver 5′-nucleotidase is sialylated and infer a hybrid type of glycosylation.

Experiments with immunoaffinity purified rat liver enzyme have shown that neuraminidase treatment whilst not affecting enzyme activity modifies the charge on electrophoresis confirming the presence of sialic acid (Chuang 1983). The enzyme is inhibited by wheat germ agglutinin and lentil lectin which bind to β-N-acetyl glucusamine and mannose residues respectively. Interestingly glycosidases specific for these residues appear to have some inhibiting effect on the enzyme (Table 1) though less than that reported for brain 5′-nucleotidase (Kreutzberg et al., this volume).

Table 1. Inhibition of rat liver ecto-5′-nucleotidase by lectins and glycosidases [a]

Addition	% Inhibition
α-Mannosidase (1.25 Uml^{-1})	24 ± 2
Lentil lectin (125 μgml^{-1})	48 ± 4
β-N-acetyl glucoseaminidase (10 $mUml^{-1}$)	17 ± 2
Wheat germ agglutinin (125 μgml^{-1})	39 ± 4

[a] Immunoaffinity purified enzyme (0.5 mU, 280 Umg^{-1} protein) was incubated with lectin or glycosidase for 30 min at 37°C prior to radioassay. Results as mean ± SEM

6.3 Lipid

There is no general agreement as to the role of lipid in 5′-nucleotidase activity. The enzyme can be prepared in a lipid-free form (Evans and Gurd 1973) so there is no absolute requirement. However the ability to release the enzyme from rat liver membranes and isolated hepatocytes with a phosphatidyl inositol-specific phospholipase c has suggested a role for this phospholipid in anchoring the enzyme in the membrane (Low and Finean 1978; Shukla et al. 1980).

The question of whether lipid modulates 5′-nucleotidase activity has been approached by examining the effect of different lipid environments on Arrhenius plots (i.e. plots of log rate vs reciprocal of absolute temperature). Some enzymes have one or more discontinuities of slope due to a change in the activation energy of the reaction at the transition temperature(s). When such a break occurs in the Arrhenius plot of intrinsic membrane enzymes it is usually attributed to a phase change in the bound lipid. When soluble proteins, e.g. trypsin, urease and fumarate hydratase exhibit a break, a different explanation is sought (Dixon and Webb 1979). Dipple et al. (1978, 1982), found that some detergents and anionic drugs at high concentration had a profound effect on the Arrhenius plots of 5′-nucleotidase. The temperature of the breakpoint varied with the detergent; values of 6.8 to 23°C were obtained. On the other hand, we found that the breakpoint temperature of 30.4°C was not affected by the amphiphile environment (Stanley and Luzio 1978) and that immunoaffinity purified rat liver enzyme retains the breakpoint (Worku 1984). Merisko et al. (1981) found that detergent solubilization of 5′-nucleotidase did not affect the breakpoints obtained but the Arrhenius plot of the purified sphingomyelin-complexed enzyme was linear, with no discontinuity. Subsitution of this complex with specific phosphatidylcholines reintroduced one or two breakpoints but the transition temperatures found bore no relation to the phase transition temperatures of the lipids. From experiments with affinity purified heart enzyme Naito and Lowenstein (1981) suggested that the breakpoint of 30.4°C found for rat liver enzyme (Stanley and Luzio 1978) is related to the changed sensitivity to magnesium ions above this temperature which they observed. The affinity purified bovine liver 5′-nucleotidase shows no break in the Arrhenius plot though a break at 28°C is present in the membrane-bound enzyme (Harb et al. 1983). Attempts to rationalise the literature data by examining Arrhenius plots of 5′-nucleotidase in membranes from rats fed diets containing different lipids have shown an effect of cholesterol (Venkatesan et al. 1983) but have otherwise failed to clarify the reported discrepancies (Alam and Alam 1983). At present the question of the effect of the lipid environment on 5′-nucleotidase remains unresolved.

6.4 Active Site

Both rat liver plasma membrane 5′-nucleotidase and the cytosolic enzyme can be inhibited by diethylpyrocarbonate (Worku et al. 1984) and the inhibition is reversible by hydroxylamine. Reversible inhibition by diethylpyrocarbonate suggests the presence in the active site of histidyl, seryl or tyrosyl residues, though in the case of ecto-5′-nucleotidase evidence against an active site tyrosyl residue was provided by the ability of the enzyme to be iodinated whithout loss of activity. Neither the ectoenzyme nor cytosolic enzyme can be inhibited by 4 m*M* phenylmethylsulphonyl fluoride showing that they are not serine esterases. Thus, the evidence suggests that both the cytosolic enzyme and ectoenzyme belong to a group of histidine phosphatases which may include other enzymes such as acid phosphatase and glucose-6-phosphatase. Inactivation by thiol reagents suggests the presence of an essential cysteine and by analogy with a catalytic mechanism proposed for cytosolic 5′-nucleotidase (Worku et al. 1984) it may be that catalysis involves nucleophilic attack by a histidine on AMP, formation of a phosphoenzyme intermediate and a role for cysteine in general acid/base catalysis to break down the histidine phosphate. Evidence that the ectoenzyme is a metalloenzyme with Mg^{2+} or Mn^{2+} as cation has been suggested for bovine liver 5′-nucleotidase (Harb et al. 1983) and such a cation may play a role in catalysis (Worku et al. 1984).

7 Molecular Cloning of 5′-Nucleotidase

A complete understanding of the structural properties of a protein requires a knowledge of its primary sequence. However, for minor membrane proteins it is difficult to purify sufficient protein to achieve a complete primary sequence by conventional protein chemistry techniques. An alternative approach is to sequence cDNA coding for the protein, though this in turn requires the availability of a probe capable of detecting the specific DNA.

There are approximately 5×10^5 monomer molecules of 5′-nucleotidase per liver cell which puts it in the class of a minor membrane component (Luzio et al. 1984). No primary sequence information has been published making it impossible to construct oligonucletide probes. In addition, no means are known of increasing the abundance of mRNA for 5′-nucleotidase which suggests that screening cDNA clones by hybrid selected translation (Harpold et al. 1978) would be inappropriate for this

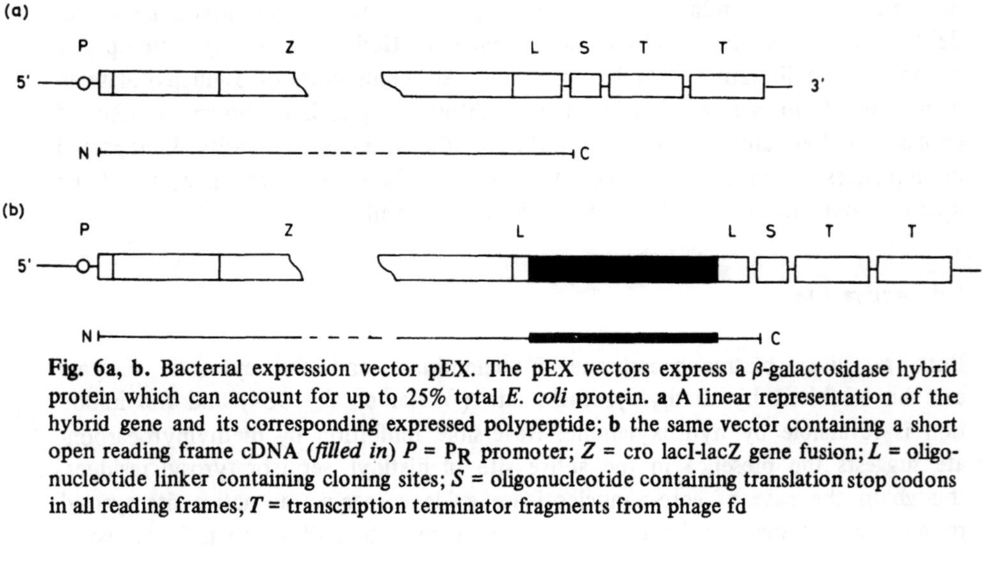

Fig. 6a, b. Bacterial expression vector pEX. The pEX vectors express a β-galactosidase hybrid protein which can account for up to 25% total *E. coli* protein. **a** A linear representation of the hybrid gene and its corresponding expressed polypeptide; **b** the same vector containing a short open reading frame cDNA (*filled in*) *P* = P_R promoter; *Z* = cro lacI-lacZ gene fusion; *L* = oligonucleotide linker containing cloning sites; *S* = oligonucleotide containing translation stop codons in all reading frames; *T* = transcription terminator fragments from phage fd

molecule. An elegant solution of overcoming screening difficulties is to use specific antibodies to identify cDNA clones.

We are currently attempting to clone a cDNA coding for 5′-nucleotidase using the pEX bacterial expression vectors (Stanley and Luzio 1984; Fig. 6). By combining an optimised expression vector with a new method for immunoblotting precipitated hybrid proteins in situ (Stanley 1983) a very good signal can be obtained from positive colonies and large numbers can be screened in a short period of time. This method has now been used successfully to isolate clones coding for rare cytoplasmic and membrane proteins, and preliminary screening, using antibodies against 5′-nucleotidase have produced several putative positive clones (Colaco, Luzio and Stanley, unpublished).

Confirmation of this, and subsequent sequencing of the cDNA inserts should provide the primary sequence of the enzyme.

There are a number of additional benefits which will accrue as a result of the expression cloning of 5′-nucleotidase. The complete cDNA can be cloned downstream of a strong promoter so that the synthesis and assembly of the enzyme may be studied in an in vitro transcription and translation system. Moreover, domains of

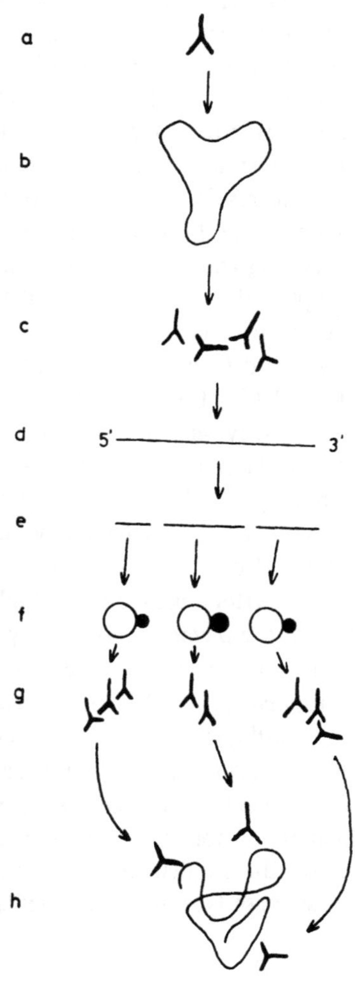

Fig. 7. a–h. Scheme for using cloned cDNA in protein topographical studies. Monoclonal antibodies (**a**) on immunoadsorbents are used to purify minor membrane proteins (**b**) which in turn are used to prepare polyclonal antisera of mono-specificity (**c**). These antisera are used to screen cDNA libraries by colony blotting to obtain a cDNA or part of a cDNA coding for the protein (**d**). This cDNA may be fragmented (**e**) and subcloned back into the expression vector system to generate hybrid proteins each containing discrete domains of the original protein (**f**). These hybrid proteins may be used for raising antibodies of predetermined specificity (**g**) or used for affinity purifying a particular specificity of antibody from the original polyclonal serum. These mono-specific antibodies can then be used in structural and functional studies on the original protein (**h**)

the protein may be expressed in vitro and used to raise antibodies of predetermined specificity. These can then be used to map protein topography with respect to the impermeability barrier of the plasma membrane and also with respect to the DNA sequence. Figure 7 shows the overall approach of using antibodies to purify a protein and identify cDNA clones which in turn are used to generate antibodies to different parts of the protein molecule for topographical studies.

8 Synthesis of 5′-Nucleotidase

In common with other cell surface glycoproteins 5′-nucleotidase is synthesised by membrane-bound ribosomes (Bergeron et al. 1975). Cytochemical staining and biochemical assay showed the presence of 5′-nucleotidase in liver microsomes and Golgi apparatus (Widnell 1972; Bergeron et al. 1973a; Farquhar et al. 1974; Little and Widnell 1975) suggesting that the Golgi apparatus was the site of glycosylation and important in routing the enzyme to the cell surface (Bergeron et al. 1973b; Evan 1976). The function of the Golgi apparatus in directing newly synthesised 5′-nucleotidase to the cell surface has been confirmed in experiments using immunogold localisation of the enzyme in the liver of rats after cycloheximide treatment to inhibit protein synthesis (Geuze et al. 1984a). These experiments showed that after 1 h cycloheximide treatment 5′-nucleotidase was absent from the Golgi apparatus but still present on the cell surface, indicating a long residence at the plasma membrane relative to transit time.

The results of some histochemical and biochemical experiments on 5′-nucleotidase in rat liver Golgi and microsomal membranes suggested that the synthesis, processing and insertion of 5′-nucleotidase into the plasma membrane may be more complex than predicted by the 'signal hypothesis' (Wickner 1979). In particular, activity was localised on the cytoplasmic side of microsomal membranes (Widnell 1972) and Golgi cisternae but the inside face of Golgi secretory vesicles (Farquhar et al. 1974; Little and Widnell 1975). This data implied that during formation of secretory vacuoles from the Golgi cisternae 5′-nucleotidase is translocated from the cytoplasmic side of the membrane to the inside. More recent experiments using immunogold labelling of the enzyme have cast doubt on this hypothesis since 5′-nucleotidase was detected on the non-cytoplasmic side of both Golgi stacks and vesicles (Geuze et al. 1984a).

A further difficulty in understanding the synthesis of liver 5′-nucleotidase is that the enzyme is not evenly distributed over the surface of the parenchymal liver cell but concentrated at the bile canalicular front. Histochemical staining (Farquhar et al. 1974), subcellular fractionation (Wisher and Evans 1975), and immunofluorescent staining (Geuze et al. 1984a) have all shown this concentration of the enzyme at the bile canalicular surface domain. The latter techniques have also shown the presence of enzyme, though in lesser amounts, on the blood sinusoidal membrane domain. Immunogold labelling with electron microscopy (Fig. 8) has shown the contiguous region to be essentially free of 5′-nucleotidase. Evidence from pulse chase labelling experiments of membrane glycoproteins in rat liver (Evans et al.

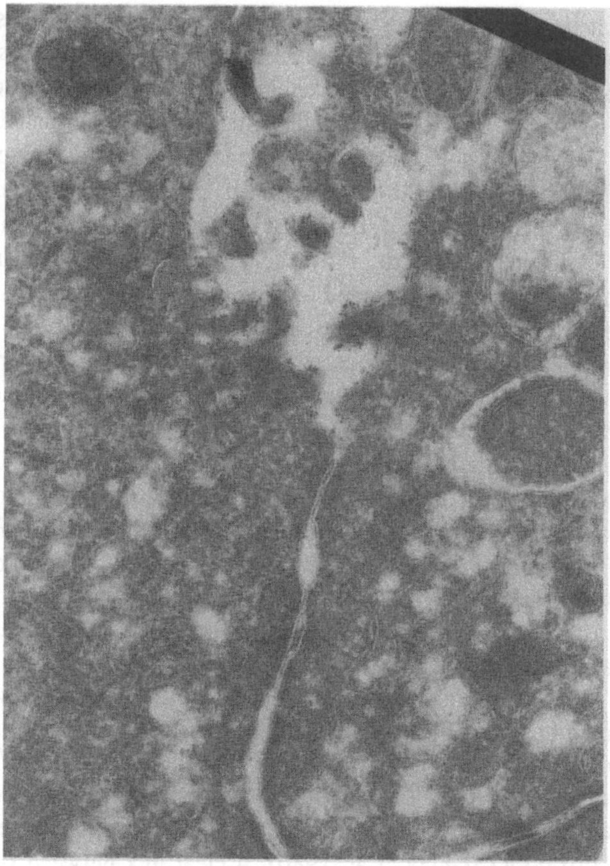

Fig. 8. Immunogold labelling of 5′-nucleotidase on the hepatocyte surface in situ. Cryosections of rat liver were treated with mouse monoclonal anti-5′-nucleotidase, rabbit anti-mouse IgG and gold-labelled protein A. Enzyme is localised on the bile canalicular surface but not on the contiguous domain of the cell surface (Gift of Dr. G. Griffiths)

1980) suggests that after processing in the Golgi apparatus they are incorporated into the blood sinusoidal surface before reaching the bile canalicular surface. Thus, newly synthesised 5′-nucleotidase may be inserted into the blood sinusoidal surface before lateral diffusion (Evans 1976) or endocytosis and trans-cellular transport to the bile canalicular surface. No evidence is available concerning a mechanism for maintaining the highest concentration of enzyme at this domain of the parenchymal liver cell surface, though tight junctions may prevent or slow down its redistribution by lateral diffusion. The presence of specific lipids or binding proteins also cannot be ruled out (Evans 1976).

9 Intracellular Localisation and Movement of 5′-Nucleotidase

Whilst it was expected that ectoenzymes would also be found in small amounts at intracellular sites on pathways of synthesis and degradation (Newby et al. 1975), the observation that 50% of 5′-nucleotidase was intracellular in collagenase-isolated rat hepatocytes was a surprise (Stanley et al. 1980). This proportion of the enzyme activity was latent in intact cells, i.e. inaccessible to substrate unless the cell surface permeability barrier was broken by severe mechanical homogenisation or detergent treatment. The non-ionic detergents (Triton X-100 and Lubrol 12A9) used to break the cells did not activate the enzyme (Stanley and Luzio 1978; Stanley et al. 1980; Bailyes 1982) and it was possible to perform mechanical homogenisation such that subsequent addition of detergent had no further effect on homogenate activity.

In an attempt to study the orientation in situ of intracellular 5′-nucleotidase, hepatocytes were subjected to progressive cell damage by mechanical homogenisation and access of substrate to intracellular 5′-nucleotidase and glucose-6-phosphatase was measured and compared with the release of the enzymes aryl sulphatase and lactate dehydrogenase (Stanley et al. 1980). Glucose-6-phosphatase activity measured by the addition of exogenous substrate closely followed the release of the cytosolic enzyme lactate dehydrogenase which was expected since the former enzyme is located on the luminal face of the endoplasmic reticulum with access to the cytosol via a specific glucose-6-phosphate carrier (Gold and Widnell 1976). In contrast, intracellular 5′-nucleotidase activity showed a slower increase correlating with the release of lysosomal aryl sulphatase, suggesting that the enzyme was inside intracellular vesicles of similar fragility to lysosomes, and that the enzyme had no access to cytosolic AMP.

Intracellular 5′-nucleotidase could be measured separately from the plasma membrane enzyme by inhibiting cell surface activity with polyclonal anti-5′-nucleotidase serum at 2°C. It was found that the cell surface 5′-nucleotidase and this intracellular activity were the same enzyme since they were both inhibited by the specific antiserum, had the same K_m for AMP and were pelleted by centrifugation at 100 000 g (Stanley et al. 1980). A clear link between the two pools of enzyme was established by incubating intact cells with polyclonal anti-5′-nucleotidase serum at 37°C for 30 min and measuring the remaining activity of whole cells and homogenate. Cell surface activity was rapidly inhibited, and intracellular activity was also inhibited though more slowly. Similar inhibition of cell surface 5′-nucleotidase and intracellular activity occurred when heptatocytes were incubated with either the lectin concanavalin A or monovalent Fab fragments of IgG from anti-5′-nucleotidase serum. When cell surface activity was inhibited by incubating hepatocytes with antiserum at 2°C and the cells then washed and resuspended at 37°C it was found that surface activity returned with time and intracellular activity decreased. Total hepatocyte activity was constant and the result was not affected by the presence of cycloheximide to inhibit protein synthesis. Thus in hepatocytes the exchange of 5′-nucleotidase between the cell surface and an intracellular membrane bound pool of the enzyme has been clearly established and a complete cycle of membrane 5′-nucleotidase circulation is probable (Stanley et al. 1980).

The location of the intracellular pool of 5′-nucleotidase in hepatocytes has been further studied by immunofluorescent microscopy (Stanley et al. 1983; Geuze et al. 1984a) using monoclonal antibodies to the ectoenzyme and specific polyclonal antibodies raised against enzyme immunoaffinity purified using a monoclonal antibody. In a cultured rat liver cell line it was found that the intracellular distribution of the enzyme was not the same as the distribution of Golgi and endoplasmic reticulum markers, being more widespread. In intact heptocytes immunogold electron microscopy has clearly shown the presence of 5′-nucleotidase in the Golgi complex (Geuze et al. 1984a). but antigenic activity is also present elsewhere. Estimates of the 5′-nucleotidase content of Golgi and lysosomes from rat liver indicate that these organelles may only account for <8% of total cell activity thus also implying the presence of other intracellular membranes containing 5′-nucleotidase activity (Luzio and Stanley 1983).

Intracellular pools of 5′-nucleotidase have been shown in several other cell types, though the size of the pool is variable (Stanley et al. 1982). Internalisation and recycling of the enzyme has been clearly demonstrated in cultured rat fibroblasts (Widnell et al. 1982) by similar experiments to those described above for hepatocytes. In addition, in fibroblasts, circulation of surface-bound IgG from polyclonal anti-(5′-nucleotidase) serum was shown and also the effects of the ionophore monensin on 5′-nucleotidase recycling have been suggested to indicate a function of the Golgi complex in the re-cycling process (Wilcox et al. 1982). Comparison of the time courses of reappearance of 5′-nucleotidase at the surface of cells pretreated with inhibitory antiserum at 0°C suggests that the rate of enzyme circulation may vary in different cell types (Widnell 1983). It does, however, appear to be an extremely useful marker for some of the membrane events of endocytosis.

10 Isolation of Intracellular Liver Vesicles Containing 5′-Nucleotidase

An understanding of the endocytic pathways and intracellular circulation of 5′-nucleotidase within the hepatocyte cannot be complete without a greater knowledge of the intracellular structures in which it resides. The inability of the Golgi complex and lysosomes to account for much of the intracellular pool of enzyme suggested that a strategy was required to identify and isolate other intracellular vesicles containing the enzyme. Such vesicles were expected to have latent 5′-nucleotidase activity that would be inhibited if cells were incubated with anti-5′-nucleotidase antibodies at 37°C prior to homogenisation, but not inhibited if the incubation of cells with antiserum was performed at 2°C. Using a (1–15% wt/vol) Ficoll density gradient to analyse post-10 000 g supernatants from collagenase-isolated hepatocyte homogenates it was found that 5′-nucleotidase with the characteristics expected of the intracellular pool enzyme was distributed throughout the gradient as was enzyme derived from the cell surface (Luzio and Stanley 1983). A peak fraction of enzyme activity derived from surface-inhibited cells prepared by incubating cells with antibody at 2°C was further purified by removal of plasma membrane onto an anti-IgG immunoadsorbent. This procedure, shown diagrammatically in Fig. 9, allowed the preparation of a small vesicle (diam 65 nm) fraction containing latent 5′-nucleotidase activity, free of lysosomal and endoplasmic reticulum markers and having a restricted phospholipid and polypeptide composition (Luzio and Stanley 1983).

A second series of experiments was designed to isolate the endocytic vesicles involved in the trans-epithelial movement of IgA across the hepatocyte from the blood sinusoidal front to the bile (Mullock and Hinton 1981; Orlans et al. 1983).

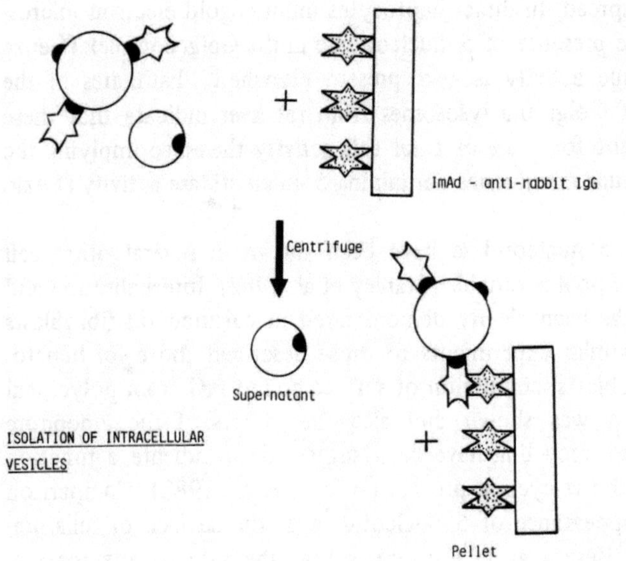

Fig. 9. The purification of intracellular vesicles containing latent 5′-nucleotidase by removal of cell surface membrane pre-labelled with antibody

Polymeric IgA becomes covalently linked to its membrane receptor (secretory component) by disulphide bonds after binding at the blood sinusoidal surface and is subsequently released into bile as a result of proteolytic cleavage of the receptor after transport across the liver cell (Kuhn and Kraehenbuhl 1982). After pulse labelling of rat liver with ^{125}I-IgA followed by homogenisation and sequential centrifugation on Ficoll (1–22% wt/vol) and metrizamide (10–30% wt/vol) density gradients a fraction which contained IgA and latent 5′-nucleotidase activity and consisted of 140 nm diam vesicles, was isolated (Mullock et al. 1983; Fig. 10). The fraction was free of endoplasmic reticulum and lysosomal markers and contained a Mg^{2+}-ATPase which could be inhibited by tributyltin, a result which by analogy with studies of other intracellular organelles (Hutton and Peshavaria 1982) indicates the presence of a proton pump.

The relationship between the two vesicle fractions isolated and their relationship with other subcellular organelles is unknown. Small amounts of galactosyl transferase found in the fraction (Luzio and Stanley 1983; Mullock et al. 1983) may indicate an involvement with the Golgi apparatus, but could simply be contamination. Whether intact vesicles have been isolated or fragments of larger vesicles is not clear, but it is tempting to speculate that those containing IgA derive from the intracellular organelle called CURL (compartment of uncoupling of receptor and ligand) and from subsequent transfer vesicles shown to be involved in IgA transport by immunogold electron microscopy (Geuze et al. 1984b). Both vesicle fractions appear to derive from an intracellular site and have the characteristics expected of vesicles derived from what has been broadly described as the endosome (Brown et al. 1983), to which endocytosed cell surface components and bound ligands are thought to be internalised before any subsequent processing by lysosomes and Golgi; or return to the cell surface.

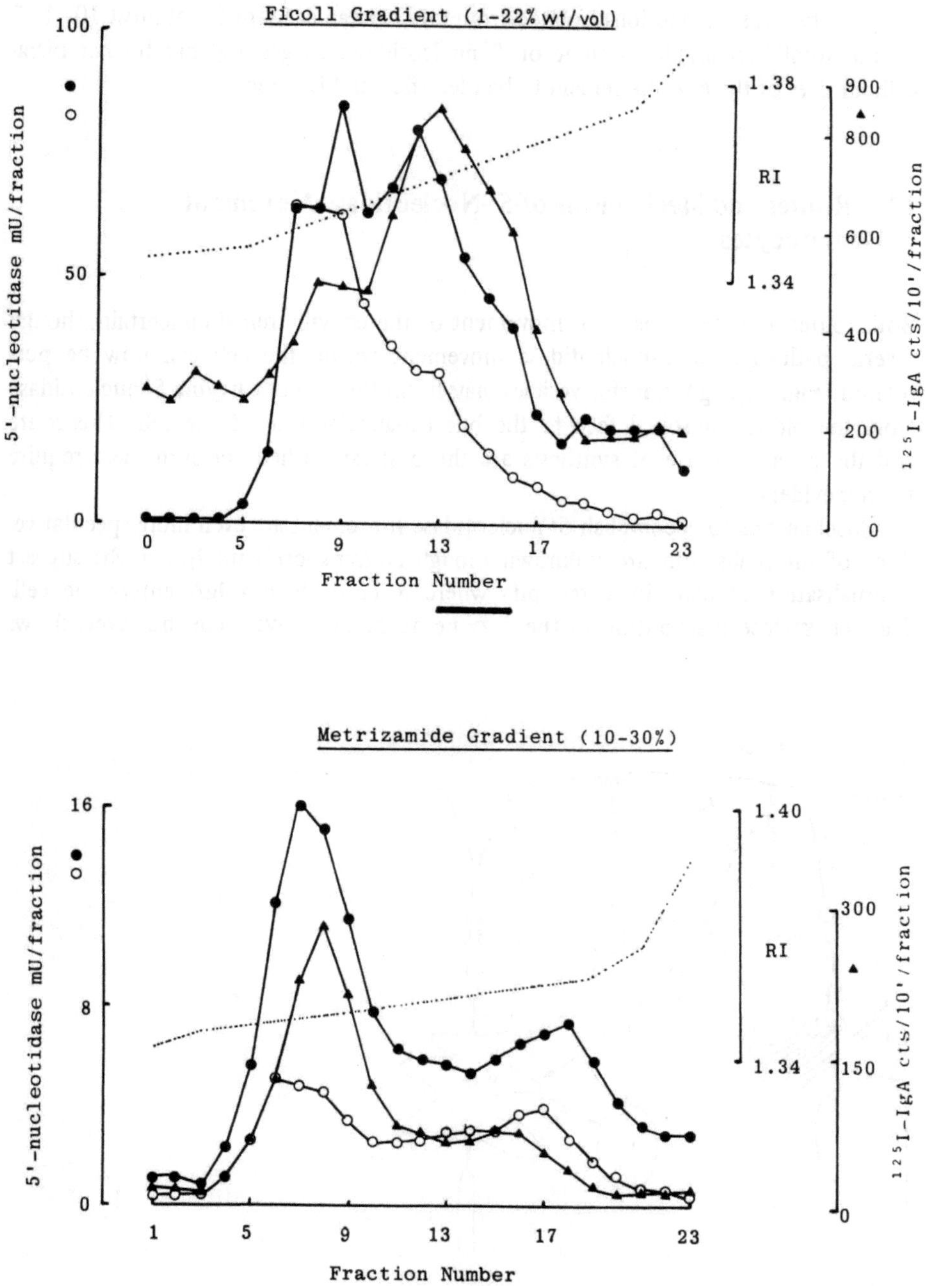

Fig. 10. The isolation of endocytic vesicles containing IgA from rat liver. A representative experiment showing the correspondence of latent 5′-nucleotidase with ^{125}I IgA on sequential Ficoll and metrizamide gradients. Rats were pulsed with ^{125}I IgA before killing. *Closed circles* represent 5′-nucleotidase assayed in the presence of 1% Triton X-100 and *open circles* in its absence. Fractions *13 to 16* from the Ficoll gradient were loaded onto the metrizamide gradient. Experimental details as in Mullock et al. (1983)

The two vesicle fractions isolated from rat liver can account for at most 10–15% of the total hepatocyte content of 5′-nucleotidase, suggesting that further intracellular sites of the enzyme remain to be identified and isolated.

11 Routes and Mechanisms of 5′-Nucleotidase Movement in Heptatocytes

Both routes and mechanisms of movement of the enzyme remain uncertain, though several pathways of 5′-nucleotidase movement within the cell can now be postulated (Fig. 11). IgA transfer vesicles may fulfill the role of carrying 5′-nucleotidase from the blood sinusoidal face to the bile canalicular face of the cell. This route and the putative route of synthesis are the best established yet even these require further evidence.

Mechanisms to accomplish 5′-nucleotidase movement are even more speculative. Sites of internalisation are unknown though co-transport with IgA might suggest internalisation at clathrin-coated pits where receptor bound IgA enters the cell. The cell surface distribution of the enzyme on cultured liver cells, however, shows

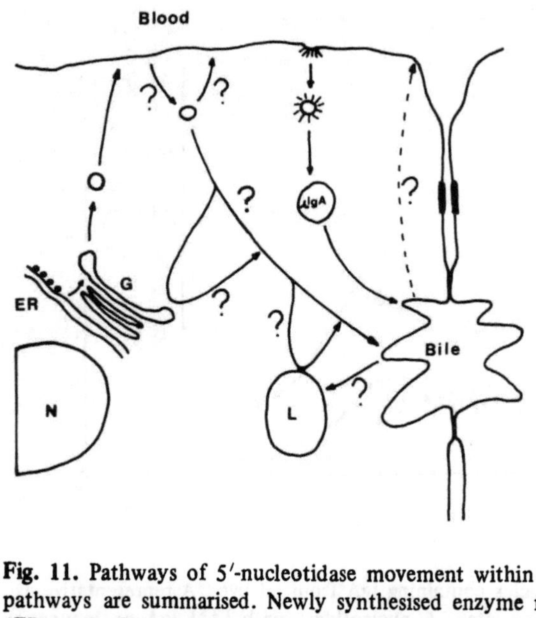

Fig. 11. Pathways of 5′-nucleotidase movement within the hepatocyte. A number of possible pathways are summarised. Newly synthesised enzyme moves from the endoplasmic reticulum (*ER*) to the Golgi apparatus (*G*) and then to the plasma membrane. Enzyme at the blood sinusoidal surface of the hepatocyte may be internalized by coated pit or non-coated vesicle mechanisms and can subsequently either recycle to the blood sinusoidal surface or be transcytosed to the bile canalicular surface. At least one route of transcytosis involves comigration with polymeric immunoglobulin A (pIgA) though other routes may exist. It is not known whether the enzyme circulates within the cell via the Golgi apparatus (*G*) or lysosome (*L*). Reverse transcytotic movement from the bile canalicular cell surface to the blood sinusoidal surface is suggested. *N*, nucleus

that it is not concentrated in the coated pits as are some receptors (Stanley et al. 1983). The low specific activity of 5′-nucleotidase in endosome-derived vesicles suggests that it might travel 'piggy back' in vesicles as a result of incomplete filtering of plasma membrane proteins during coated pit formation. Inhibition of the intracellular pool of 5′-nucleotidase in collagenase-isolated hepatocytes by polyclonal antisera cannot be prevented by cytochalasin B and colchicine, drugs disrupting microtubules and microfilaments, nor by decreasing the ATP content to 2% of normal by treatment with oligomycin and 2-deoxy-glucose (Stanley et al. 1980). These data and the inhibition of intracellular enzyme by antibody Fab fragments distinguish 5′-nucleotidase circulation from the phenomenon of patching and capping though they cannot rule out the possibility of cytoskeletal involvement in the correct routing of the enzyme.

A variety of experiments have led to the suggestion that an interaction between actin and 5′-nucleotidase might be involved in the enzyme's intracellular distribution and movement (Carraway et al. 1979; Mescher et al. 1981; Crawford and Stewart 1984). The most direct observation is that various preparations of 5′nucleotidase cause the dissociation of actin-DNAase I complexes (Mannherz and Rohr 1978; Rohr and Mannherz 1979; and Dieckhoff this volume). We have observed that sulphobetaine 14 solubilised, immunoaffinity purified rat liver 5′-nucleotidase preparations can also prevent rabbit skeletal muscle G actin inhibition of bovine pancreatic DNAase I. However this can be explained by the presence of an ATPase as minor contaminant (0.0002% of the 5′-nucleotidase) which can inactivate the actin (Therien and Gruda 1983). In other experiments we found no evidence for actin binding to the enzyme when using cross-linking agents, nor binding of 5′-nucleotidase to an actin-sepharose column nor binding of radiolabelled G actin to enzyme subjected to electrophoresis using an overlay technique (Snabes et al. 1981). Although the purified 5′-nucleotidase preparations affected the polymerisation of the fluorescent actin derivative N-pyrenyliodoacetamide-actin this was due to the action of sulphobetaine 14. It remains possible that this detergent destroys the enzyme binding site to actin or affects the actin in such a way as to prevent it interacting with 5′-nucleotidase.

12 The Fate of Liver 5′-Nucleotidase

The eventual fate of the enzyme in hepatocytes may be degradation in lysosomes or shedding from the cell surface into bile, though the relative quantitative importance of the two routes has not been assessed. Isolated rat liver lysosomes have been shown to contain 5′-nucleotidase activity that is the same as the ectoenzyme by the criteria of inhibition by antibodies and by $\alpha\beta$-methylene ADP. This enzyme which can account for 3% of total cell activity was found to be >90% latent, and in immunoblotting experiments was shown to contain the α and β subunits, but also other immunoreactive species of intermediate molecular weight (Maguire and Luzio 1985). These results are consistent with lysosomes being a site of proteolytic degradation of the ectoenzyme.

A well-established alternative fate of the hepatocyte ectoenzyme involves shedding of membrane fragments into bile, without cell lysis, under the influence of conjugated bile salts (Billington et al. 1980; Barnwell et al. 1983). In perfused rat liver, with bile salts present in the perfusate it has been estimated that the liver sheds 5′-nucleotidase at the rate of 0.004% of total activity per minute, which is equivalent to 6% per 24 h (P.J. Lowe, personal communication). Ligation of the common bile duct in the rat (Mullock et al. 1977) or cholestatic disease in Man causes a rise in serum 5′-nucleotidase concentration and measurement of the raised level of the enzyme in serum has proved a useful clinical indicator of cholestatic disease (Ellis et al. 1978). It is thought that 5′-nucleotidase enters the bloodstream by bile regurgitation when the bile duct is obstructed (Bergan et al. 1975; Mullock et al. 1977) though the possibility of complex mechanisms such as reversed endocytic flow through the hepatocyte has also been suggested (Desmet 1977).

In studies of the bile and serum 5′-nucleotidases of rat and Man it was found that much of the activity in bile and cholestatic serum was present as high molecular weight aggregates (De Broe et al. 1975; Brocklehurst et al. 1976; Chuang et al. 1984). Solubilisation with zwitterionic detergent and polyacrylamide gel electrophoresis under non-denaturing conditions showed the presence of enzyme with the

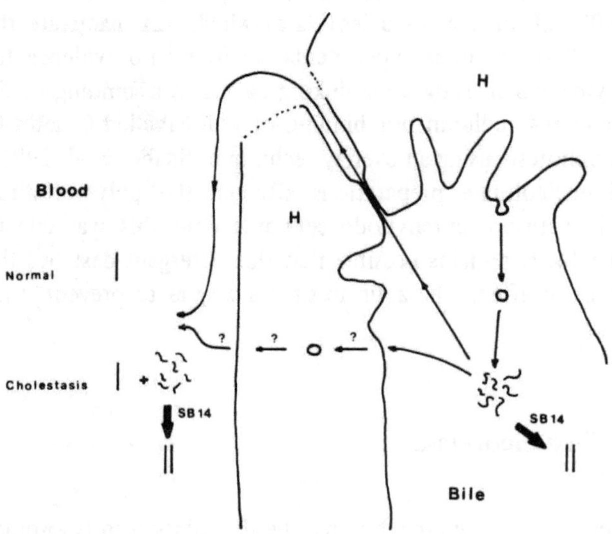

Fig. 12. 5′-nucleotidase in blood and bile. 5′-Nucleotidase can be released into the bile as a component of membrane vesicles shed from the bile canalicular surface of the hepatocyte. Such membrane vesicles may break down to produce micelles or polymeric aggregates containing the enzyme, which all appears to be the normal membrane dimer when treated with the detergent sulphobetaine 14 (SB14). The enzyme can pass from bile to the blood stream by leaking between hepatocytes (*H*) or possibly via an endocytic route. Leakage or regurgitation increases during cholestasis. Normally the enzyme in the blood stream is a soluble form with the characteristics of the enzyme monomer. In cholestasis the concentration of this form of the enzyme in the blood stream is increased and in addition aggregates and micelles containing the enzyme are also present. Sulphobetaine 14 (SB 14) treatment of the latter shows them to contain the normal membrane dimer

properties of the $\alpha\alpha$ homodimer. Cholestatic serum also contained a soluble form of the enzyme with the electrophoretic properties of α monomer. Bile did not contain this form of the enzyme, but it was the only form present in normal serum and in rat was shown also to have the immunological properties of α monomer (Chuang et al. 1984). Although this form of the enzyme may derive from a tissue other than liver, this is unlikely since its serum concentration is raised during cholestasis. Possible routes of release of liver 5′-nucleotidase into bile and serum are summarized in Fig. 12.

13 Function of Liver 5′-Nucleotidase

Despite much effort the function of 5′-nucleotidase remains obscure. Its ubiquitous distribution as an ectoenzyme has led to suggestions of involvement in nutrition, nucleotide scavenging or adenosine production, the latter to act subsequently as a 'local hormone' (Stanley et al. 1982). In hepatocytes there is evidence against its involvement in the vectorial production of adenosine (Sasaki et al. 1983), and its cellular orientation, substrate K_m and inhibition by intracellular nucleotide concentrations argue against its participation in a substrate cycle between AMP and adenosine that has been proposed to permit a rapid increase in the concentration of adenosine which can be transported into the extracellular medium to act as a local hormone (Arch and Newsholme 1978).

The possibility that in liver the function of 5′-nucleotidase is linked to its selective localisation at one domain of the cell surface or to its intracellular location and endocytic route cannot be excluded. It is of interest that a circadian rhythm in the activity of the enzyme at different sites in the cell has been reported (Von Mayersbach and Klaushofer 1979). A further hypothetical function is an involvement in the control of cell growth and differentiation. Fluctuations in 5′-nucleotidase content have been observed in different stages of the cell cycle and during cell maturation in various cell types (Lelievre et al. 1971; Sivak et al. 1972; Rodan et al. 1977; Graham 1979; Bastian et al. 1984). It is intriguing to note the observation that in rat liver plasma membrane preparations, 5′-nucleotidase can act in concert with other membrane enzymes to degrade RNA (Yannarell and Aronson 1977) in view of suggestions that exogenous nucleic acid can control differentiation (Niu et al. 1974). Thus the absence of 5′-nucleotidase and related enzymes in immature cells may allow them to be more susceptible to environmental influences during development.

14 Summary

Liver 5′-nucleotidase is an ectoenzyme predominantly located at the bile canalicular front of the hepatocyte. The enzyme exists mainly as an active dimer of M_r 70 000 subunits. Half of the total hepatocyte content of 5′-nucleotidase exists at intracellular sites on pathways of synthesis, membrane circulation and degradation. It can be expected that future research efforts will be directed to understanding its synthesis and the identification and quantitation of amounts of the enzyme on different membrane traffic routes within the cell. The function of the enzyme remains unknown.

Acknowledgements. We are indebted to all our collaborators and colleagues and especially thank Prof. C.N. Hales, Prof. L. Herman, Dr. A.C. Newby, Dr. P.J. Richardson, Dr. M.R. Edwards, Dr. C.A.L.S. Colaco, Dr. R.H. Hinton, Dr. G. Griffiths, Dr. A. Weeds, Dr. P. Jackson, Ms. M. Soos, Miss N.-N. Chuang, Mr. Y. Worku and Miss L. Shaw for many valuable discussions about, and experiments with 5′-nucleotidase. Some of the work reported was supported by the Medical Research Council, and the European Molecular Biology Organisation. K.S. is a Wellcome Lecturer.

References

Alam SQ, Alam BS (1983) Acyl group composition of lipids and the activities of (Na^+ and K^+)-ATPase 5′-nucleotidase and γ-glutamyltranspeptidase in salivary glands and kidneys of rats fed diets containing different dietary fats. Biochim Biophys Acta 758: 1–9

Arch, JRS, Newsholme EA (1978) Activities and some properties of 5′-nucleotidase, adenosine kinase and adenosine deaminase in tissues from vertebrates and invertebrates in relation to the control of the concentration and the physiological role of adenosine. Biochem J 174: 965–977

Arkesteijn, CL (1976) A kinetic method for serum 5′-nucleotidase using stabilized glutamate dehydrogenase. J Clin Chem Clin Biochem 14:155–158

Avruch J, Wallach DFH (1971) Preparation and properties of plasma membrane and endoplasmic reticulum fragments from isolated rat fat cells. Biochim Biophys Acta 233:334–347

Bailyes EM (1982) Purification and structure of the plasma membrane enzyme 5′-Nucleotidase. Ph D Thesis, University of Cambridge

Bailyes EM, Newby AC, Siddle K, Luzio JP (1982) Solubilisation and purification of rat liver 5′-nucleotidase by use of a zwitterionic detergent and a monoclonal antibody immunoadsorbent. Biochem J 203:245–251

Bailyes EM, Soos M, Jackson P, Newby AC, Siddle K, Luzio JP (1984) The existence and properties of two dimers of rat liver ecto-5′-nucleotidase. Biochem J 221:369–377

Barnwell SG, Godfrey PP, Lowe PJ, Coleman R (1983) Biliary protein output by isolated perfused rat livers. Effects of bile salts. Biochem J 210:549–557

Bastian, JF, Ruedi JM, MacPherson GA, Golombesky HE, O'Connor RD, Thompson LF (1984) Lymphocyte ecto-5′-nucleotidase activity in infancy. Increasing activity in peripheral blood B cells precedes their ability to synthesize IgG in vitro. J Immunol 132:1767–1772

Bergan A, Taksdal S, Sander J (1975) Transport and conjugation of 14C-bilirubin during acute and chronic cholestasis in the cholecystectomized dog. Eur Surg Res 7:355–365

Bergeron JJM, Ehrenreich JH, Siekevitz P, Palade GE (1973a) Golgi fractions prepared from rat liver homogenates II. Biochemical characterization. J Cell Biol 59:73–78

Bergeron JJM, Evans WH, Geschwind II (1973b) Insulin binding to rat liver Golgi fractions. J Cell Biol 59:771–776

Bergeron JJM, Berridge M, Evans WH (1975) Biogenesis of plasmalemma glycoproteins. Intracellular site of synthesis of mouse liver plasmalemma 5′-nucleotidase as determined by the subcellular location of messenger RNA coding for 5′-nucleotidase. Biochim Biophys 407: 325–337

Billington D, Evans, CE, Godfrey PP, Coleman R (1980) Effects of bile salts on the plasma membrane of isolated rat hepatocytes. Biochem J 188:321–327

Brocklehurst D, Lathe GH, Aparicio SR (1976) Serum alkaline phosphatase, nucleotide pyrophosphatase, 5′-nucleotidase and lipoprotein-X in cholestasis. Clin Chim Acta 67:269–279

Brown MS, Anderson RGW, Goldstein JL (1983) Recycling receptors: the round-trip itinerary of migrant membrane proteins. Cell 32:663–667

Carraway KL, Doss RC, Huggins JW, Chesnut RW, Carraway CAC (1979) Effects of cytoskeletal perturbant drugs on ecto-5′nucleotidase, a concanavalin A receptor. J Cell Biol 83:529–543

Chuang N-N (1983) Human serum 5′-nucleotidase: multiplicity and origin. Ph D Thesis, University of Cambridge

Chuang N-N, Newby AC, Luzio JP (1984) Characterisation of different molecular forms of 5′-nucleotidase in normal serum and in serum from cholestatic patients and bile duct ligated rats. Biochem J 224, 689–695

Crawford N, Stewart DIH (1984) Redistribution of 5′-nucleotidase and assembly of actin on the membrane of rabbit peritoneal polymorphonuclear leucocytes during phagocytosis. J Submicrosc Cytol 16:69–71

De Broe ME, Borgers, M, Wieme RJ (1975) The separation and characterisation of liver plasma membrane fragments circulating in the blood of patients with cholestasis. Clin Chim Acta 59:369–372

De Pierre JW, Karnovsky ML (1974a). Ectoenzymes of the guinea pig polymorphonuclear leukocyte I. Evidence for an ecto-adenosine monophosphatase, adenosine triphosphatase and p -nitrophenyl phosphatase. J Biol Chem 249:7111–7120

De Pierre JW, Karnovsky ML (1974b) Ectoenzymes of the guinea pig polymorphonuclear leukocyte II. Properties and suitability as markers for the plasma membrane. J Biol Chem 249:7121–7129

Desmet VJ (1977) In: Gentilini P, Teodori U, Gorini S, Popper H (eds). Intrahepatic cholestasis. Raven, New York, pp 7–23

Dipple I, Elliott KRF, Houslay MD (1978) Detergents modify the form of Arrhenius plots for 5′-nucleotidase activity. FEBS Lett 89:153–156

Dipple I, Gordon LM, Houslay MD (1982) The activity of 5′-nucleotidase in liver plasma membranes is affected by the increase in bilayer fluidity achieved by anionic drugs but not by cationic drugs. J Biol Chem 257:1811–1815

Dixon M, Webb EC (1979) Enzymes, 3rd edn. Longmans, London, pp 175–181

Dornand J, Bonnafous JC, Mani JC (1978) Purification and properties of 5′-nucleotidase from lymphocyte plasma membranes. Eur J Biochem 87:459–465

Edwards MT, Maguire MH (1970) Purification and properties of rat heart 5′-nucleotidase. Mol Pharmacol 6:641–648

El-Aaser AA, Fitzsimons JTR, Hinton RH, Norris KA, Reid E (1973) Electron microscopic examination of hepatic subcellular fractions for phosphatases. Histochem J 5:199–223

Ellis G, Goldberg DM, Spooner RJ, Ward AM (1978) Serum enzyme tests in diseases of the liver and biliary tree. Am J Clin Pathol 70:248–258

Engelhardt WA (1957) Enzymes as structural elements of physiological mechanisms. Proc Int Symp Enzyme Chemistry (Tokyo). IUB Symposium vol 2, pp 163–166

Essner E, Novikoff AB, Masek B (1958) Adenosine triphosphatase and 5′-nucleotidase activities in the plasma membrane of liver cells as revealed by electron microscopy. J Biophys Biochem Cytol 4:711–730

Evans WH (1976) The liver plasma membrane as a functional mosaic. Biochem Soc Trans 4: 1007–1011

Evans WH (1978) Preparation and characterisation of mammalian plasma membranes. North-Holland, Amsterdam

Evans WH, Gurd JW (1973) Properties of a 5′-nucleotidase purified from mouse liver plasma membranes. Biochem J 133:189–199

Evans WH, Flint N, Vischer P (1980) Biogenesis of hepatocyte plasma membrane domains. Incorporation of [^{3}H] fucose into plasma membrane and Golgi apparatus glycoproteins. Biochem J 192:903–910

Farquhar MG, Bergeron JJM, Palade GE (1974) Cytochemistry of Golgi fractions prepared from rat liver. J Cell Biol 60:8–25

Fukui H, Schichi H (1981) 5′-nucleotidases of retinal rod membranes. Arch Biochem Biophys 212:78–87

Geuze HJ, Slot JW, Strous GJ, Luzio JP, Schwartz AL (1984a) A cycloheximide resistant pool of receptors for asialoglycoproteins and mannose-6-phosphate residues in the Golgi complex of hepatocytes. EMBO J 3:2687–2694

Geuze HJ, Slot JW, Strous GJAM, Peppard J, von Figura K, Hasilik A, Schwartz AL (1984b) Intracellular receptor sorting during endocytosis: comparative immunoelectronmicroscopy of multiple receptors in rat liver. Cell 37:195–204

Gold G, Widnell CC (1976) Relationship between microsomal membrane permeability and the inhibition of hepatic glucose-6-phosphatase by pyridoxal phosphate. J Biol Chem 251: 1035–1041

Graham JM (1979) In: Hynes RO (ed) Surfaces of normal and malignant cells. Wiley, Chichester, pp 199–246

Harb J, Meflah K, Duflos Y, Bernard S (1983) Purification and properties of bovine liver plasma membrane 5′-nucleotidase. Eur J Biochem 137:131–138

Harb J, Meflah K, Duflos Y, Bernard S (1984) Evidence for a metalloprotein structure of plasma membrane 5′-nucleotidase. FEBS Lett 171:215–220

Harpold MM, Dobner PR, Evans RM, Bancroft FC (1978) Construction and identification by positive hybridization-translation of a bacterial plasmid containing a rat growth hormone structural gene sequence. Nucleic Acids Res 5:2039–2053

Hutton JC, Peshavaria M (1982) Proton-translocating Mg^{2+}-dependent ATPase activity in insulin secretory granules. Biochem J 204:161–170

Ipata PL (1968) Sheep brain 5′-nucleotidase. Some enzymic properties and allosteric inhibition by nucleotide triphosphates. Biochemistry 7:507–515

Itoh R (1981) Purification and some properties of cytosol 5′-nucleotidase from rat liver. Biochim Biophys Acta 659:31–37

Kenny AJ, Maroux S (1982) Topoplogy of microvillar membrane hydrolases of kidney and intestine. Physiol Rev 62:91–128

Kenny AJ, Maroux S (1982) Topology of microvillar membrane hydrolases of kidney and inbrane. Reconstitution of endopeptidase in liposomes shows that it is a short-stalked protein. Biochem J 211:755–762

Krebs HA, Cornell NW, Lund P, Hems R (1974) In: Lundquist F, Tygstrup N (eds). Regulation of hepatic metabolism. Academic, New York

Kuhn LC, Kraehenbuhl J-P (1982) The sacrificial receptor-translocation of polymeric IgA across epithelia. Trends Biochem Sci 7:299–302

Kummer U, Mysliwietz J, Gutensohn W, Buschette S, Jahn J, Neuser D, Munker R (1984) Development and properties of a monoclonal antibody specific for human ecto-5′-nucleotidase. Immunobiology 166:203–211

Lelievre L, Prigent B, Paraf A (1971) Contact inhibition – plasma membranes enzymatic activities in cultured cell lines. Biochem Biophys Res Commun 45:637–643

Little JA, Widnell CC (1975) Evidence for the translocation of 5′-nucleotidase across hepatic membranes in vivo. Proc Natl Acad Sci USA 72:4013–4017

Low MG, Finean JB (1978) Specific release of plasma membrane enzymes by a phosphatidyl inositol specific phospholipase C. Biochim Biophys Acta 508:565–570

Luzio JP, Stanley KK (1983) The isolation of endosome derived vesicles from rat hepatocytes. Biochem J 216:27–36

Luzio JP, Bailyes EM, Daw RA, Newby AC, Siddle K (1984) Approaches to isolating minor membrane components using monoclonal antibodies, as in purifying 5′-nucleotidase. In: Reid E, Cook GMW, Morre DJ (eds) Investigation of Membrane-Located Receptors Plenum, New York, pp 95–106

Maguire GA, Luzio JP (1985) The preseence and orientation of "ecto-5′-nucleotidase" in rat liver lysosomes. FEBS Lett

Mannherz HG, Rohr G (1978) 5′-Nucleotidase reverses the inhibitory action of actin on pancreatic deoxyribonuclease. I. FEBS Lett 95:284–289

Meflah K, Harb J, Duflos Y, Bernard J (1984a) 5′-Nucleotidase from bovine caudate nucleus synaptic plasma membranes: specificity for substrates and cations; study of the carbohydrate moiety by glycosidases. J Neurochem 42:1107–1115

Meflah K, Bernard S, Massoulie J (1984b) Interactions with lectins indicate differences in the carbohydrate composition of the membrane bound enzymes acetylcholinesterase and 5′-nucleotidase in different cell types. Biochim (Paris) 66:59–60

Merisko EM, Ojakian GK, Widnell CC (1981) The effects of phospholipids on properties of hepatic 5′-nucleotidase. J Biol Chem 256:1983–1993

Mescher MF, Jose MJL, Balk SP (1981) Actin-containing matrix associated with the plasma membrane of murine tumour and lymphoid cells. Nature (Lond) 289:139–144

Mullock BM, Hinton RH (1981) Transport of proteins from blood to bile. Trends Biochem Sci 6:188–191

Mullock BM, Issa FS, Hinton RH (1977) Bile 5′-nucleotidase in the serum of jaundiced rats. Clin Chim Acta 79:129–140

Mullock BM, Luzio JP, Hinton RH (1983) Preparation of a low density species of endocytic vesicles containing IgA. Biochem J 214:823–827

Naito Y, Lowenstein JM (1981) 5′-Nucleotidase from rat heart. Biochemistry 20:5188–5194

Nakamura S (1976) Effect of sodium deoxycholate on 5′-nucleotidase. Biochim Biophys Acta 426:338–347

Newby AC (1980) Role of adenosine deaminase, ecto-(5′-nucleotidase) and ecto-(non-specific phosphatase) in cyanide-induced adenosine monophosphate catabolism in rat polymorphonuclear leucocytes. Biochem J 186:907–918

Newby AC, Luzio JP, Hales CN (1975) The properties and extracellular location of 5′-nucleotidase of the rat fat cell plasma membrane. Biochem J 146:625–633

Newby AC, Rodbell M, Chrambach A (1978) Adenylate cyclase in polyacrylamide gel electrophoresis: solubilized but active. Arch Biochem Biophys 190:109–117

Niu MC, Deshpande AK, Niu LC (1974) Poly(A)-attached RNA as activator in embryonic differentiation. Proc Soc Exp Biol Med 147:318–322

Orlans E, Peppard JV, Payne AWR, Fitzharris BM, Mullock BM, Hinton RH, Hall JG (1983) Comparative aspects of the hepatobiliary transport of IgA. Ann NY Acad Sci 409:411–426

Reis J (1934) La nucleotidase et sa relation avec le desamination des nucleotides dans le coeur et dans le muscle. Bull Soc Chim Biol 16:385–399

Reis JL (1951) The specificity of phosphomonoesterases in human tissues Biochem J 48:548–551

Riordan JR, Slavik M (1974) Interactions of lectins with membrane glycoproteins. Effects of concanavalin A on 5′-nucleotidase. Biochim Biophys Acta 373:336–360

Rodan GA, Bourret LA, Cutler LS (1977) Membrane changes during cartilage maturation. Increase in 5′-nucleotidase in adenosine inhibition of adenylate cyclase. J Cell Biol 72: 493–501

Rohr G, Mannherz HG (1979) The activation of actin: DNAase I complex with rat liver plasma membranes. The possible role of 5′-nucleotidase. FEBS Lett 99:351–356

Sasaki T, Abe A, Sakagami T (1983) Ecto-5′-nucleotidase does not catalyse vectorial production of adenosine in the perfused rat liver. J Biol Chem 258:6947–6951

Segal HL, Brenner BM (1960) 5′-Nucleotidase of rat liver microsomes. J Biol Chem 235:471–474

Shukla SD, Coleman R, Finean JB, Michell RH (1980) Selective release of plasma membrane enzymes from rat hepatocytes by a phosphatidyl inositol specific phospholipase C. Biochem J 187:277–280

Siddle K, Bailyes EM, Luzio JP (1981) A monoclonal antibody inhibiting rat liver 5′-nucleotidase. FEBS Lett 128:103–107

Sivak A, Mossman BT, Van Duuren BL (1972) Activation of cell membrane enzymes in the stimulation of cell division. Biochem Biophys Res Commun 46:605–609

Slavik M, Kartner N, Riordan JR (1977) Lectin induced inhibition of plasma membrane 5′-nucleotidase: sensitivity of purified enzyme. Biochem Biophys Res Commun 75:342–349

Snabes MC, Boyd AE, Bryan J (1981) Detection of actin binding proteins in human platelets by ^{125}I-actin overlay of polyacrylamide gels. J Cell Biol 90:809–812

Song CS, Bodansky O (1966) Purification of 5′-nucleotidase from human liver. Biochem J 101:5c–6c

Song CS, Bodansky O (1967) Subcellular localization and properties of 5′-nucleotidase in the rat liver. J Biol Chem 242:694–699

Stanley KK (1983) Solubilization and immune detection of beta galactosidase hybrid proteins carring foreign antigenic determinants. Nucleic Acids Res 11:4077–4092

Stanley KK, Luzio JP (1978) The Arrhenius plot behaviour of rat liver 5′-nucleotidase in different lipid environments. Biochim Biophys Acta 514:198–205

Stanley KK, Luzio JP (1984) Construction of a new family of high efficiency bacterial expression vectors: identification of cDNA clones coding for human liver proteins. EMBO J 3:1429–1434

Stanley KK, Edwards MR, Luzio JP (1980) Subcellular distribution and movement of 5′-nucleotidase in rat cells. Biochem J 186:59–69

Stanley KK, Newby AC, Luzio JP (1982) What do ectoenzymes do? Trends Biochem Sci 7: 145–147

Stanley KK, Burke B, Pitt T, Siddle K, Luzio JP (1983) Localisation of 5′-nucleotidase in a rat liver cell line using a monoclonal antibody and indirect immunofluorescent labelling. Exp Cell Res 144:39–46

Therien HM Gruda J (1983) Effect of liver plasma membranes on G-actin II. Fate of actin-bound nucleotide. Can J Biochem Cell Biol 61:93–98

Trams EG, Lauter CJ (1974) On the sidedness of plasma membrane enzymes. Biochim Biophys Acta 345:180–197

Uusitalo RJ, Karnovsky MJ (1977) Surface localization of 5′-nucleotidase on the mouse lymphocyte. J Histochem Cytochem 25:87–96

Venkatsean S, Gallagher JJ, Mitropoulos KS (1983) Modulation of the activity of 5′nucleotidase by the transfer of non esterified cholesterol to rat liver microsomal fraction and evidence for regulation of the concentration of non esterified cholesterol in plasma membranes in vivo. Biochim Biophys Acta 756:72–82

Von Mayersbach H, Klaushafer K (1979) Circadian variation of 5′-nucleotidase activity in rat liver. Cell Mol Biol 24:73–79

Wickner W (1979) The assembly of proteins into biological membranes: the membrane trigger hypothesis. Annu Rev Biochem 48:23–45

Widnell CC (1972) Cytochemical localization of 5′-nucleotidase in subcellular fractions isolated from rat liver. I. The origin of 5′-nucleotidase activity in microsomes. J Cell Biol 52:542–558

Widnell CC (1975) Purification of rat liver 5′-nucleotidase as a complex with sphingomyelin. Methods Enzymol 32:368–374

Widnell CC (1983) Use of antibody to 5′-nucleotidase as a marker to study membrane flow during pinocytosis. Methods Enzymol 98:395–404

Widnell CC, Little JS (1977) Membranes involved in exocytosis and endocytosis: a cytochemical approach. In: Reid E (ed). Membranous elements and movement of molecules. Horwood, Chichester, pp 149–162

Widnell CC, Schneider Y-J, Pierre B, Baudhuin P, Trouet A (1982) Evidence for a continual exchange of 5′-nucleotidase between the cell surface and cytoplasmic membranes in cultured rat fibroblasts. Cell 28:61–70

Wilcox DK, Kitson RP, Widnell CC (1982) Inhibition of pinocytosis in rat embryo fibroblasts treated with monensin. J Cell Biol 92:859–864

Wisher MH, Evans WH (1975) Functional polarity of the rat hepatocyte surface membrane. Isolation and characterisation of plasma membrane subfractions from the blood sinusoidal, bile canalicular and contiguous surfaces of the hepatocyte. Biochem J 146:375–388

Worku Y (1984) The properties and physiological role of cytosolic 5′-nucleotidase. Ph D Thesis, University of Cambridge

Worku Y, Newby AC (1983) The mechanisms of adenosine production in rat polymorphonuclear leucocytes. Biochem J 214:325–330

Worku Y, Luzio JP, Newby AC (1984) Identification of histidyl and cysteinyl residues essential for catalysis by 5′-nucleotidase. FEBS Lett 167:235–240

Yannarell A, Aronson NN (1977) Localisation of an RNA-degrading system of enzymes on the rat liver plasma membrane. Biochim Biophys Acta 311:191–204

Zachowski A, Evans WH, Paraf A (1981) Immunological evidence that plasma membrane 5′-nucleotidase is a transmembrane protein. Biochim Biophys Acta 644:121–126

Ecto-5′-Nucleotidase as a Probe for the Analysis of Membrane Flow During Fluid Phase Pinocytosis

C. C. Widnell, R. P. Kitson, and D. K. Wilcox [1]

1 Introduction

Much of the research on ectoenzymes is concentrated on understanding their physiological functions. However, the fact that these proteins are located on the cell surface, and therefore accessible from the external medium, means that they provide ideal markers for attempts to analyze the fate of membrane during endocytosis. The work to be described here explores the use of 5′-nucleotidase as a probe for the recycling of the plasma membrane during pinocytosis.

1.1 Membrane Flow

It has been evident for some time (Steinman et al. 1976) that membrane from the cell surface, which is internalized during pinocytosis, is reutilized, instead of being degraded completely. This has been studied in most detail in the case of receptor-mediated pinocytosis (Brown et al. 1983). In this process, ligands bind to receptors located either in coated pits, or more uniformly over the cell surface; in the latter case, the receptor ligand complex moves rapidly into a coated pit. The complex is then internalized as a vesicle which loses its clathrin coat rapidly; the uncoated vesicle then fuses with an intermediate compartment in the transport to lysosomes, now called the endosome. In the endosome, the ligand and receptor may be separated, with the receptor returning to the cell surface and the ligand being delivered to lysosomes, or the ligand-receptor complex can remain intact and be transported to lysosomes, where both ligand and receptor are degraded (see Steinman et al. 1983).

In all cases in which receptors recycle, the time required for a complete cycle is of the order of a few minutes. There is also evidence that the coated vesicles which enter the cell may not contain a representative group of plasma membrane proteins (see Kerjaschki et al. 1984). In contrast, the rate of plasma membrane recycling as a consequence of fluid phase pinocytosis is much slower (Steinman et al. 1976) and vesicles which contain fluid phase markers seem to contain a representative group of plasma membrane proteins (Mellman et al. 1980). Moreover, an undetermined fraction of this membrane has been shown to enter secondary lysosomes,

[1] Department of Anatomy and Cell Biology, University of Pittsburgh School of Medicine, Pittsburgh, Pennsylvania 15261, USA

from which it returns to the cell surface (Schneider et al. 1979; Muller et al. 1980). While it seems likely that a major fraction of fluid phase pinocytosis occurs via coated vesicles, it is also possible that uptake occurs at uncoated regions of the membrane (see Steinman et al. 1983). Once internalized, however, the content of the vesicles is generally delivered to lysosomes, probably in much the same way as ligands taken up by receptor-mediated pinocytosis.

In addition to pinocytosis, which is a constitutive activity of most cells, specialized cells (e.g., leukocytes) can also internalize appreciable amounts of their cell surface during phagocytosis. The general features of membrane flow in this type of uptake are similar to pinocytosis, i.e., the formation of phagocytic vacuoles which then fuse with lysosomes. The major difference is that phagocytosis can result in the incorporation of a major fraction of the plasma membrane into secondary lysosomes, from which it can recycle to the cell surface (Muller et al. 1980).

A further complication is caused by the fact that not all the material (membrane or content) which enters endosomes is automatically transported to lysosomes. Both ligands bound to receptors, and fluid phase markers, may be exocytosed directly, or, particularly in endothelia or transporting epithelia, may pass across the the cell and be released into the extracellular space at the opposite side of the cell (see Steinman et al. 1983). In the case of transferrin (see Brown et al. 1983), iron is released from the molecule in endosomes, while it is still bound to its receptor, after which the receptor-ligand complex returns to the cell surface, and the apo-transferrin is released from the cell.

In secretory cells there is yet another aspect to the movement of membrane from the cell surface to the interior (Farquhar 1981). The membrane of secretion granules or vacuoles is retrived following exocytosis of the content, and this mem-

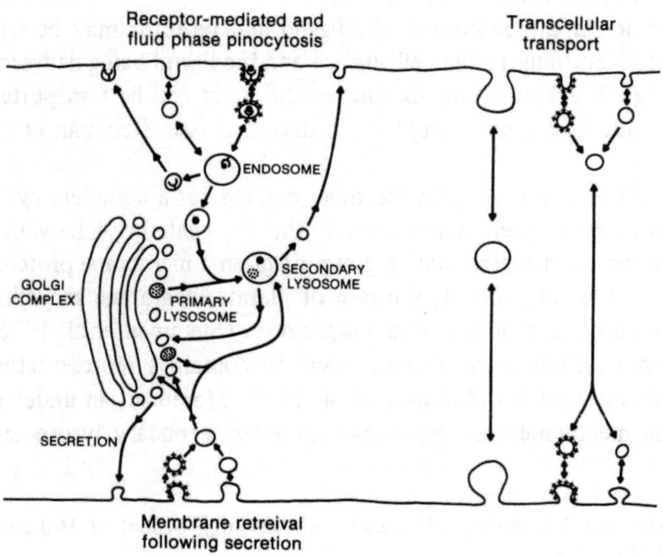

Fig. 1. The various patterns of plasma membrane flow which have been identified in mammalian cells

brane can move directly to the Golgi complex without passing through lysosomes. At present, it is not clear to what extent other plasma membrane constituents follow this pathway.

In the last decade, then, there has been a remarkable increase in the understanding of the flow of membrane between the cell surface and cytoplasmic organelles. This is summarized in Fig. 1. With the exception of a few well-studied receptors, much of this information reflects a qualitative analysis of traffic patterns, so that quantitative information concerning the significance of individual pathways is clearly needed.

1.2 Plasma Membrane 5′-Nucleotidase

As discussed elsewhere in this volume, 5′-nucleotidase activity is present in many cells, and concentrated in the plasma membrane. The active site of the enzyme is disposed externally and, in cells which are not part of organized epithelium, the enzyme is distributed relatively uniformly on the cell surface (Widnell et al. 1982). In heptaocytes the enzyme is distributed over the entire cell surface, but concentrated in the microvilli of the bile canaliculi (Matsuura et al. 1984). In principle, then, the enzyme represents an interesting marker both for analyzing membrane flow in pinocytosis, and for determining the mechanisms involved in maintaining plasma membrane differentiation in epithelia.

However, in order to use the enzyme to probe these events, it is necessary to label the protein. An ideal labeling technique would be specific, and would permit the labeling of the enzyme either at the cell surface or at intracellular locations. In addition, the label incorporated into protein should have no effect on either its subcellular location or stability, or on the ability of the enzyme to follow the normal pathways involved in membrane flow.

This ideal has yet to be realized in full. The analysis of membrane recycling requires labeling at a specific site (usually the cell surface) and the subsequent demonstration that the labeled membrane protein is internalized and then returns to the cell surface. Biogenetic labeling, which results in no modification of the protein, does not permit the distinction between surface and internal locations for the molecule. Other labeling techniques can provoke events which would not occur during the normal lifetime of the molecule. For example, iodinated proteins may exhibit a rapid breakdown (Muller et al. 1980) and the binding of antibodies may provoke the internalization of membrane proteins (see Schneider et al. 1979). It is thus of considerable importance to demonstrate that the labeling technique employed for an individual protein does not result in a series of events which are unrelated to normal membrane flow.

The technique of utilizing anti-plasma membrane antibodies to probe membrane recycling was first exploited successfully by Schneider et al. (1979). These experiments prompted the analyses of 5′-nucleotidase and its antibody which are described here.

2 Experimental Results

2.1 5′-Nuleotidase Activity in Fibroblasts

Although in some cells most of the 5′-nucleotidase activity is accessible to the external medium (DePierre and Karnovsky 1974), this is not the situation in all cases. Less than 50% of the 5′-nucleotidase activity in cultured rat embryo fibroblasts was accessible to the surface (Widnell et al. 1982), and ~90% of the surface activity was inhibited when cells were incubated at 4° with a polyclonal anti-5′-nucleotidase antibody. This antibody inhibited the enzyme almost completely when incubated with cells solubilized in detergent.

This distribution of the enzyme between the cell surface and the interior was supported by cytochemical studies, which demonstrated reaction product from the enzyme both on the cell surface and at internal sites, the most prominent of which were numerous cytoplasmic vesicles and vacuoles, together with the Golgi complex. When cells were treated with anti-5′-nucleotidase antibody at 4°, and then processed for cytochemistry, only the surface activity was inhibited. It should be emphasized that the inhibition of surface activity was virtually complete, so that antibody and substrate were accessible to the enzyme both at the surface of the cells in contact with the culture dish, and the surface in direct contact with the medium (Widnell et al. 1982).

2.2 Effect of Binding Anti-5′-Nucleotidase Antibody to Fibroblasts

Anti-5′-nucleotidase was bound to fibroblasts at 4°, to inhibit the surface activity, and the cells were returned to culture. The amount of antibody remaining with the cells was analyzed by the use of [^{3}H]-antibody, and the amount on the cell surface was analyzed, in parallel experiments, by the subsequent binding of a [^{3}H]-sheep anti-rabbit $F(ab')_2$. The use of a second antibody directed against $F(ab')_2$ fragments of IgG is essential, since it has been shown that the Fc region of the IgG molecule is particularly susceptible to degradation by lysosomal hydrolases (Schneider et al. 1981). The distribution of 5′-nucleotidase activity between the cell surface and the interior of the cell was also studied.

The results are shown in Table 1. After reculture for 9 h ~75% of the anti-5′-nucleotidase antibody remained associated with the cells, and most of the release of radioactivity which occurred was detected during the first 3 h. At the same time, ~65% of the anti-5′-nucleotidase was internalized, and the rate of this internalization was similar to the rate at which 5′-nucleotidase activity was transferred from the cell interior to the surface. The findings suggested that an exchange occurred between the cytoplasmic and surface locations of the enzyme.

To determine the subcellular fate of the enzyme and its antibody, cells were analyzed by isopycnic density gradient centrifugation. In untreated cells, 5′-nucleotidase activity exhibited a major peak with a density of ~1.14, with a shoulder extending to a density of ~1.22 (Fig. 2). Anti-5′-nucleotidase bound to the cells at 4° exhibited a symmetrical peak at $\rho = 1.14$, as did sheep anti-rabbit $F(ab')_2$,

Table 1. Exchange of 5′-nucleotidase between the cell surface and cytoplasmic membranes in rat fibrolasts [a]

Treatment	Reculture	Cell-bound [^{3}H]-anti-5′-nucleotidase (μgmg^{-1} protein)	5′-Nucleotidase activity (nmolmin^{-1} mg^{-1} protein) Surface	Total	Internal	Surface anti-5′-nucleotidase (μg sheep anti-rabbit F(ab′)$_2$ boundmg^{-1} protein)
None	None	None	31	72	41	None
Anti-5′-nucleotidase	None	0.86	5	41	36	1.33
Anti-5′-nucleotidase	3 h	0.71	11	43	32	0.87
Anti-5′-nucleotidase	6 h	0.66	15	44	29	0.61
Anti-5′-nucleotidase	9 h	0.64	18	42	24	0.54

[a] Experimental procedures were as described by Widnell et al. (1982), i.e., cells were treated with anti-5′-nucleotidase antibody at 2° and then recultured for the times indicated. [^{3}H]-anti-5′-nucleotidase F(ab′)$_2$ was bound to one set of dishes to determine the fraction of the antibody which remained associated with the cells during the course of the experiment. Nonradioactive anti-5′-nucleotidase was bound to another set of dishes which were then used to analyze surface and internal (calculated as the differences between total and surface) 5′-nucleotidase activity, as well as surface anti-5′-nucleotidase !analyzed by the binding of [^{3}H]-sheep anti-rabbit F(ab′)$_2$]

also bound at 4°. The binding of anti-5′-nucleotidase at this temperature also inhibited the enzyme activity recovered in the low density region of the gradient, but had no effect on the activity recovered in the denser region of the gradient.

When the cells were treated with anti-5′-nucleotidase and then recultured for 16 h, there was no accumulation of antibody in the region of the gradient corresponding to lysosomes, and no change in the distribution of the lysosomal marker, acid phosphatase. However, the distribution of the anti-5′-nucleotidase antibody was essentially the same as that of the enzyme in untreated cells, indicating that antibody had gained access to the same compartments occupied normally by the enzyme.

Since the membranes of the Golgi complex, as judged by the distribution of galactosyl transferase, were recovered in a similar region of the gradient, it was desirable to investigate the fraction of the activity in this compartment. By using the digitonin-shift procedure (Amar-Costesec et al. 1974), the membranes of the Golgi complex were partly resolved from the majority of the 5′-nucleotidase (not shown). This confirmed for fibroblasts the results obtained with hepatocyte fractions by Ravoet et al. (1981), and indicated that most of the internal 5′-nucleotidase was located in membrane with a cholesterol content similar to that of the cell surface.

These results showed that binding of anti-5′-nucleotidase antibody to the enzyme had no obvious effect on its subcellular distribution, and did not result in the complete internalization of the antigen-antibody complex. However, as shown by

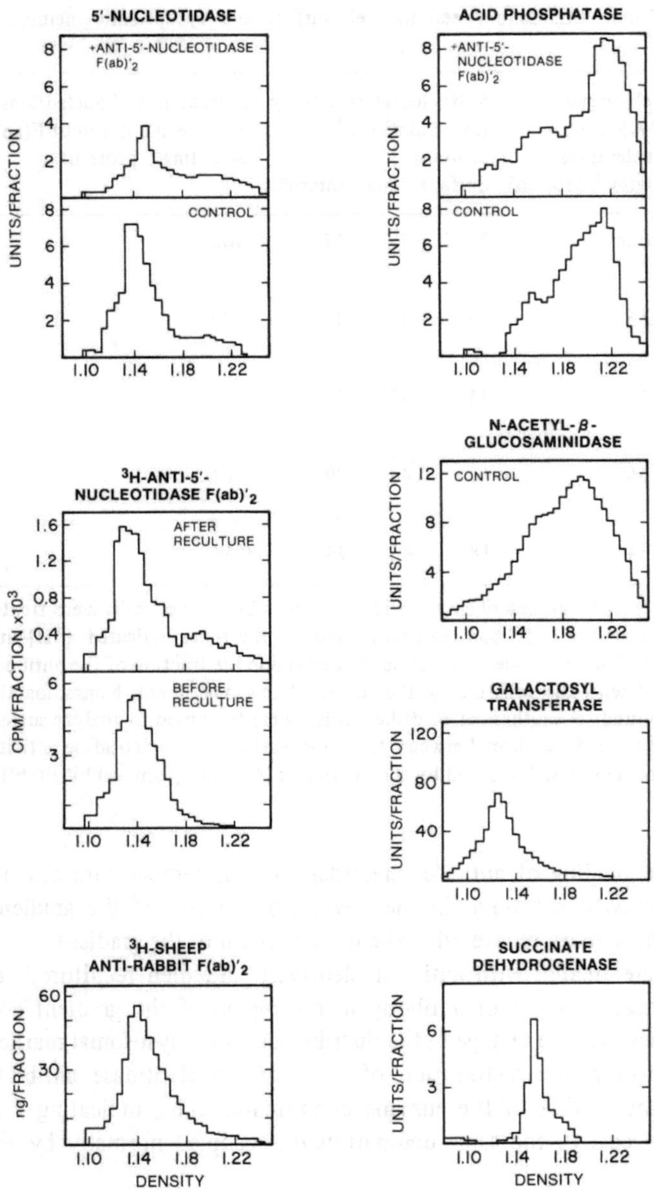

Fig. 2. Isopycnic density gradient analysis of postnuclear supernatant fractions from homogenates of rat embryo fibroblasts. The distribution of 5′-nucleotidase, acid phosphatase, N-acetyl-β-glucosaminidase, galactosyl transferase, and succinate dehydrogenase is shown for control (untreated) cells. The distribution of 5′-nucleotidase and [^{3}H]-sheep anti-rabbit F(ab′)$_2$ is shown for cells incubated at 2° first with anti-5′-nucleotidase F(ab′)$_2$ and then with [^{3}H]-sheep anti-rabbit F(ab′)$_2$. The distribution of [^{3}H]-anti-5′-nucleotidase is shown for cells treated with the antibody at 2° both before and after reculture for 15 h; for the latter cells, the distribution of acid phosphatase is also shown

Huet et al. (1980) for the H_2 antigen, binding a second [sheep anti-rabbit $F(ab')_2$] antibody to the anti-5′-nucleotidase did result in the complete internalization of the resultant immunocomplex (Berman et al. 1982).

2.3 Exchange of 5′-Nucleotidase, Pinocytosis, and Recycling

While the previous experiments suggested that binding anti-5′-nucleotidase antibody to cells did not result in gross alterations in the subcellular distribution of the enzyme, it was still possible that the exchange was unrelated to any major aspect of plasma membrane flow. Since the enzyme represents a fraction of a percent of the plasma membrane protein (Widnell 1974; Bailyes et al. 1982), it would be possible for the cell to achieve an exchange of the enzyme with minimal exchange of the total area of the cell surface membrane. The experiments also did not establish whether the anti-5′-nucleotidase returned to the surface following internalization, as was shown for anti-plasma membrane IgG by Schneider et al. (1979) and for iodinated phagolysosomal membrane protein by Muller et al. (1980).

To determine the relationship between the exchange of 5′-nucleotidase and pinocytosis, the effect was studied under conditions in which pinocytosis was inhibited. When rat fibroblasts were treated with low concentrations (1 μM) of monensin for 12 h or more, the rate of fluid phase uptake was decreased by 70–80% (Wilcox et al. 1982). Under the same conditions, the exchange of 5′-nucleotidase was also inhibited, almost completely (Fig. 3). The effect of monensin was reversible (Fig. 3), and seemed to be the consequence of the accumulation of membrane in vacuoles in the cells (Wilcox et al. 1982).

The exchange of 5′-nucleotidase also correlated with the rate of pinocytosis in mouse peritoneal macrophages. Macrophages have been shown to internalize

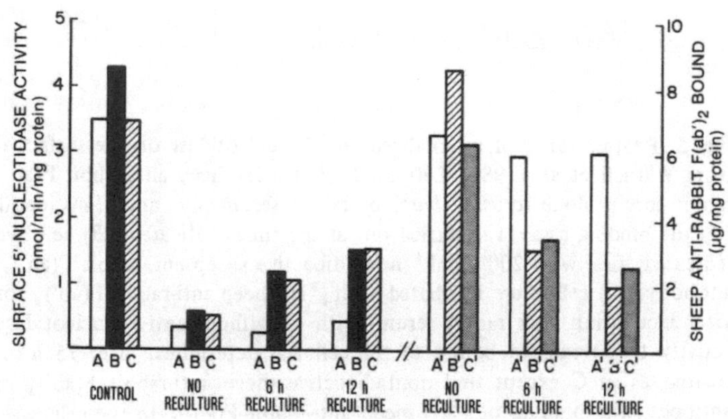

Fig. 3. Effect of monensin treatment on the exchange of 5′-nucleotidase, reprinted from Wilcox et al. (1982). The experiment was carried out essentially as described in Table 1, and was performed with cells treated for 18 h with 1 μM monensin (*A*), control (untreated) cells (*B*), and cells treated for 18 h with 1 μM monensin followed by 12 h in fresh medium (*C*). Results in the *left panel* describe surface 5′-nucleotidase activity, and in the *right*, the binding of [^{3}H]-sheep anti-rabbit $F(ab')_2$ at 2°, at the beginning of the experiment and following reculture for the times indicated

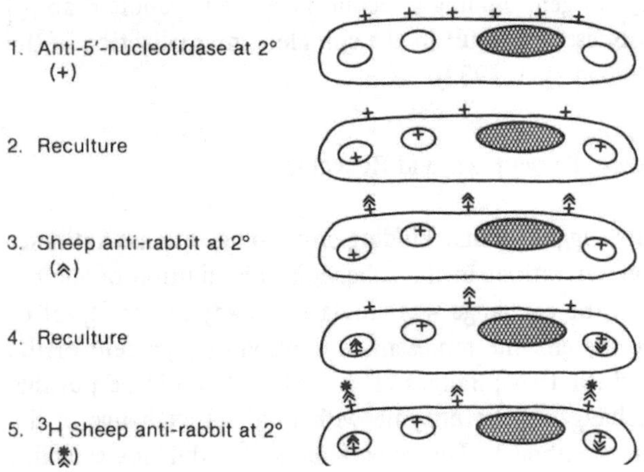

Fig. 4. The experimental approach used to study the recycling of anti-5′-nucleotidase antibody in rat fibroblasts

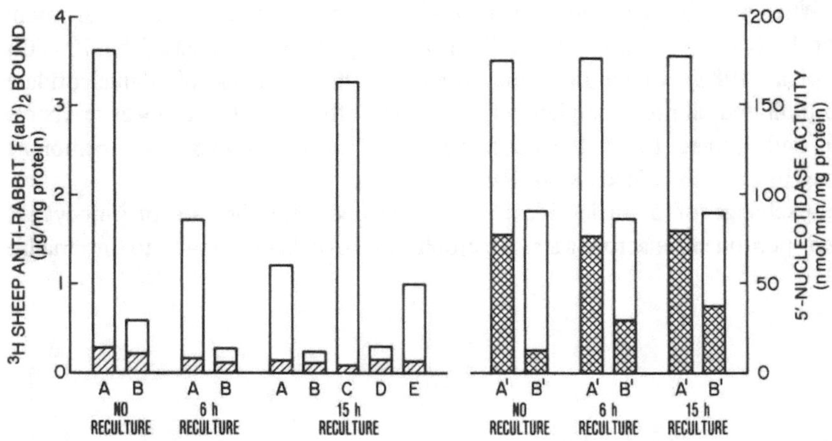

Fig. 5. Reappearance of internalized anti-5′-nucleotidase on the surface of fibroblasts, reprinted from Widnell et al. (1982) **Left** Binding of [^{3}H]-sheep anti-rabbit F(ab′)$_2$ to cells treated with rabbit serum alone (*crossed bars*) or rabbit serum plus anti-5′-nucleotidase serum (*open bars*); (*A*) the binding assay was carried out at the times indicated; (*B*) cells were treated, at the times indicated, first with 200 µgml^{-1} nonradioactive sheep anti-rabbit F(ab′)$_2$, then with radioactive antibody; (*C*) cells were incubated with [^{3}H]-sheep anti-rabbit F(ab′)$_2$ for 2 h at 2° immediately after incubation with rabbit serum, with or without anti-5′-nucleotidase serum, and the radioactivity that remained bound to the cell was determined after 15 h in culture; (*D*) cells were treated as in *C* except that nonradioactive sheep anti-rabbit F(ab′)$_2$ replaced the radioactive antibody; the binding of [^{3}H]-sheep anti-rabbit F(ab′)$_2$ to the cells was determined after 15 h in culture; (*E*) cells were treated with nonradioactive sheep anti-rabbit F(ab′)$_2$ after 6 h in culture, and binding of the radioactive antibody was determined after a further 9 h in culture. **Right** Total (*open bars*) and cell surface (*hatched bars*) 5′-nucleotidase activity; (*A*′) cells treated with rabbit serum; (*B*) cells treated with rabbit serum plus anti-5′nucleotidase serum

an amount of membrane corresponding to their entire cell surface in ~30 min as a result of fluid phase pinocytosis, whereas in L cells this extent of internalization required ~2 h (Steinmann et al. 1975). The exchange of 5′-nucleotidase in macrophages was essentially complete after 90 min (Hitchener and Widnell 1980), in contrast to the ~8 h required for rat fibrobolasts. In other respects, however, the properties of the exchange in macrophages was very similar to that found in rat fibroblasts.

The recycling of anti-5′-nucleotidase was also studied in rat fibroblasts. The experimental approach, which is shown in outline in Fig, 4, involved the use of the second antibody, sheep anti-rabbit $F(ab')_2$. Anti-5′-nucleotidase was bound to cells at 4°. The cells were returned to culture, and the amount of free anti-5′-nucleotidase (i.e., uncomplexed with the second antibody) on the cell surface was analyzed by the binding of [^{3}H]-sheep anti-rabbit $F(ab')_2$. The results are shown in Fig. 5.

5′-Nucleotidase activity exhibited the expected exchange between the cell interior and the cell surface, and the antibody binding studies showed: (1) that prior binding of nonradioactive sheep anti-rabbit antibody inhibited the binding of the radioactive molecule by at least 90%; (2) that very little sheep antibody was released from the cells during the course of the experiment, as judget by the loss of [^{3}H]-sheep anti-rabbit $F(ab')_2$ bound to anti-5′-nucleotidase at the beginning; and (3) that there was no appearance of [^{3}H]-sheep anti-rabbit antibody binding sites in cells in which anti-5′-nucleotidase was saturated with nonradioactive antibody at zero time.

However, the binding studies also indicated that a third of the anti-5′-nucleotidase which was internalized during the first period of reculture, returned to the cell surface during the second, i.e., that the antibody recycled completely. In addition, the rate at which anti-5′-nucleotidase was transferred to the cell surface during the second reculture in this experiment was similar to the rate at which 5′-nucleotidase activity was transferred from the internal compartment to the cell surface in cells treated with anti-5′-nucleotidase alone.

2.4 Intracellular Events in the Recycling of 5′-Nucleotidase

The experiments described in the previous sections have treated the cell as a black box, in which the surface and the interior were distinguished, but in which the possibility of compartmentation in the cytoplasm was ignored. Since 5′-nucleotidase activity has been detected in several different cytoplasmic membranes, it is clearly important to determine which of these organelles participate in recycling of the plasma membrane.

To achieve this, it would obviously be helpful if site-specific inhibitors were available to block recycling at particular sites along the pathway. The Golgi complex represents a site of obvious potential interest: this organelle is involved in the segregation of plasma membrane proteins from lysosomal matrix proteins following their synthesis (Rothman 1981) and in the recycling of secretory granule membrane from the cell surface (Farquhar 1981). Low concentrations of the ionophore, monensin, have been shown to inhibit secretion by interfering with the function of the Golgi complex (Tartakoff and Vassali 1978) and higher concentrations have been

Table 2. Effects of monensin treatment on pinocytosis and sodium and potassium ion concentrations in rat embryo fibroblasts

Treatment	HRP Uptake ($ngmin^{-1} mg^{-1}$ protein)	Na^+/K^+
None	2.13	0.09
1 μM Monensin 3 h	1.87	0.11
1 μM Monensin 6 h	0.93	0.10
1 μM Monensin 12 h	0.62	0.12
25 μM Monensin 1 h	1.34	0.54
25 μM Monensin 3 h	0.59	0.57
25 μM Monensin 6 h	0.43	0.62

shown to inhibit the recycling of the LDL receptor (Basu et al. 1981). It, therefore, seemed worthwhile to study the effect of this compound on fluid phase pinocytosis (Wilcox et al. 1982).

Monensin treatment, in the concentration range 1–25 μM, resulted in an inhibition of pinocytosis. However, the time course of the inhibition was concentration dependent (Table 2), so that inhibition was detected at 1 h in the presence of 25 μM monensin, whereas there was no effect until 3 h in the presence of 1 μM monensin. This difference did not seem to be caused by a slow uptake of the ionophore by the cells, since 25 μM monensin caused a rapid change (within 15 min) in the ratio of the concentration of Na^+ and K^+, whereas the ratio was not affected by 1 μM monensin, even after 15 h (Table 2). In addition, experiments using [^{14}C]-monensin provided no evidence for a significant, time-dependent, accumulation of monensin by the cells.

There were also qualitative differences in the nature of the inhibition. In control cells, and in cells treated with 1 μM monensin, most of the fluid phase marker was transported to lysosomes. In cells treated with 25 μM monensin, the marker accumulated in a vesicle fraction with a low density, exhibiting a distribution expected for endosomes (Fig. 6); only a trace was recovered in lysosomes. This failure to transport markers to lysosomes in the presence of 25 μM monensin was confirmed by morphological studies analyzing the uptake of fluorescein dextran (using the light microscope) and HRP or cationic ferritin (using the electron microscope).

The results were consistent with the findings of others (see Harford et al. 1983), which indiate that high concentrations of monensin, which alter the pH within lysosomes and endosomes (Maxfield 1982), inhibit pinocytosis by blocking the transport of material from endosomes to lysosomes. In contrast, this stage in the pathway did not seem to be altered in cells treated with 1 μM monensin, suggesting that the inhibition was the consequence of an effect on membrane recycling at a stage during the return of membrane to the cell surface either from lysosomes, or from a pre-lysosomal or endosome compartment.

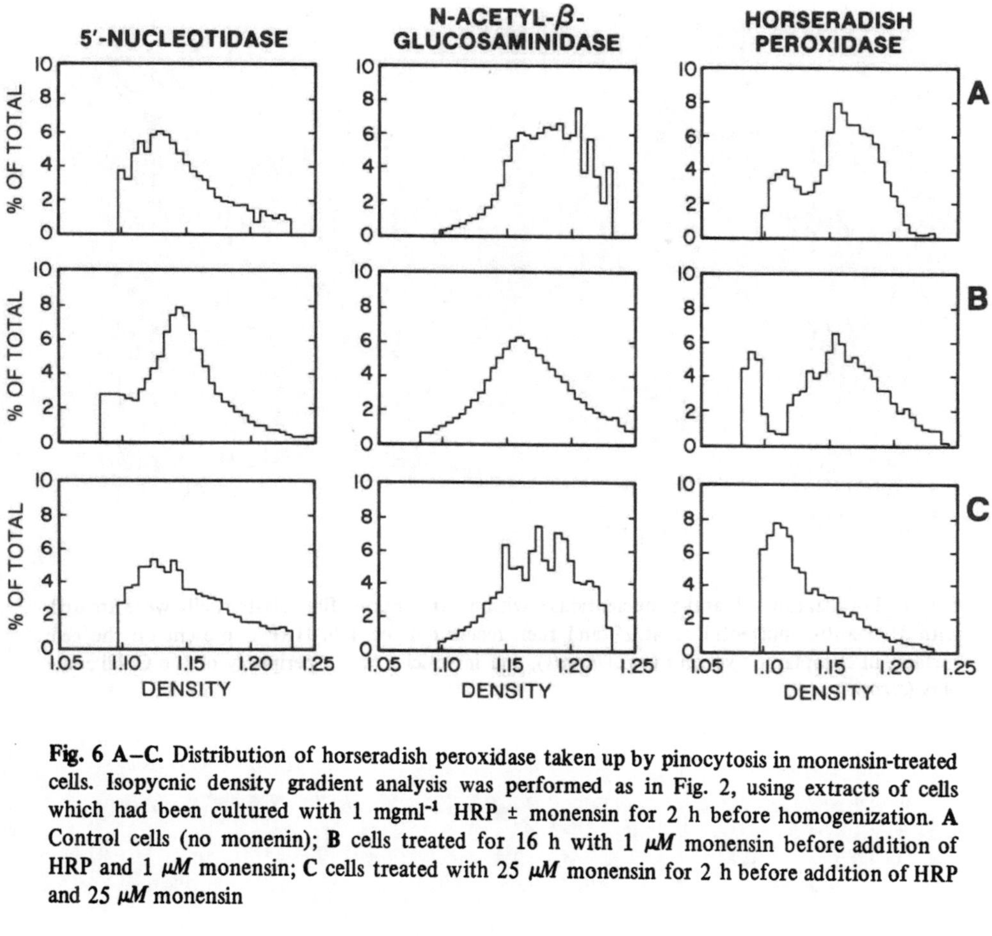

Fig. 6 A–C. Distribution of horseradish peroxidase taken up by pinocytosis in monensin-treated cells. Isopycnic density gradient analysis was performed as in Fig. 2, using extracts of cells which had been cultured with 1 mgml^{-1} HRP ± monensin for 2 h before homogenization. **A** Control cells (no monenin); **B** cells treated for 16 h with 1 *μM* monensin before addition of HRP and 1 *μM* monensin; **C** cells treated with 25 *μM* monensin for 2 h before addition of HRP and 25 *μM* monensin

To investigate the possible role of the Golgi complex in the return of membrane to the cell surface during plasma membrane recycling, the experiments of Schneider et al. (1979) were repeated at the morphological level, using anti-5′-nucleotidase instead of anti-plasma membrane antibodies. In the first experiments, an anti-5′-nucleotidase IgG-HRP conjugate was bound to cells at 4^{0}, and the cells were then recultured. Analytical isopycnic density gradient analysis showed that the HRP-conjugated antibody exhibited a similar distribution to that shown for the [^{3}H]-antibody in Fig. 2. When cells were studied by electron microscopy after reculture, HRP was localized on the cell surface, in numerous cytoplasmic vesicles and vacuoles, and in vesicles at the periphery of the Golgi complex (Fig. 7). In contrast to the cytochemical findings (Widnell et al. 1982), very little HRP was detected in the Golgi stacks, suggesting that the 5′-nucleotidase activity detected in the stacks might represent enzyme in the process of biosynthesis, as opposed to enzyme engaged in recycling.

In the next experiments, cells were incubated with a sheep anti-rabbit F(ab′)$_2$ IgG-HRP conjugate, to load lysosomes with this second antibody as a result of

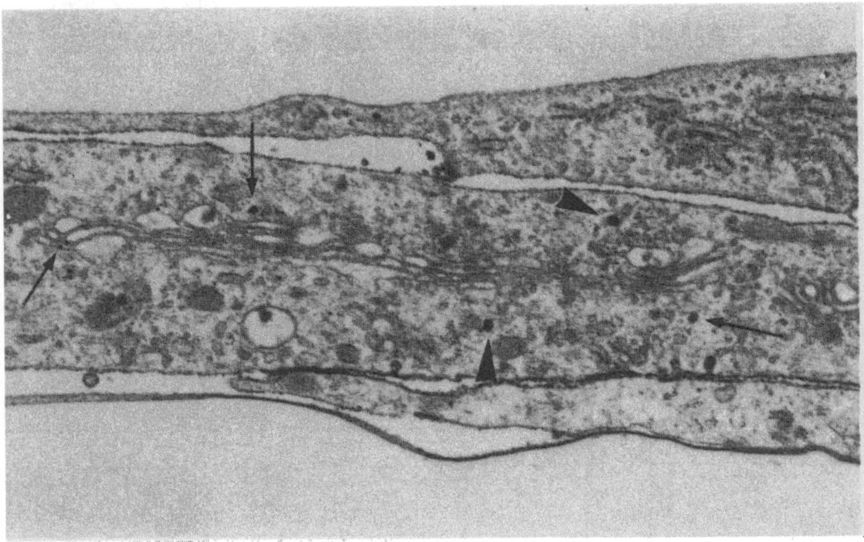

Fig. 7. Localization of anti-5′-nucleotidase within rat embryo fibroblasts. Cells were treated with HRP-anti-5′-nucleotidase at 2° and then recultured for 4 h. HRP is present on the cell surface, in cytoplasmic vesicles (*arrowheads*), and in vesicles at the periphery of the Golgi complex (*arrows*)

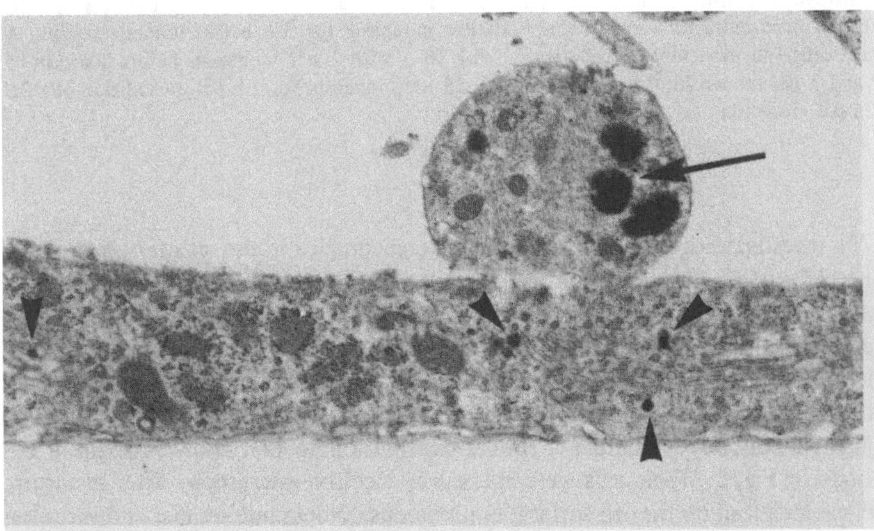

Fig. 8. Localization of sheep anti-rabbit $F(ab')_2$ transported from lysosomes by anti-5′-nucleotidase. Lysosomes were loaded with HRP-sheep anti-rabbit $F(ab')_2$, using the technique of Schneider et al. (1979), treated with anti-5′-nucleotidase at 2° and then recultured for 4 h. HRP is located in lysosomes (*arrows*) and in small vesicles at the periphery of the Golgi complex (*arrowheads*)

pinocytosis. After this, cells were treated with control or anti-5′-nucleotidase IgG at 4^{0} and finally recultured. Cells treated with control igG showed HRP restricted to lysosomes (not shown), whereas cells treated with anti-5′-nucleotidase IgG contained numerous cytoplasmic vesicles and vacuoles, including many at the periphery of the Golgi complex, which stained for HRP (Fig. 8).

3 Discussion

These experiments have been based on the hypothesis that antibody to ecto-5′-nucleotidase could be used to follow the intracellular events in plasma membrane recycling as a result of fluid phase pinocytosis. For this approach to be useful, it is essential that binding the antibody to the enzyme should not affect the normal flow of the protein within the cell.

Several results were consistent with this being the case. First, antibody remained bound to the cells and the enzyme remained inhibited as a result of this binding for several hours (Table 1). Second, when antibody was bound to the cell surface, and the cells were then then recultured, the distribution of antibody in isopycnic density gradients was very similar to that of the enzyme itself in preparations from untreated cells (Fig. 2). Third, the rate of transport of the enzyme-antibody complex from the interior of the cell to the surface was very similar to that of the enzyme when it was free of antibody (Fig.5).

These results provide additional support for the conclusions of Schneider et al. (1979), who were the first to use antibodies to plasma membrane constituents for studies of this type. It must be emphasized that none of the results to date have determined the precise stability of the antigen-antibody complex within the cell. Binding analyses (Berman et al. 1982) suggest that the complexes both form and are stable in the range pH 4.5–7.5, which corresponds to what might be expected in intracellular compartments (Okhuma and Poole 1978). A major question which remain to be answered is whether during recycling, the antibody remains bound to the same molecule of 5′-nucleotidase with which it associated originally at the cell surface. If this is not the case, the system will still provide useful information on the pathway(s) followed by membrane; to obtain quantitative information on the flow of plasma membrane, however, it will be essential to resolve this problem.

The earlier studies of Schneider et al. (1979) which indicated that plasma membrane constituents can pass through lysosomes before returning to the cell surface have been confirmed for ecto-5′-nucleotidase (Fig. 8). Since the amount of 5′-nucleotidase activity recovered in the same region as lysosomes is very small (Fig. 2), and since the protein composition of the plasma membrane and lysosome membrane can be distinguished (see Storrie et al. 1984), the results indicate that the transport time of 5′-nucleotidase through lysosomes must be relatively short. For future quantitative studies, it will be necessary to show that the 5′-nucleotidase-antibody complex resides in lysosomes long enough for it to react completely with a second antibody.

It is well established that the Golgi complex and its peripheral vesicles play a major role in the segregation of newly synthesized plasma membrane, secretory protein, and lysosomal enzymes (see Rothman 1981; Brown et al. 1984; Gabel and Kornfeld 1984; Green and Shields 1984). The results suggest that the peripheral vesicles are also involved in the recycling of plasma membrane from lysosomes to the cell surface. This might provide an opportunity to resegregate any lysosomal enzymes which were transported with the plasma membrane from lysosomes, and thus avoid their exocytosis from the cell. In addition, a role for small vesicles at the periphery of the Golgi complex in recycling is consistent with the effects of 1 μM monensin. At this concentration, the compound causes an extensive vacuolation of the Golgi complex (Tartakoff and Vassalli 1978); if this were to involve fusion of peripheral vesicles with the stacks, plasma membrane could become trapped in the vacuoles and fail to return to the cell surface. Preliminary immunocytochemical studies in the laboratory are consistent with this interpretation. Again, it is essential to emphasize the current lack of quantitative information. Future studies must establish what fraction of the 5′-nucleotidase, which enters the cell as pinocytic vesicles during a brief pinocytic pulse, actually reaches the Golgi complex via lysosomes.

Finally, it is perhaps worth emphasizing that in many cells a substantial fraction of the ectoenzyme population may reside temporarily within the cell as membranes engaged in plasma membrane recycling. This situation provides the potential for the enzymes to derive their substrates from, and release their products into, the cell cytoplasm as a result of transport across the vesicle membrane.

4 Summary

Ecto-5′-nucleotidase represents a useful marker protein for studies on plasma membrane flow during fluid phase pinocytosis. The experimental approach is based on binding a polyclonal antibody, which inhibits enzymatic activity, to the surface of rat embryo fibroblasts, and following both the transport of the enzyme from cytoplasmic membrane to the cell surface, and the intracellular pathway followed by the antibody during its internalization and subsequent return to the cell surface. Evidence is presented that a fraction of the antibody enters lysosomes and the the pathway from lysosomes back to the cell surface involves vesicles at the periphery of the Golgi complex.

References

Amar-Costesec A, Wibo M, Thines-Sempoux D, Beaufay H, Berthet J (1974) Analytical study of microsomes and isolated subcellular membranes from rat liver. IV. Biochemical, physical, and morphological modifications of microsomal components induced by digitonin, EDTA, and pyrophosphate. J Cell Biol 62:717–745

Bailyes EM, Newby AC, Siddle K, Luzio JP (1982) Solubilization and purification of rat liver 5′-nucletidase by use of a zwitterionic detergent and a monoclonal-antibody immunoadsorbent. Biochem J 203:245–251
Basu SK, Goldstein JL, Anderson RGW, Brown MS (1981) Monensin interrupts the recycling of low density lipoprotein receptors in human fibroblasts. Cell 24:493–502
Berman CL, Hilderson HK, Widnell CC (1982) Stability and fate of membrane-bound antibody-antigen complexes. J Cell Biol 95:438a
Brown MS, Anderson RGW, Goldstein JL (1983) Recycling receptors: the round-trip itinerary of migrant membrane proteins. Cell 32:663–667
Brown WJ, Constantinescu E, Farquhar MG (1984) Redistribution of mannose-6-phosphate receptors induced by tunicamycin and chloroquine. J Cell Biol 99:320–326
DePierre JW, Karnovsky ML (1974) Ecto-enzymes of the guinea pig polymorphonuclear leucocyte. I. Evidence for an ecto-adenosine monophosphatase, -adenosine triphosphatase, and -p-nitrophenyl phosphatase. J Biol Chem 249:7111–7120
Gabel CA, Kornfeld S (1984) Targeting of β-glucuronidase to lysosomes in mannose 6-phosphate receptor-deficient MOPC 315 cells. J Cell Biol 99:296–305
Green R, Shields D (1984) Somatostatin discriminates between the intracellular pathways of secretory and membrane proteins. J Cell Biol 99:97–104
Harford J, Wolkoff AW, Ashwell G, Klausner R (1983) Monensin inhibits intracellular dissociation of asialoglycoproteins from their receptor. J Cell Biol 96:1824–1828
Hitchener WR, Widnell CC (1980) Membrane exchange between the plasma membrane and cytoplasm during endocytosis by mouse macrophages. Eur J Cell Biol 22:206
Huet D, Ash JF, Singer SJ (1980) The antibody-induced clustering and endocytosis of HLA antigens on cultured human fibroblasts. Cell 21:429–438
Kerjaschki D, Noronha-Blob L, Sacktor B, Farquhar MG (1984) Microdomains of distinctive glycoprotein composition in the kidney proximal tubule brush border. J Cell Biol 98: 1505–1513
Matsuura S, Eto S, Kato K, Tashiro Y (1984) Ferritin immunoelectron microscopic localization of 5′-nucleotidase on rat liver cell surface. J Cell Biol 99:166–173
Maxfield FR (1982) Weak bases and ionophores rapidly and reversibly raise the pH of endocytic vesicles in cultured mouse fibroblasts. J Cell Biol 95:676–681
Mellman IS, Steinman RM, Unkeless JC, Cohn ZA (1980) Selective iodination and polypeptide composition of pinocytic vesicles. J Cell Biol 86:712–722
Muller WA, Steinman RM, Cohn ZA (1980) The membrane proteins of the vacuolar system II. Bi-directional flow between secondary lysosomes and plasma membrane. J Cell Biol 86:304–314
Okhuma S, Poole B (1978) Fluorescence probe measurement of the intralysosomal pH in living cells and the perturbation of pH by various agents. Proc Nat Acad Sci USA 75:3327–3331
Ravoet A-M, Amar-Coestesec A, Godelaine D, Beaufay H (1981) Quantiative assay and subcellular distribution of enzymes acting on dolichyl phosphate in rat liver. J Cell Biol 91: 679–688
Rothman JE (1981) The Golgi apparatus: two organelles in tandem. Science (Wash DC) 213: 1212–1219
Schneider Y-J, Tulkens P, deDuve D, Trouet A (1979) Fate of plasma membrane during endocytosis II. Evidence for recycling (shuttle) of plasma membrane constituents. J Cell Biol 82:466–474
Schneider Y-J, deDuve C, Trouet A (1981) Fate of plasma membrane furing endocytosis III. Evidence for incomplete breakdown of immunoglubulins in lysosomes of cultured fibroblasts. J Cell Biol 88:380–387
Steinman RM, Brodie SE, Cohn ZA (1976) Membrane flow during pinocytosis. J Cell Biol 68:665–687
Steinman RM, Mellman IS, Muller WA, Cohn ZA (1983) Endocytosis and the recycling of plasma membrane. J Cell Biol 96:1–27
Storrie B, Pool RR Jr, Sachdeva M, Maurey KM, Oliver C (1984) Evidence for both prelysosomal and lysosomal intermediates in endocytic pathways. J Cell Biol 98:108–115

Tartakoff A, Vassalli P (1978) Comparative studies of intracellular transport of secretory proteins. J Cell Biol 79:694–707

Widnell CC (1974) Purification of rat liver 5′-nucleotidase as a complex with spingomyelin. In: Fleischer S, Packer L, Estabrook RW (eds). Biomembranes, part B. A Volume in methods in enzymology, vol 32. Academic, New York, pp 368–374

Widnell CC, Schneider Y-J, Peirre B, Baudhuin P, Trouet A (1982) Evidence for a continual exchange of 5′-nucleotidase between the cell surface and membranes of the cytoplasm in cultured rat fibroblasts. Cell 28:61–70

Wilcox DK, Kitson RP, Widnell CC (1982) Inhibition of pinocytosis in rat embryo fibroblasts treated with monensin. J Cell Biol 92:859–864

Isolation and Characterization of 5′-Nucleotidase from Avian Muscle Sources and its Interaction with Filamentous Actin

J. Dieckhoff, M. Heidemann, R. Lietzke, and H. G. Mannherz [1]

1 Introduction

5′-Nucleotidase has been known for a long time to be an ectoenzyme concentrated in the plasma membrane of many eukaryotic cells. DePierre and Karnovsky (1974) first demonstrated its nature as an ectoenzyme and since then a vast amount of literature has been compiled concerning its enzymatic properties and presumed functional significance. In spite of efforts in many laboratories it has been impossible to assign a definitive function to this enzyme. It has been found, however, that adenosine, the product formed by its enzymatic action on 5′-AMP, can act as a local modulator of cellular functions in a number of different tissues, such as the vascular and central nervous systems (for a review see Arch and Newsholme 1978).

Our interest in 5′-nucleotidase arose from results we obtained with 5′-nucleotidase isolated from snake venom (Crotalus adamanteous) and rat bile which indicated its ability to interact with the cytosekeletal component actin. These results showed that the otherwise very stable 1:1 complex of actin and bovine pancreatic DNase I is separated into its protein constituents in a time-dependent manner when incubated with 5′-nucleotidase either from snake venom or rat bile (Mannherz and Rohr 1978; Rohr and Mannherz 1979). A possible interaction of 5′-nucleotidase and actin has also been inferred from results obtained using living cells. These experiments showed that microfilament disturbing drugs, such as cytochalasin B, altered the response of the cellular ecto-AMPase to concanavalin A (Carraway et al. 1979) which ordinarily inhibits the AMPase activity of isolated 5′-nucleotidase (Slavik et al. 1977). Since 5′-nucleotidase is concentrated in the plasma membrane of most eurkaryotic cells, we speculated that this enzyme might participate in mechanisms that anchor actin filaments to the cytoplasmic face of the plasma membrane. This hypothesis is supported by experimental results indicating 5′-nucleotidase to be a spanning protein with domains at the extracellular and cytoplasmic faces of the plasma membrane (Zachowski et al. 1981). In order to test this hypothesis it was considered necessary to study the interaction of both proteins after their purification in vitro and to assess their spatial relationship in vivo by morphological techniques. However, since 5′-nucleotidase from snake venom or rat bile can be regarded as a secretory isoform with possible structural and functional differences, we chose as starting material for purification the chicken gizzard, skeletal and heart muscle where it is assumed to occur solely incorporated into the membrane.

[1] Institut für Anatomie und Zellbiologie, Robert-Koch-Str. 6, 3550 Marburg, FRG

2 Purification of 5′-Nucleotidase from Avian Muscle Sources

The isolation of 5′-nucleotidase from various avian muscle sources was attempted according to the methods of Dornand et al. (1978) This procedure involves two affinity chromatographic steps using immobilized lens culinaris lectin (LcL) and 5′-AMP-Sepharose 4B. After a number of modifications had been introduced, a very pure 5′-nucleotidase preparation was obtained from chicken gizzard, as judged by SDS-PAGE and the silver stain procedure (Fig. 1). Our purification scheme will be outlined briefly. 5′-Nucleotidase (chicken gizzard) was found to be optimally solubilized with Chaps [3- (3-cholamidopropyl)- dimethylammonio -1- propanesulfate] (Heidemann et al., manuscript in preparation). Triton X-100 at 2% (wt/vol) did not solubilize this enzyme from chicken gizzard. Therefore, the crude muscle homogenate was pretreated with Triton X-100 to remove any readily soluble membrane constituents other than the 5′-nucleotidase. After centrifugation, the residual pellet was treated with 5% (wt/vol) Chaps at 4°C for 1 h. The solubilized crude 5′-nucleotidase was then sequentially subjected to affinity chromatography on AMP-Sepharose and LcL-Sepharose. The enriched 5′-nucleotidase fraction thus obtained was further chromatographed over a DEAE-cellulose column which was designed to retain any protein contaminants still present, while pure 5′-nucleotidase was not adsorbed. The purified enzyme was taken up in 5 m*M* Hepes, pH 7.4, 0.1 m*M* $CaCl_2$, 0.5 m*M* NaN_3, and 0.1% Triton X-100 (buffer A) and could be stored frozen

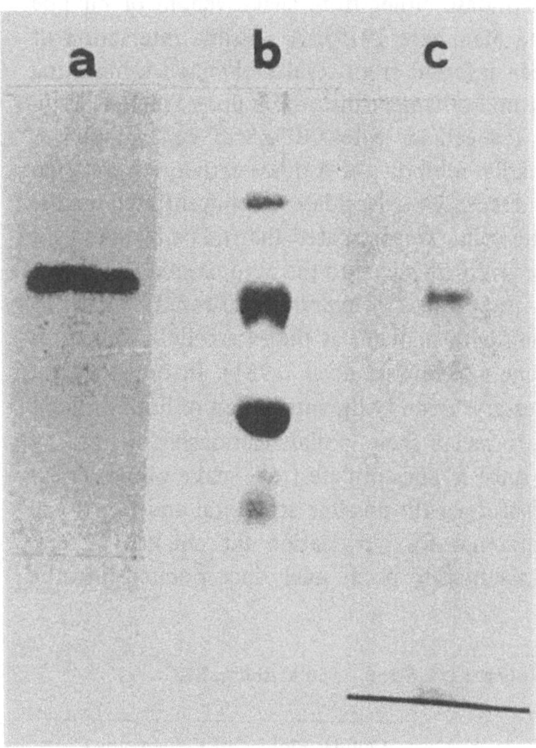

Fig. 1 a–c. SDS-PAGE of purified 5′-nucleotidase from chicken gizzard stained with the silver stain procedure. **a** Purified 5′-nucleotidase and **b** molecular mass marker proteins from *top to bottom;* α-actinin, bovine serum albumin, actin, and bovine pancreatic DNase I. **c** 5′-Nucleotidase purified from rat liver

at -20°C without any noticable loss of enzymatic activity for up to 3 months. Using this procedure 5′-nucleotidase from rat liver could also be purified (Fig. 1). This enzyme appears be of lower molecular mass.

3 Properties of 5′-Nucleotidase Purified from Chicken Gizzard

3.1 Enzymatic Properties

According to SDS-PAGE, purified 5′-nucleotidase appears as a single polypeptide of 79 000 molecular mass. After gel filtration of the purified enzyme on Sephacryl S300 the AMPase activity appeared at an elution volume corresponding to a molecular mass of 152 000, indicating that the native enzyme might exist as a homodimer in the presence of 0.1 *M* NaCl and 0.6% (wt/vol) sodium deoxycholate (Fig. 2). Using other buffer conditions higher molecular mass aggregates were observed with this technique, indicating a strong tendency of purified 5′-nucleotidase for self-aggregation (Heidemann et al. to be published).

Hydrolysis of 5′-AMP was found to follow Michaelis-Menten kinetics up to 0.1 m*M* with a K_m of 6.2 ± 1 μ*M* and a V_{max} of 2.7 x 10^{-6} *M* AMP hydrolyzed per minute per mg enzyme. Setting the rate of AMP-hydrolysis at 100% the rate of breakdown of 5′-GMP, 5′-UMP, 5′CMP, and 5′-IMP was found to be 17%, 12%, 16%, and 20%, respectively. 5′-Nucleotidase is competitively inhibited by ATP, ADP, and APCP with K_i-values of 15 μ*M*, 2.9 μ*M*, and 19 n*M*, respectively. In agreement with Burger and Lowenstein (1975) $MgCl_2$ was found to lower the inhibitory capacity of these nucleotides (data not shown). The purified 5′-nucleotidase was found to be free of contaminating phosphatase activities. However, the presence of an unidentified ATPase activity to a variable degree was noted, but in most cases this constitutes less than 1% of its AMPase activity.

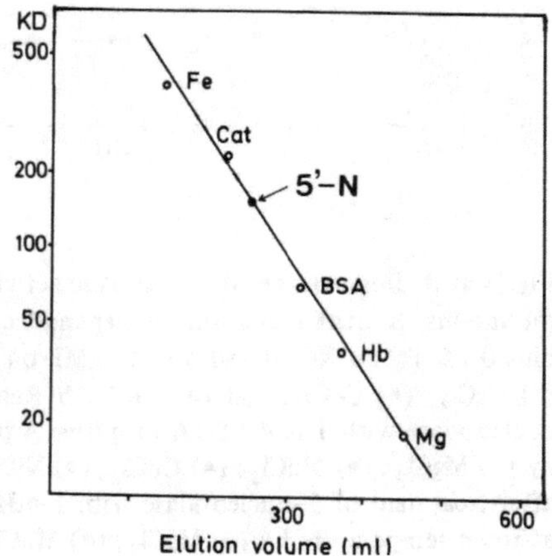

Fig. 2. Determination of the native molecular mass of purified chicken gizzard 5′-nucleotidase by gel filtration on Sephacryl S300. The column was calibrated using ferritin (*Fe*), catalase (*cat*), bovine serum albumin (*BSA*); hemoglobin (*Hb*) and myoglobin (*Mg*). *Arrow* indicates elution volume of purified 5′-nucleotidase which was determined by its enzymatic activity

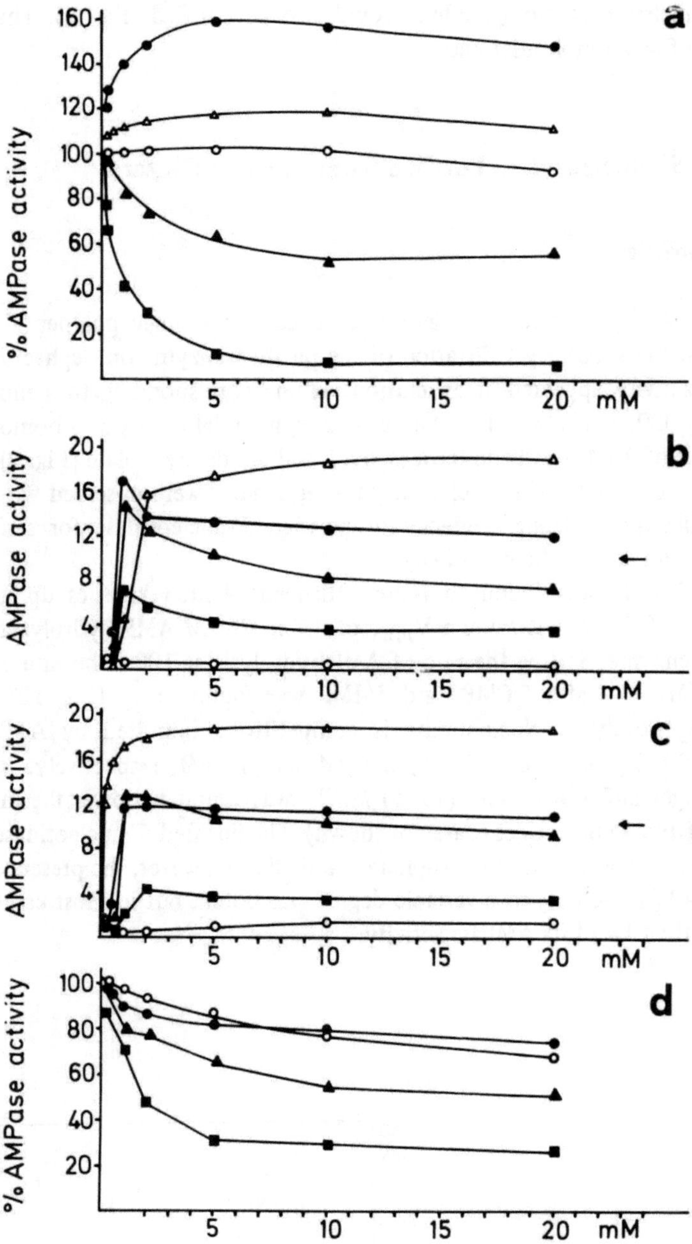

Fig. 3 a–d. Dependence of the AMPase activity of 50 μg purified chicken gizzard 5′-nucleotidase on various divalent metal ions. **a** Dependence of relative activity of 5′-nucleotidase in buffer A plus 0.1% Triton X-100 and 50 μ*M* AMP on increasing concentrations of (●) $MnCl_2$; (△) $MgCl_2$; (○) $CaCl_2$; (▲) $CoCl_2$; and (■) $NiCl_2$. **b** Reactivation of AMPase-activity after treatment of 5′-nucleotidase with 1 m*M* EDTA in buffer A plus 0.1% Triton X-100 for 3 h of room temperature by (△) $MgCl_2$; (●) $MnCl_2$; (▲) $CoCl_2$; (■) $NiCl_2$; and (○) $CaCl_2$. **c** Reactivation of AMPase activity after treatment of 5′-nucleotidase with 1 m*M* EGTA in buffer A plus 0.1% Triton X-100 for 3 h at room temperature by (△) $MgCl_2$; (●) $MnCl_2$: (▲) $CoCl_2$; (■) $NiCl_2$; and (○) $CaCl_2$. *Arrows* in

3.2 Effect of Divalent Cations

The effect of various divalent metal ions on the enzymatic activity of 5′-nucleotidase isolated from chicken gizzard was tested under a number of conditions. Figure 3a gives the effect of increasing concentrations of $MnCl_2$, $MgCl_2$, $CaCl_2$, $CoCl_2$, and $NiCl_2$ on the AMPase activity of purified enzyme isolated from chicken gizzard. It can be seen that its enzymatic activity in buffer A is further increased with increasing Mn^{2+} and Mg^{2+} concentrations and little effect is exhibited by increasing the Ca^{2+} concentration, whereas both Co^{2+} and Ni^{2+} strongly inhibit AMPase activity. When purified 5′-nucleotidase from gizzard is incubated at room temperature with either 1 m*M* EDTA or 1 m*M* EGTA its enzymatic activity is either completely inhibited or reduced to about 5% after 3 h. The activity could, however, be restored by the addition of divalent metal ions (Fig. 3b, c). It can be seen that this restoring or stimulating effect is highest for Mg^{2+} and decreases in the order Mn^{2+}, Co^{2+}, Ni^{2+}, and Ca^{2+}. After incubation with EDTA, increasing the Mn^{2+}, Co^{2+}, or Ni^{2+} concentration produces a biphasic dependence of the AMPase activity: a stimulating effect up to 1 m*M* and at higher concentrations an inhibitory action. The biphasic hehaviour is not observed after treatment with EGTA (Fig. 3c). Therefore, after treatment with EDTA this behaviour might be attributable to either contamination of the metal ion solutions with Ca^{2+} or a successive liberation of Ca^{2+} from EDTA when titrating with the other metal ions. Saturation of the enzyme with Mg^{2+} was achieved by treating 5′-nucleotidase with 1 m*M* EGTA in 5 m*M* $MgCl_2$, 0.5 m*M* NaN_3, and 5 m*M* Hepes, pH 7.4. Under these conditions all the divalent cations tested were found to exert an inhibitory effect on its AMPase activity (Fig. 3d) with increasing potency in the order Mn^{2+}, Ca^{2+}, Co^{2+}, and Ni^{2+}. Since both EDTA and EGTA lead to an inactivation of the enzyme which can only be fully restored after addition of Mg^{2+} or Mn^{2+}, it has to be assumed that 5′-nucleotidase in buffer B still contained a bound divalent metal ion other than Ca^{2+}, most probably Mg^{2+}. It appears, therefore, that 5′-nucleotidase is inactive in the absence of any divalent cations.

3 Inhibition of the Enzymatic Activity of 5 -Nucleotidase by Lectins

The inhibitory effect of lectins specific for mannose and glucose residues on 5′-nucleotidase activity is well documented (Jarett and Smith 1974; Riordan and Slavik 1974, Carraway and Carraway 1975).

---

b and **c** indicate level of AMPase activity before treatment. AMPase activity is expressed in △ OD 265 per min (Mannherz and Rohr 1978). **d** Relative dependence of the 5′-nucleotidase activity after treatment with 1 m*M* EDTA in buffer A plus 0.1% Triton X-100 for 3 h at room temperature and subsequent addition of 5 m*M* $MgCl_2$ on increasing concentrations of (●) $MnCl_2$; (○) $CaCl_2$; (▲) $CoCl_2$; and (■) $NiCl_2$

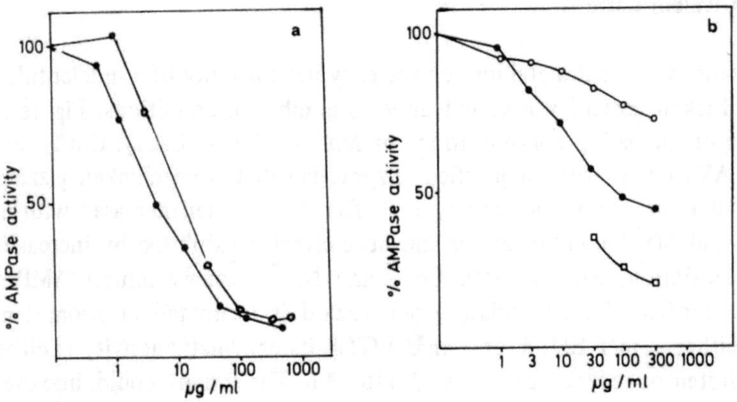

Fig. 4a, b. Inhibition of the AMPase activity of 50 μg purified chicken gizzard 5′-nucleotidase by increasing concentrations of lectins. **a** Increasing concentrations of Con A (•) and LcL (○). **b** Increasing concentrations of succinyl-Con A (•), succinyl-ConA plus 25 m*M* methylmannopyranoside (○), and after addition of 44 mgml⁻¹ (final concentration) of anti-Con A-antiserum (□)

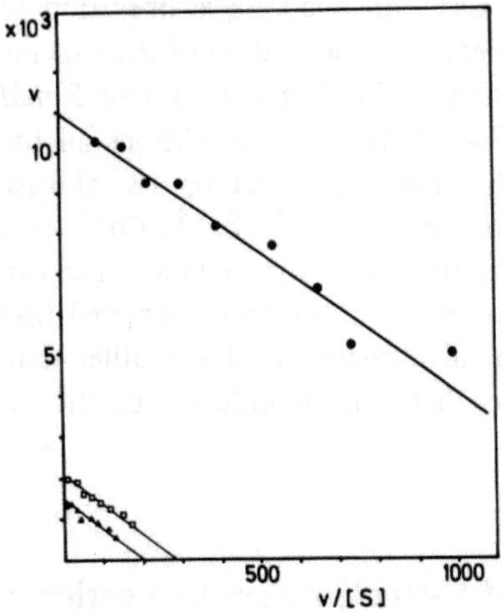

Fig. 5. Eadi-Hofstee plot of the inhibition of the AMPase activity of ca. 50 μg purified chicken gizzard 5′-nucleotidase by Con A. Without Con A (•); plus 30 μgml⁻¹ Con A (□); and plus 100 μgml⁻¹ Con A (▲)

5′-Nucleotidase from all three avian muscular sources employed is also strongly inhibited by concanavalin A (Con A) and lens culinaris lectin (Fig. 4a). The tetrameric structure of Con A is not a necessary prerequisite, since succinyl-Con A [mostly dimeric (Gunther et al. 1973)] also exhibits an inhibitory effect on its AMPase activity, although to a much lesser extent (Fig. 4b). Addition of anti-Con A IgG further increases the inhibitory effect of dimeric Con A to a value almost identical to that of tetrameric Con A (Fig. 4b). The mode of inhibition of Con A is non-competitive as illustrated in Fig. 5, which gives the data according to Eadie (1942) and Hofstee (1952). It can be seen that straight lines are obtained for two inhibitor concentrations (Con A at 30 μgml⁻¹ and 100 μgml⁻¹) which run strictly parallel to each other.

From the intercepts with the y-axis K_i-values of 17.6×10^{-8} *M* and 7.6×10^{-8} *M* can be calculated for the inhibitor concentrations of 1 μM and 0.3 μM, respectively.

4 Antibodies Against 5′-Nucleotidase from Chicken Gizzard

Antisera against 5′-nucleotidase were elicited in rabbits. These antisera were used to detect 5′-nucleotidase on "Western blots" after SDS-PAGE of crude (Chaps) extracts of chicken muscular tissues and of purified enzyme (Fig. 6). In all three crude extracts only a single band arount 80 kD is stained using the procedure of Towbin et al. (1979). Figure 6 also gives the staining of a transblot of purified 5′-nucleotidase after SDS-PAGE with concanavalin A according to Hawkes (1982), clearly corroborating its glycoprotein nature. Antibodies to chicken gizzard 5′-nucleotidase were also raised in mice followed by hybridoma techniques in order to select clones producing specific monoclonal antibodies. A number of such clones were obtained, half of which inhibit the AMPase activity of 5′-nucleotidase (Fig. 7). All clones selected, however, bind purified 5′-nucleotidase when immobilized on Sepharose 4B (data not shown).

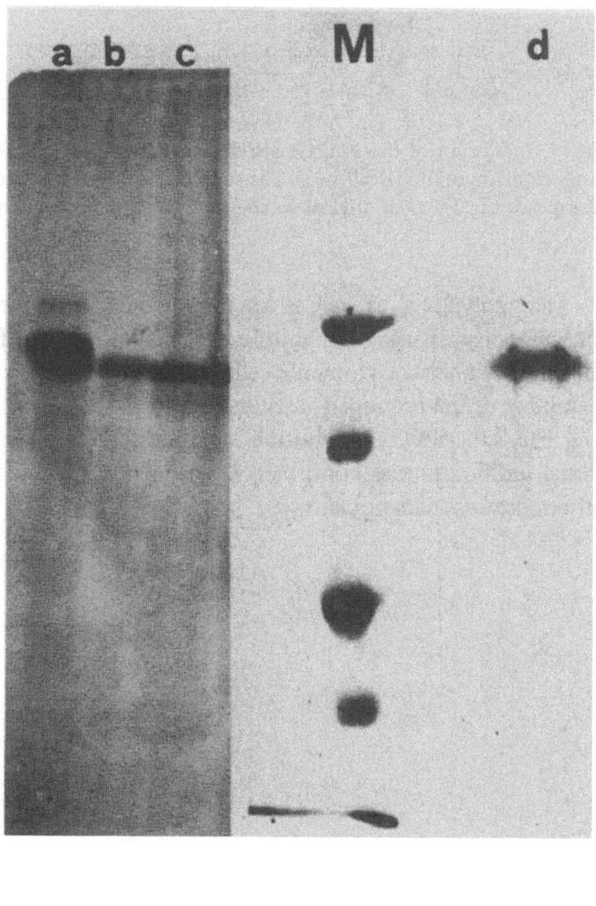

Fig. 6a, d. Western blots of solubilized 5′-nucleotidase (200 000 g supernatant after extraction with Chaps) stained with anti-5′-nucleotidase (from chicken gizzard) IgG at 0.05 mgml^{-1} final concentration according to Towbin et al. (1979). **a** Chicken heart muscle; **b** chicken skeletal; and **c** gizzard. **d** Stain of a Western blot of purified chicken gizzard 5′-nucleotidase with Con A for carbohydrates according to Hawkes (1982). *M* = transferred marker proteins as in Fig. 1

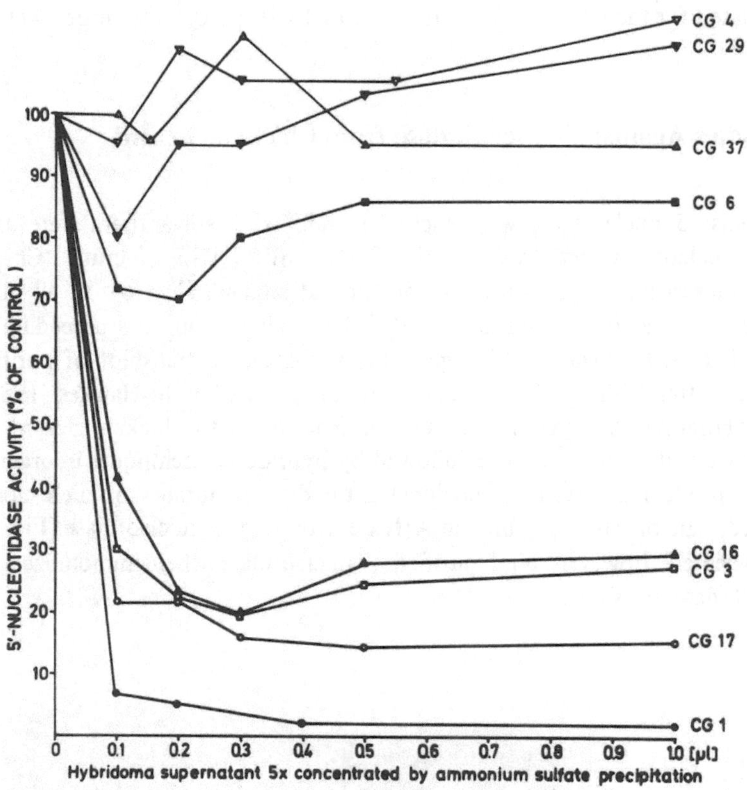

Fig. 7. Inhibition of the AMPase activity of chicken gizzard 5′-nucleotidase by supernatants of monoclonal hybridoma cell lines. The supernatants were precipitated by 50% ammonium sulfate, resuspended in 1/5 their original volume and dialyzed against phosphate buffered saline

The polyclonal as well as the monoclonal antibodies were used to stain chicken muscular tissues using immunofluorescent techniques (Fig. 8a, b, c). It can be seen that the periphery of all muscle cells is stained. In cardiac muscular tissue a prominent staining of the presumed vascular endothelial cells also becomes evident (Fig. 8c).

The subcellular distribution of 5′-nucleotidase as revealed by using the polyclonal antibodies was confirmed by using the monoclonal antibodies either singly or in mixtures (data not shown).

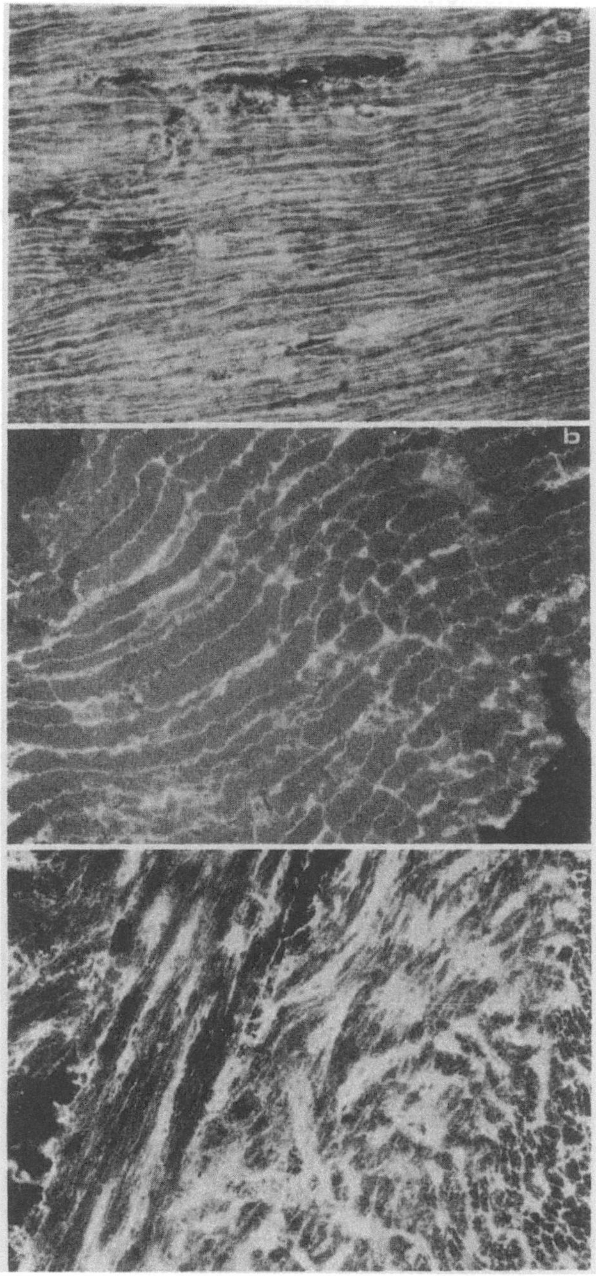

Fig. 8 a–c. Immunolocalization of 5'-nucleotidase. Frozen tissue sections were incubated at room temperature with an IgG-fraction against chicken gizzard 5'-nucleotidase at 0.05mgml^{-1} in phosphate buffered saline for 30 min. Further staining followed the protocol given by Drenckhahn and Mannherz (1983). **a** Chicken gizzard; **b** chicken pectoralis muscle; note staining of the cell peripheries. **c** Chicken heart muscle; note faint staining of muscle cell borders and prominent staining of the capillaries

5 Interaction of Purified 5′-Nucleotidase with Rabbit Skeletal Muscle Actin

The interaction of purified chicken gizzard 5′-nucleotidase with rabbit skeletal muscle actin was studied in vitro. The contractile protein actin can exist in two forms of supramolecular organization, G-actin, the monomeric, or F-actin, the polymeric form. The latter is the physiologically relevant form. Actin contains the stoichiometrically bound nucleotide, G-actin ATP, which is hydrolyzed into ADP and inorganic phosphate during polymerization (after increasing the ionic strength). The ADP formed remains firmly attached to F-acting (for a review see Mannherz and Goody 1976).

Prior to experimentation G-actin was subjected to gel filtration over a Sephadex G-100 column in order to remove any unbound nucleotide. F-actin was obtained from the thus treated G-actin by addition of 1 m*M* $MgCl_2$. Addition of F-actin to purified 5′-nucleotidase instantaneously inhibits its AMPase activity. An inhibitory effect is not exhibited by actin-DNase I complex and to an only lesser degree by G-actin (Fig. 9; Dieckhoff et al. 1984).

Furthermore, a time-dependent transformation of F-actin by 5′-nucleotidase led to a DNase I-resistent form, i.e., F-actin lost its capacity to inhibit DNase I. Since this effect occurred in both a time- and a 5′-nucleotidase concentration-dependent manner, a catalytic rather than stoichometric underlying mechanism must be assumed. At present, we are unable to explain fully its nature. However, it was found that the desensitization of F-actin can be correlated to a transformation of the actin-bound nucleotide into adenosine (unpublished observations). This process occurs faster when F-actin instead of G-actin is incubated with 5′-nucleotidase and corresponds to the observed faster rate of transformation of F-actin (data not shown). Since the rate of desensitization was found to be faster in 5′-nucleotidase preparations with a higher percentage of contaminating ATPase, an involvement of this enzymatic entity must be assumed, although a contributing role of 5′-nucleotidase in this process cannot be excluded.

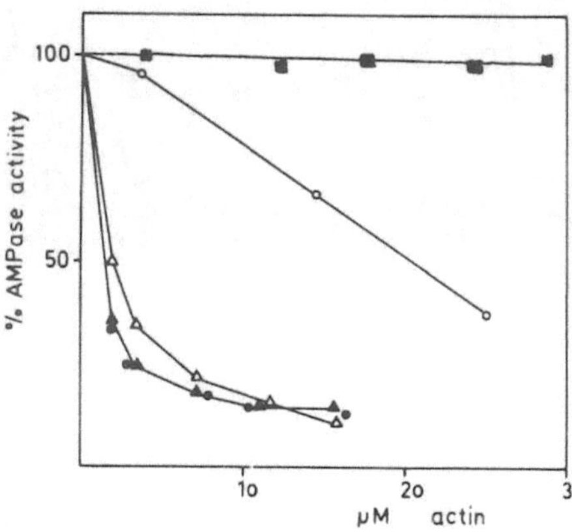

Fig. 9. Inhibition of the AMPase activity of chicken gizzard 5′-nucleotidase by increasing concentrations of different forms of rabbit skeletal muscle actin. F-actin (△), F-actin stabilized by equimolar phalloidin (▲), G-actin (○), and actin-DNase I complex (■)

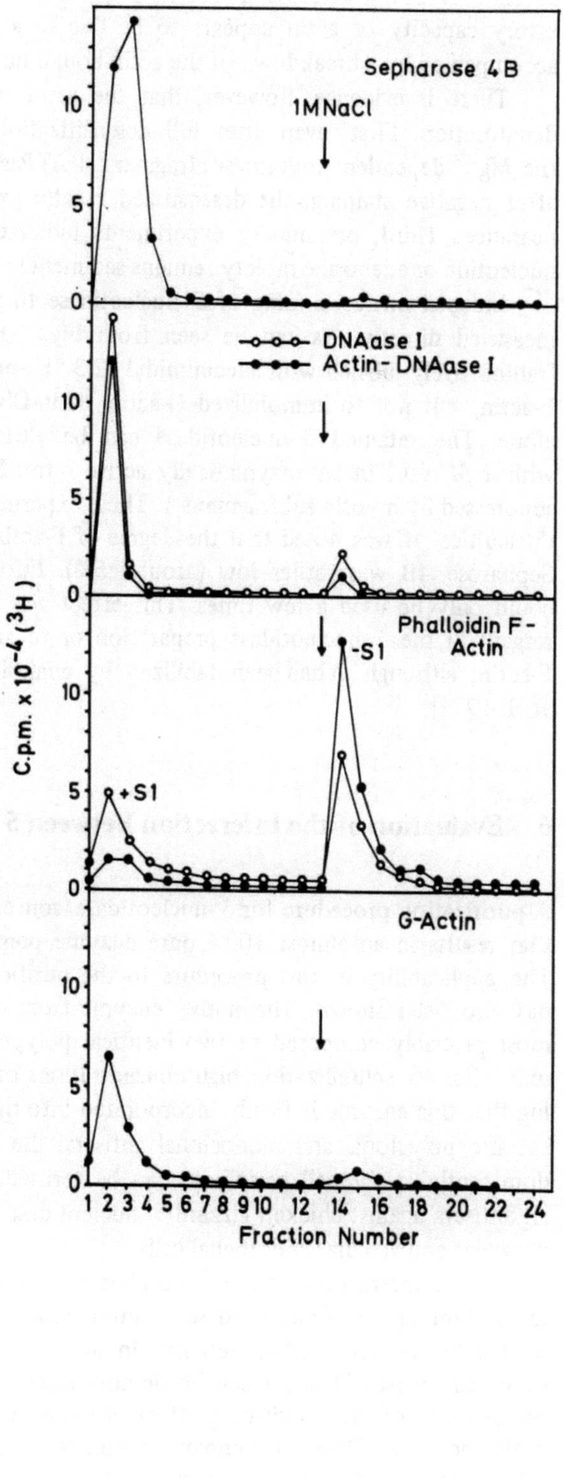

Fig. 10. Binding of chicken gizzard [^{3}H]-5′-nucleotidase to different forms of immobilized actin [^{3}H]-5′-nucleotidase (200 μl) in buffer A was applied to the four different columns (ca. 1.5 ml bed volume each) indicated and eluted with buffer until radioactivity had returned to zero. Then the columns were cluted with buffer A sup-Fraction size: 0.5 ml; the fractions collected were analyzed for radioactivity. Retention of 5′-nucleotidase was only observed to F-actin stabilized by phalloidin (●), this retention could be suppressed by simultaneous application of myosin subfragment 1 and 5′-nucleotidase (○)

A densensitization of either F- and G-actin against DNase I also occurs in the absence of 5′-nucleotidase at room temperature at a slow rate. This loss of inhibitory capacity of actin appears to be due to a denaturation process and is not accompanied by a breakdown of the actin-bound nucleotide to adenosine.

There is evidence, however, that the actin desensitization is not due to its denaturation. First, even after full desensitization, F-actin was found to activate the Mg^{2+}-dependent myosin subfragment 1-ATPase as untreated F-actin Second, after negative staining, the desensitized F-actin exhibits a normal filamentous appearance. Third, preliminary experiments indicate that after transformation, the nucleotide or adenosine moiety remains sedimentable with F-actin.

Independently, binding of 5′-nucleotidase to immobilized F- and G-actin was measured directly. As can be seen from Fig. 10 chicken gizzard 5′-nucleotidase [radioactively labeled with succinimidyl (2,3-^{3}H)-proprionate] binds to immobilized F-actin, but not to immobilized G-actin, actin-DNase I complex or Sepharose 4B alone. The retained 5′-nucleotidase can be eluted with buffer A supplemented with 1 *M* NaCl in an enzymatically active form. Binding of 5′-nucleotidase can be suppressed by myosin subfragment 1. These experiments were hampered by technical difficulties. It was noted that the degree of F-actin coupling to cyanogen bromide-Sepharose 4B was rather low (around 5%). Furthermore, the columns obtained could only be used a few times. This effect was attributed to the presence of detergent in the 5′-nucleotidase preparation or to an action of the enzyme itself on F-actin, although it had been stabilized by phalloidin prior to coupling (Lengsfeld et al. 1974).

6 Evaluation of the Interaction Between 5′-Nucleotidase and Actin

A purification procedure for 5′-nucleotidase from chicken gizzard has been developed that results in an almost 100% pure enzyme preparation as judged by SDS-PAGE. The applicability of this procedure to the purification of rat liver 5′-nucleotidase has also been shown. The native enzyme from chicken gizzard is a glycoprotein most probably composed of two identical polypeptide chains of 79 000 molecular mass. For its solubilization high concentrations of detergent are necessary, indicating that this enzyme is firmly incorporated into the plasma membrane. Using monospecific polyclonal and monoclonal antisera, the localization of this enzyme, predominantly at the cell periphery, can be verified. Judged by immunofluorescence, all antisera against chicken gizzard 5′-nucleotidase where shown to specifically stain muscular and vascular endothelial cells.

Direct interaction of 5′-nucleotidase with actin is demonstrated in vitro, i.e., direct binding of 5′-nucleotidase to immobilized F-actin and an inhibitory action of F-actin on its AMPase activity. In addition a time-dependent desensitization of F-actin versus DNase I can be demonstrated which is in agreement with earlier observations of the reactivating effect of snake venom or rat bile 5′-nucleotidase on preformed actin-DNase I complex (Mannherz and Rohr 1978; Rohr and Mannherz 1979). This latter effect appears to be of a catalytic rather than stoichiometric nature and might be due to a transformation of F-actin-bound nucleotide (ADP) to adenosine. Since desensitized F-actin appears to retain its ability to activate the Mg^{2+}-dependent myosinsubfragment 1 ATPase, it is concluded that its inability to inhibit DNase I is not due to its denaturation.

The study of the interaction of purified 5′-nucleotidase with actin is, however, hampered experimentally. Purified 5′-nucleotidase is only soluble in the presence of detergents, such as Triton X-100. Detergents modify or inhibit the actin-actin interaction, thus, reducing the rate and extent of actin polymerization (unpublished observations). Therefore, reliable measurements of the influence of 5′-nucleotidase on the actin polymerization process were impossible so far. The previously used procedure to relate the ATP hydrolysis of actin-bound nucleotide to the rate of its polymerization in the presence of 5′-nucleotidase (Mannherz and Magener 1979) needs further evaluation in view of the results that preparations of purified 5′-nucleotide are able to transform actin-bound nucleotide to adenosine.

Certainly, the difficulties encountered during our study of actin-5′-nucleotidase interaction represent a general technical problem encountered with purified intrinsic membrane proteins. The results obtained may indicate that there are membrane components which are able to interact with actin directly without the interposition of other linking components, although they do not exclude the possibility that such components may be used in vivo to modify such an interaction.

Given the known confinement of actin to the intracellular space it inevitably follows that the functional domain of 5′-nucleotidase able to interact with actin should be separated from its AMPase center by the lipid bilayer. Unfortunately, there is so far no experimental evidence for this model.

Abbreviations: Hepes, 4- (2-hydroxyethyl)-1-piperazineethenesulfonic acid; EDTA, ethylene-diamine tetraacetic acid; EGTA, ethyleneglycol-bis-(2-aminoethylether) N,N′-tetraacetic acid; SDS-PAGE, polyacrylamide electrophoresis in the presence of sodium dodecylsulfate; APCP, α, β-methylene-ADP; DNase I or deoxyribonuclease I (E.C. 3.1.21.1.); LcL, lens culinaris lectin; ConA, concanavalin A

Acknowledgment. It is a pleasure for us to acknowledge the expert technical assistance by Mrs. U. Krauskopf, Mr. J. Koch, and Mrs. A. Hunsicker for editorial help. The work reported was supported by the Deutsche Forschungsgemeinschaft.

References

Arch JRS, Newsholme EA (1978) The control of the metabolism and the hormonal role of adenosine. Essays Biochem 14:82–123

Burger RM, Lowenstein JM (1970) Preparation and properties of 5′-nucleotidase from smooth muscle of small intestine. J Biol Chem 245:6274–6279

Carraway KL, Carraway CA (1975) Cooperative effects in the perturbation of membrane enzymes by concanavalin A. Biochem Biophys Res Commun 67:1301–1306

Carraway KL, Doss RC, Huggins JW, Chesnut RW, Carraway CAC (1979) Effect of cytoskeletal perturbant drugs on ecto-5′-nucleotidase, a concanavalin A receptor. J Cell Biol 83:529–543

De Pierre JW, Karnovsky ML (1974) Ecto-enzymes of granulocytes: 5′-nucleotidase. Science (Wash DC) 183:1096–1098

Dieckhoff J, Heidemann M, Mannherz HG (1984) The isolation of 5′-nucleotidase from avian sources and its interaction with polymeric actin. J Submicrosc Cytol 16:33–34

Dornand J, Bonnafous JC, Mani JC (1978) Purification and properties of 5′-nucleotidase from lymphocyte plasma membranes. Eur J Biochem 87:459–465

Drenckhahn D, Mannherz HG (1983) Distribution of actin and the actin-associated proteins myosin, tropomyosin, alpha-actinin, vinculin, and villin in rat and bovine exocrine glands. Eur J Cell Biol 30:167–176

Eadie CS (1942) The inhibition of cholesterase by physostigmine and prostigmine. J Biol Chem 146:85–93

Gunther GR, Wong JL, Yakara I, Cunningham BA, Edelman GM (1973) Concanavalin A derivatives with altered biological activities. Proc Natl Acad Sci USA 70:1012–1016

Jarrett L, Smith RM (1974) The stimulation of adipocyte plasma membrane magnesium ion-stimulated adenosine triphosphatase by insulin and concanavalin A. J Biol Chem 249: 5195–5199

Hawkes R (1982) Identification of concanavalin A binding proteins after sodium dodecylsulfate-gelelectrophoresis and transblotting. Anal Biochem 123:143–146

Hoffstee BHJ (1952) Specificity of esterases I. Identification of two pancreatic aliesterases. B Biol Chem 199:357–364

Lengsfeld AM, Löw I, Wieland T, Danker P, Hasselbach W (1974) Interaction of phalloidin with actin. Proc Natl. Acad Sci USA 71:2803–2807

Mannherz HG, Goody RS (1976) Proteins of contractile systems. Annu Rev Biochem 45:427–465

Mannherz HG, Magener M (1979) Concanavalin A inhibits the interaction of snake venom 5′-nucleotidase and actin. FEBS Lett 103:77–80

Mannherz HG, Rohr G (1978) 5′-Nucleotidase reverses the inhibitory action of actin on pancreatic deoxyribonuclease I. FEBS Lett 95:284–289

Riordan JR, Slavik M (1974) Interactions of lectins with membrane glycoproteins. Effects of concanavalin A on 5′-nucleotidase. Biochim Biophys Acta 373:356–360

Rohr G, Mannherz HG (1979) The activation of actin-DNAse I complex with rat liver plasma membranes, the possible role of 5′-nucleotidase. FEBS Lett 99:351–356

Slavik M, Kartner N, Riordan JR (1977) Lectin-induced inhibition of plasma membrane 5′-nucleotidase. Sensitivity of purified enzyme. Biochem Biophys Res Commun 75:342–349

Towbin H, Staehlin T, Fordin J (1979) Electrophoretic transfer of proteins from polyacrylamide gels to nitrocellulose sheets: procedure and some applications. Proc Natl Acad Sci USA 76:4350–4354

Zachowski A, Evans HW, Paraf A (1981) Immunological evidence that plasma-membrane 5′-nucleotidase is a transmembrane protein. Biochim Biophys Acta 664:121–126

5′-Nucleotidase in the Nervous System

G. W. KREUTZBERG, D. HEYMANN, and M. REDDINGTON[1]

1 Introduction

Ectoenzymes represent important functional components of the extracellular space. They are of particular interest in the central nervous system since, due to the strict control of substances entering and leaving the brain via the blood-brain barrier, the extracellular space is effectively a closed system. Metabolic interactions in the extracellular compartment are therefore extremely important both for the effective recycling of cerebral metabolites and in the processes of intercellular communication. Our knowledge of the physiology of the extracellular space in the central nervous system is rather restricted. Indeed, even its geometry is still a matter for debate (Cragg 1979). It is to be hoped that a broader knowledge of the enzymes having their catalytic sites exposed to the extracellular compartment will lead to a better understanding of the factors controlling molecular interactions and thus the composition of the extracellular space in the brain.

An ectoenzyme *par excellence* which is widely distributed in the central nervous system is 5′-nucleotidase (EC.3.1.3.5). As will be evident from several contributions to this volume, 5′-nucleotidase is thought to play an important physiological role in several cells and tissues due to its key position in the breakdown of purine and pyrimidine nucleotides to their corresponding nucleosides. Thus, nucleotides released into the extracellular space, which are generally unable to cross cell membranes, are converted by 5′-nucleotidase into a membrane-permeable form and can be retrieved and reutilized by the various cellular components of the brain. In addition, at least in the case where 5′AMP can act as a substrate, the reaction product, adenosine, could exert a potent modulatory action on neuronal activity via an interaction with extracellular receptors (see Schubert et al. 1983; Williams 1974 for recent reviews). 5′-Nucleotidase is, therefore, an important element to consider in both processes of metabolic recycling and intercellular communication in central nervous tissue. Knowledge of the fine structural localization of the enzyme is of obvious importance to the elucidation of its exact physiological significance in the brain.

In the present chapter we describe the localization of 5′-nucleotidase in central tissues using electron microscopical (EM) cytochemistry. Due to a certain amount of controversy in this field, we have also attempted to face the challenge of identify-

[1] Max Planck Institute for Psychiatry, 8033 Martinsried n. Munich, FRG

ing the enzyme detected by such methods as a true 5′-nucleotidase rather than a nonspecific phosphatase. We, therefore, present data on the biochemical characterization of AMPase activity in brain tissue. Conversely, we have applied EM-cytochemistry to subcellular fractions from rat brain in order to bridge the gap between in situ and in vitro approaches to enzyme localization and, finally, consider functional aspects of AMP hydrolysis in the central nervous system.

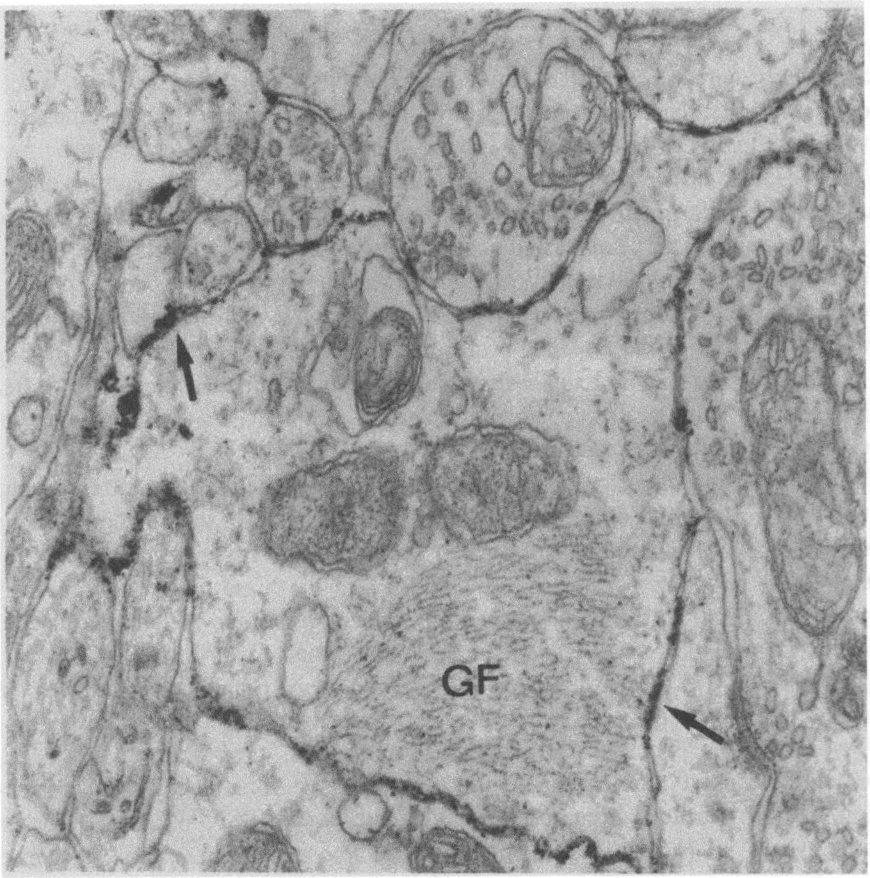

Fig. 1. 5′-Nucleotidase in an astrocyte, characterized by the presence of glial filaments (*GF*), in the rat dentate gyrus. Electron dense reaction product shows activity on the plasmalemma (*arrows*). Magnification 1:62 000

2 Cytochemical Localization of 5′-Nucleotidase in Situ

Light microscopical histochemistry using lead salt techniques has shown 5′-nucleotidase to be differentially localized in certain brain regions (Scott 1967; Schubert et al. 1979), but does not allow resolution at the cellular level. To achieve a cellular resolution EM-cytochemical techniques based on the formation of electron-dense heavy metal salts have been employed (Scott 1967; Suran 1974; Marani 1977; Bernstein et al. 1978; Kreutzberg et al. 1978). Using 5′AMP as a substrate in the presence of lead nitrate, our first studies showed deposits of lead phosphate to be associated predominantly with glial cells (Fig. 1). Astrocytes, Bergmann glia of the cerebellum, and the Müller cells of the retina clearly show reaction product at the plasmalemma, usually on the external surface (i.e., at an "ecto" position) and often filling the extracellular space. In the case of astrocytic end feet associated with the cerebral vasculature, reaction product is found at the basal lamina (Kreutzberg et al. 1978). No activity is found at intracellular sites in glial cells.

Light microscopy has revealed a high activity of 5′-nucleotidase in the cerebral white matter, leading to the conclusion that the enzyme is associated with myelin (Cammer et al. 1980). At the ultrastructural level activity is indeed present in between the myelin lamellae, in the periaxonal and the inner mesaxonal spaces (Fig. 2).

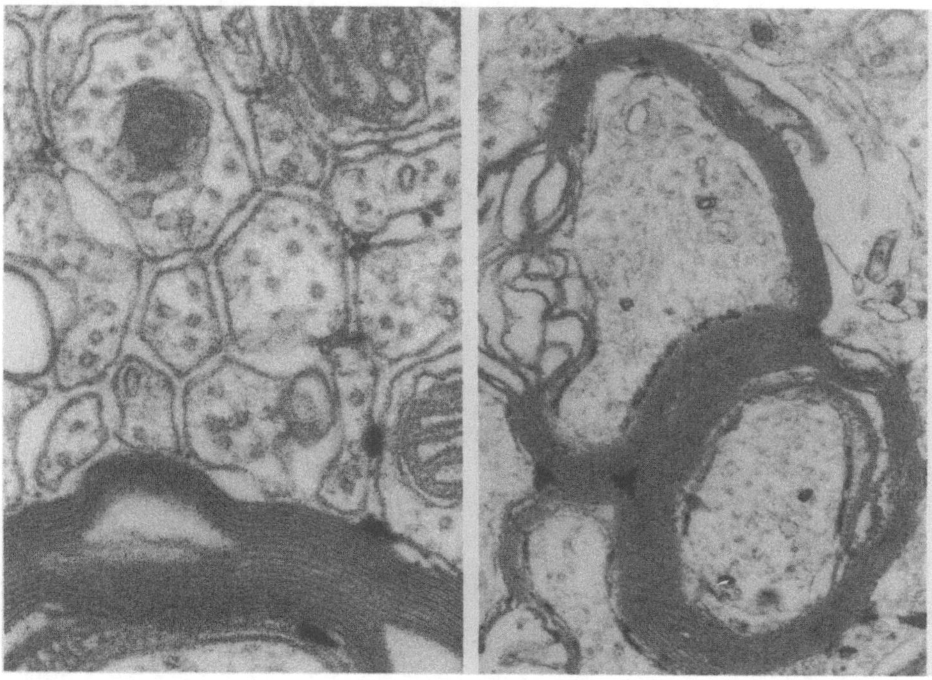

Fig. 2. Activity of 5′-nucleotidase is seen associated with myelinated fibers of the rat brain. Patchy activity is also seen in the periaxonal spaces of myelinated and unmyelinated fibers. The deposits in the axons represent nonspecific staining. Magnification *left* 1:100 000; *right* 1:50 000

Oligodendrocytes also show plasmalemma activity (Fig. 3). These findings have recently been confirmed using immunocytochemical methods (Cammer et al. 1985).

A third class of glial cells, the microglia, also shows 5′-nucleotidase reaction product in an ecto-position, at least when these cells are induced to proliferate (Fig. 4). This is the case, for instance, in a motor nucleus, such as that of the facial nerve, after peripheral transection of the nerve (Kreutzberg and Barron 1978). Under these conditions resting microglial cells become proliferative and cover with abundant processes the surface of the somata and the stem dendrites of the chromatolytic (i.e., regenerating) motoneurons. They thus become involved in the displacement or stripping of synapses from the neuronal surface (Blinzinger and Kreutzberg 1968). The significance of the high activity of 5′-nucleotidase on the

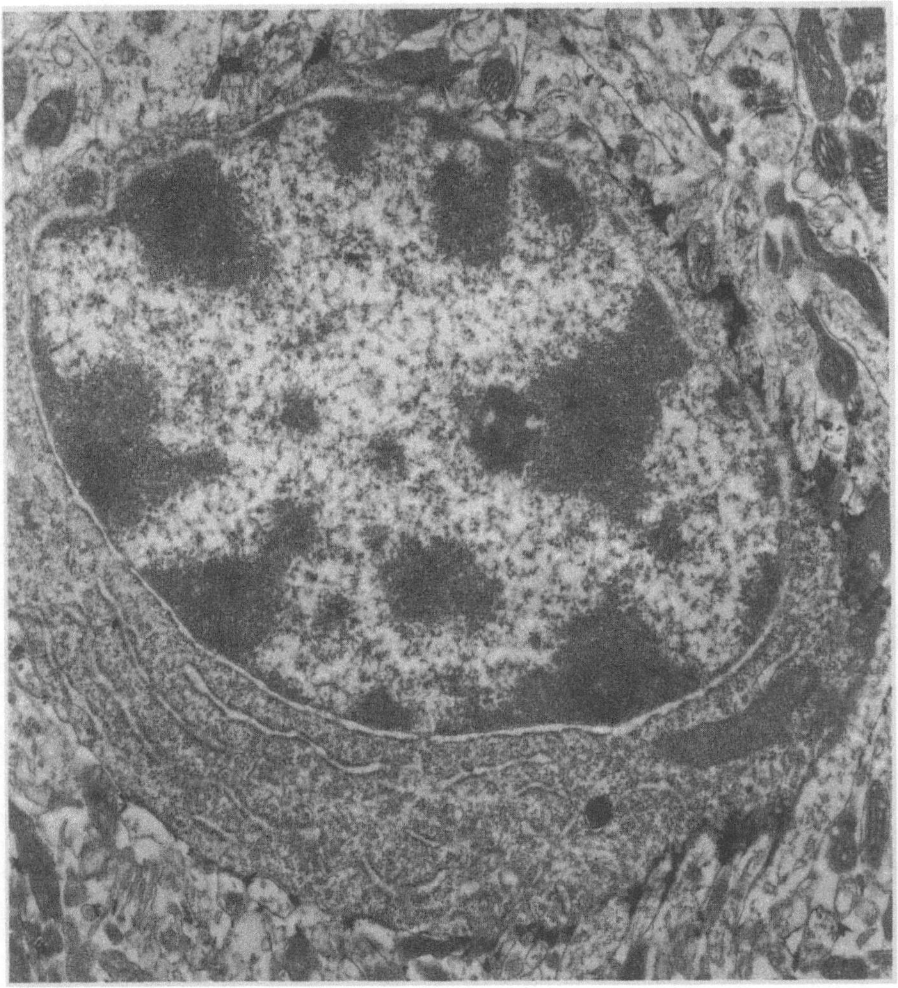

Fig. 3. Oligodendroglial cell from the rat caudate nucleus shows 5′-nucleotidase activity on the plasmalemma. Magnification 1:20 000

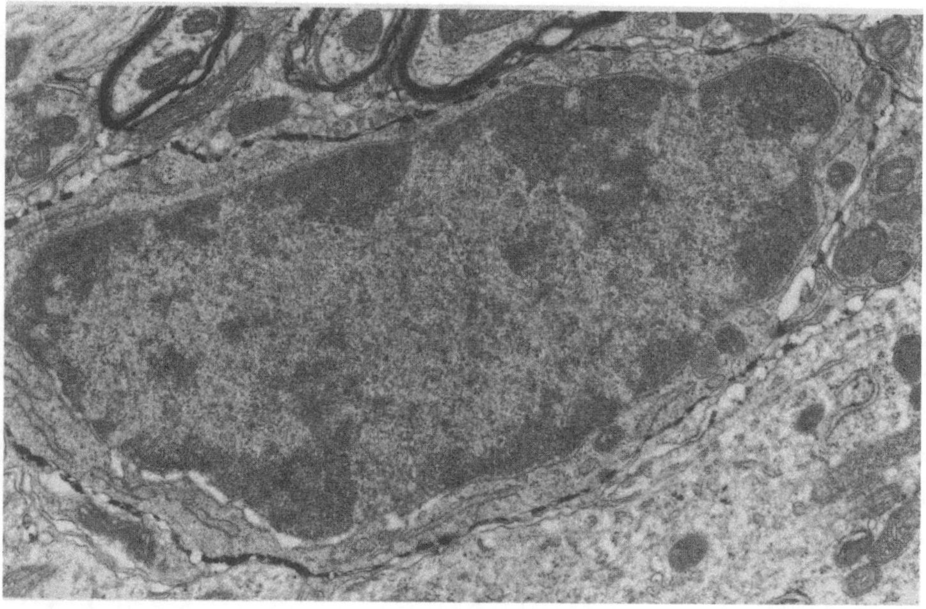

Fig. 4. Reactive microglial cell in the rat facial nucleus 6 days after transection of the seventh nerve. High activity is seen on the plasmalemma. Magnification 1:25 000

proliferating microglial cells is unclear. However, two possible functions might be considered. First, the enzyme might hydrolyze nucleotides which are lost from the displaced boutons, thus making nucleosides available to the proliferating microglia to help meet their need for increased synthesis of DNA and RNA. Secondly, adenine nucleotides could be hydrolyzed to adenosine which might play a role in increasing blood flow to supply the increased demand for substrates, such as glucose, required in the regenerating nucleus (Kreutzberg and Emmert 1980).

In more recent studies of the distribution of 5′-nucleotidase in the mammalian brain we have found clear evidence for the association of the enzyme with neuronal membranes. It had previously been shown by Marani (1977) that cytochemical recation product can be seen in subsynaptic cisternae of Purkinje cell dendrites in the mouse cerebellum. We have been able to confirm this finding in dendrites of CA3 pyramidal cells of the hippocampus where intracellular activity is seen in spine apparatus cisternae underlying synapses. In the hippocampus, 5′-nucleotidase is also seen to be associated with the mossy fiber bundles. Although species differences exist in the quantity of the enzyme associated with this fiber bundle, reaction product can be clearly seen using electron microscopy (Fig. 5). Since this nerve bundle is unmyelinated and shows no other intercalated glial profiles, and since the axons lie close together it is clear that the 5′-nucleotidase of the hippocampal mossy fiber bundle is associated with the neuronal (axonal) membranes.

A further region characterized by high 5′-nucleotidase activity at the light microscopical level is the striatum. In this case, however, it is unclear, even using EM-cytochemistry, as to which cellular elements reaction product should be attributed, largely due to intermingling of dense neuropil with astrocytic processes.

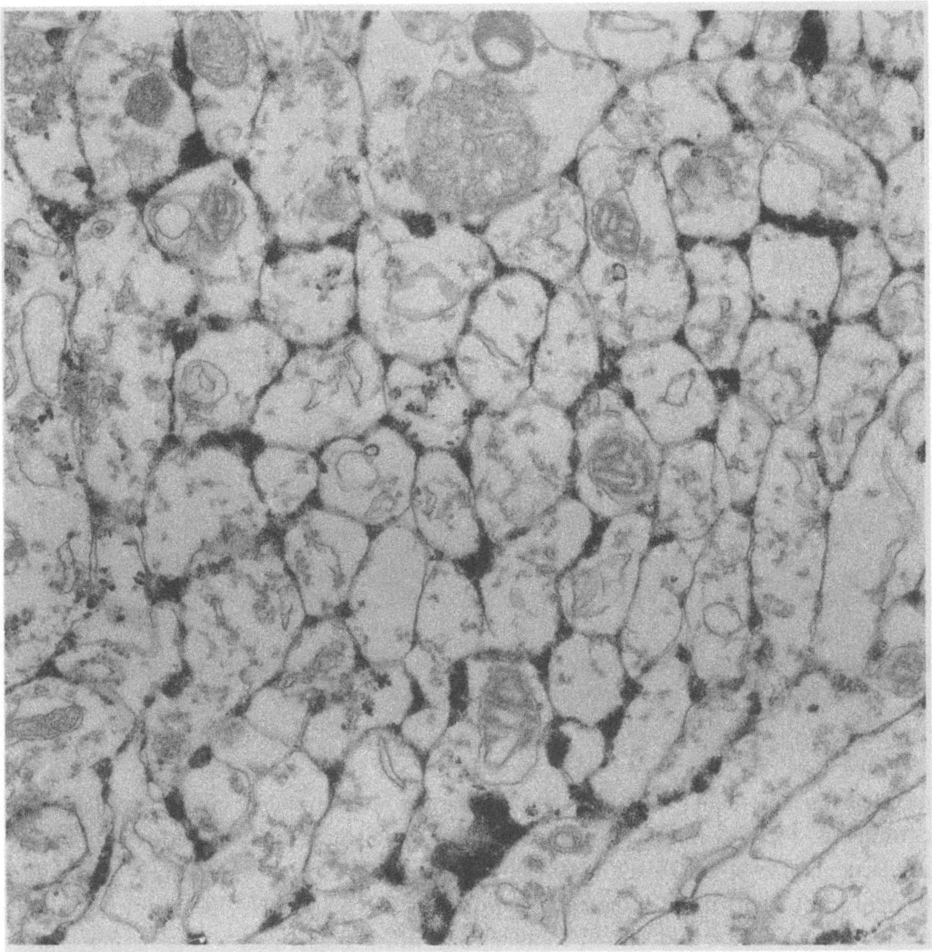

Fig. 5. A mossy fiber in the hippocampus of a golden hamster shows high activity of 5′-nucleotidase on the axons. Magnification 1:41 000

It does nevertheless appear that the 5′-nucleotidase is associated with the membranes of neuronal processes in the striatal neuropil as well as in the hippocampus. The distribution of reaction product in these areas is currently being examined using a statistical analysis of serial EM sections (see also Fig. 2, left).

The distribution of 5′-nucleotidase, predominantly glial but also neuronal has also been confirmed in the retina. In this tissue the cellular equivalent of the neuroglia of the brain is the Müller cell. This cell extends throughout the different layers of the retina thus exhibiting different domains of intercellular contact. Interestingly, 5′-nucleotidase on the surface of Müller cells is only found on the processes running through the external plexiform and external nuclear layers (Fig. 6: Kreutzberg and Hussain 1982). Evidently this glial plasma membrane shows a remarkable cytochemical heterogeneity.

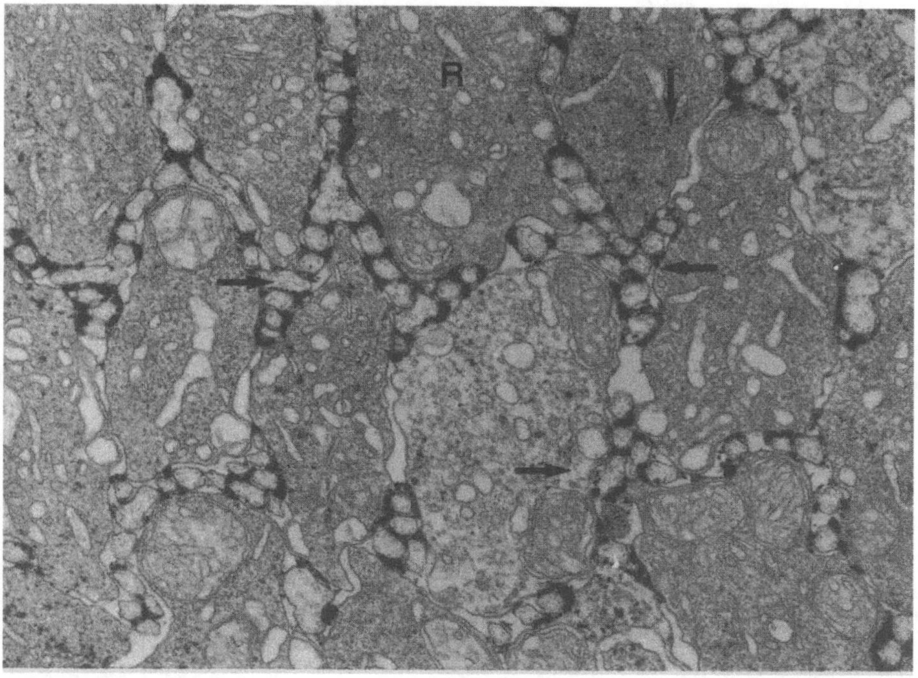

Fig. 6. Rat retina receptor layer. 5′-Nucleotidase is seen on the villous processes of the Müller cells (*arrows*) forming fiber baskets around the photoreceptor inner segment (*R*) where also intracellular activity is evident. Magnification 1:41 000

Further to its localization on retinal glial cells, 5′-nucleotidase has been found associated with the complex synapses of the external plexiform layer formed by the terminals of the photoreceptor fibers and of the horizontal and bipolar cells (Kreutzberg and Hussain 1984). Reaction product is clearly seen in the synaptic clefts. When the receptor fibers are followed in a further external direction their axolemma is also 5′-nucleotidase positive. On the basis of the observed localization of the reaction product, the enzyme is here clearly in an ecto-position.

The cell bodies of the receptor cells occur in the external nuclear layer. These, together with the so-called myoid and ellipsoid, form the rod inner segment. In contrast to other regions, reaction product in this region is clearly seen inside the cells, being scattered over the cytoplasm but not over vesicles, the cisternae of the endoplasmic reticulum, and the mitochondria. It would seem that in this region enzyme is detected while undergoing transport from its site of synthesis to its final destination, namely, the disks of the rod outer segment. Here, in the photoreceptor membranes, 5′-nucleotidase appears to be inserted in a somewhat surprising position, i.e., with its catalytic site facing the cytoplasmic side of the rod outer segment membranes (Fig. 7). No reaction product can be demonstrated in the intradiskal spaces corresponding to the ecto-side of the membranes.

Finally, the tips of the photoreceptor rod outer segments are covered by the apical processes of the pigment epithelial cells, which are involved in the constant

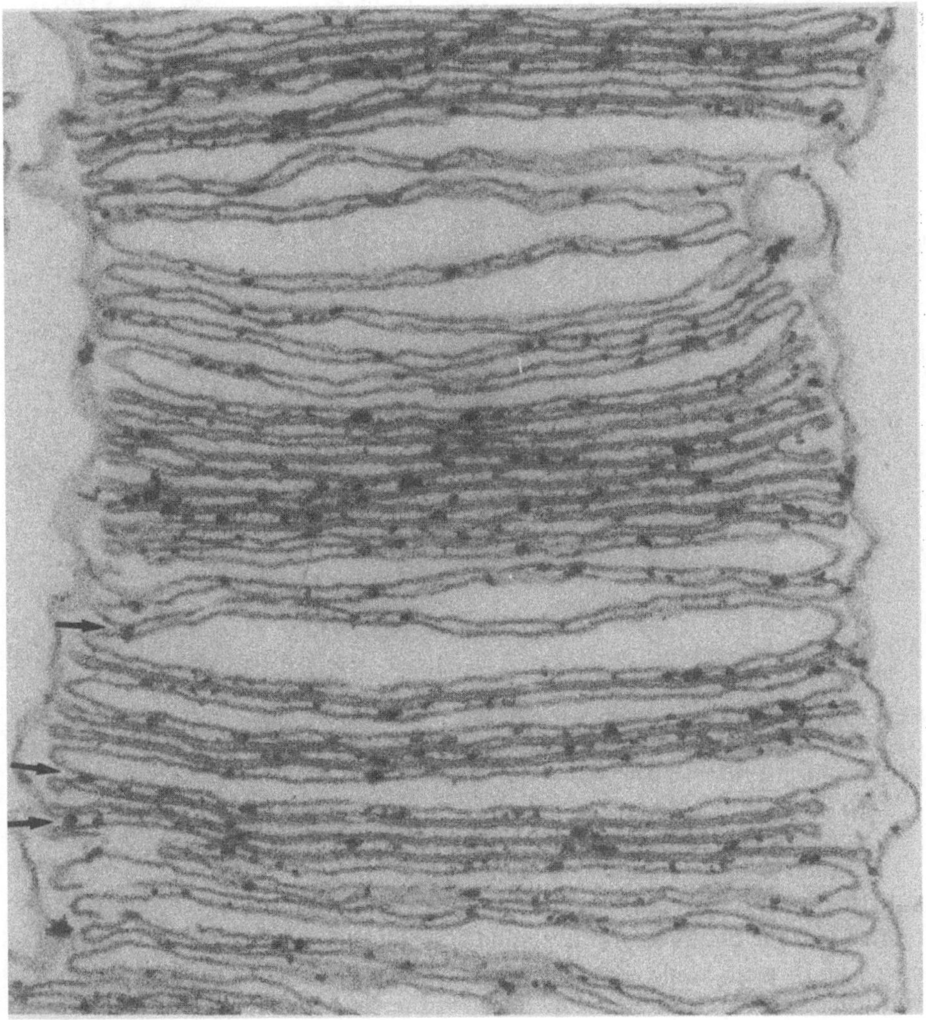

Fig. 7. Rat retina rod outer segment. 5′-Nucleotidase is seen on the cytoplasmic side of the disk membrane (*arrows*), not in the intradiskal space which corresponds to the ecto-side. Magnification 1:160 000

phagocytosis of the rod outer segments. These pigment cells are highly positive for 5′-nucleotidase with the reaction product on the ecto-side of the plasmalemma or associated with phagocytic organelles (Fig. 8).

In summary, cytochemical studies on the ultrastructural localization of 5′-nucleotidase in situ show this enzyme to be localized mainly on the external side of the plasma membranes of glial cells and neuronal processes, confirming the ecto-nature of 5′-nucleotidase in the central nervous system. It is evident, however, that the enzyme can also be found in some specialized sites in an intracellular position as in the spine apparatus of the dendrites of certain neurons or dispersed in the cytoplasm of retinal photoreceptor cells.

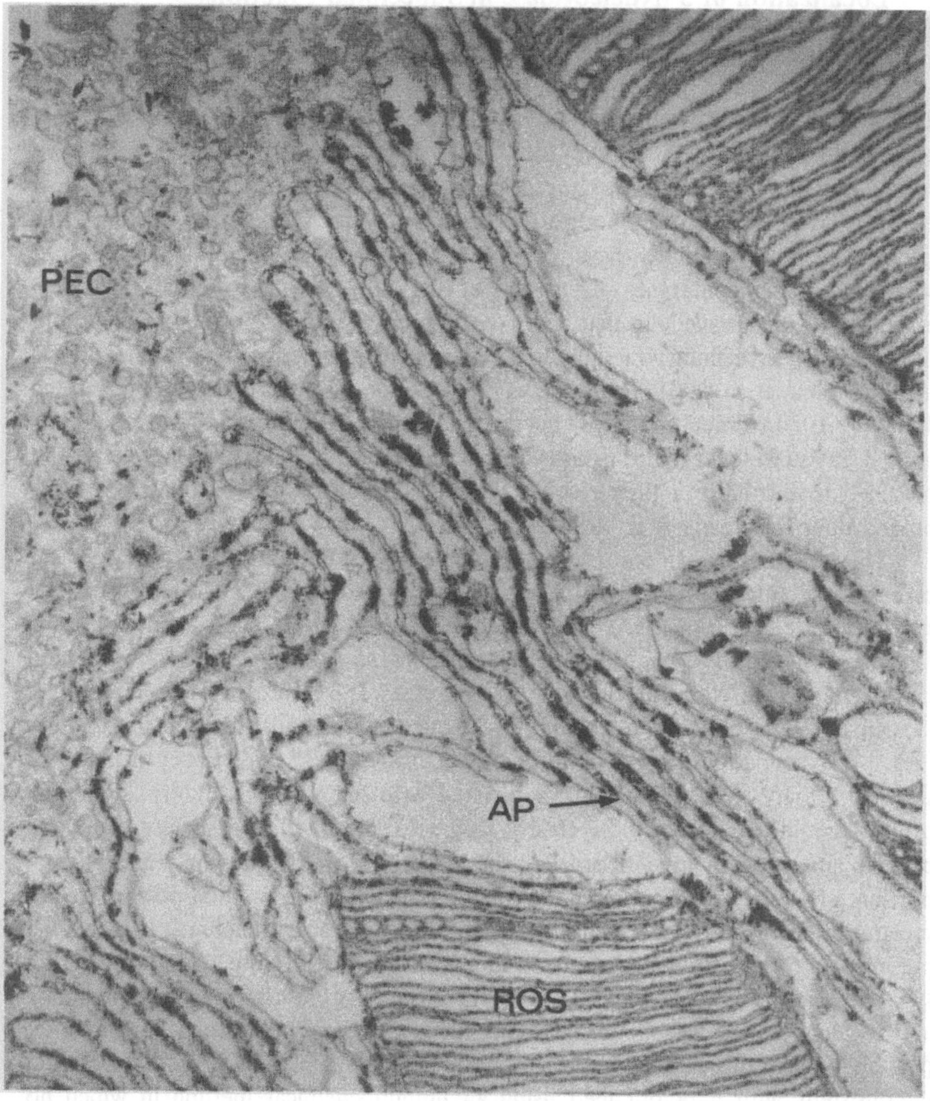

Fig. 8. Ecto-5′-nucleotidase activity is seen on the apical processes (*AP*) of retinal pigment epithelial cells (*PEC*) as they cover the tips of the rod outer segments (*ROS*) for phagocytosis. Mangification 1:63 000

3 Localization of 5′-Nucleotidase in Subcellular Fractions from Rat Brain

The problem of the localization of 5′-nucleotidase in brain tissue has not only been approached using in situ cytochemical methods, but also biochemically by subcellular fractionation. These approaches have, however, often led to conflicting views as to the subcellular localization of the enzyme. Thus, while authors using EM-cytochemical methods have emphasized the association of 5′-nucleotidase with glial plasma membranes (Kreutzberg et al. 1978), synaptic structures (Bernstein et al. 1978), or intracellular neuronal compartments (Marani 1977), interpretations of data from subcellular fractionation experiments have stressed a localization in myelin (Cammer et al. 1980; Bourre et al. 1982) or in synaptosomes (Pilcher and Jones 1970). In order to bridge the gap between these approaches we, therefore, carried out subcellular fractionations with cytochemical examination of the fractions obtained to introduce a finer resolution than is possible with the biochemical approach alone (Heymann et al. 1984).

The distribution of 5′-nucleotidase in various subcellular fractions from rat cerebrum or cerebellum was similar to that described by other authors in several species (Kai et al. 1966; Israel and Frachon-Mastour 1970; Phillips and Newsolme 1979; Cammer and Zimmermann 1981; Pilcher and Jones 1970; Harwood and Hawthorne 1969; Withy et al. 1982; Nagata et al. 1984). Most of the activity was particulate, being enriched in the myelin, synaptosomal and microsomal fractions (Fig. 9). Examination of these fractions for 5′-nucleotidase using EM-cytochemistry showed activity in all cases to be associated with membrane structures. Reaction product could, for instance, be clearly seen associated with the surface of myelin profiles and between myelin lamellae. In synaptosomal fractions only a small proportion of synaptosomes showed reaction product, but it was not possible to associate this unequivocally with the synaptosomal plasma membrane due to the presence of other, unidentifiable membrane fragments in close apposition to the synaptosomal membrane (Fig. 10). It is, therefore, possible that the 5′-nucleotidase activity found in such fractions is associated with nonsynaptosomal membranes, possibly of glial origin, adhering to the synaptosomal membrane. This interpretation is consistent with data obtained using an immunochemical method in which no evidence was obtained for synaptosomal 5′-nucleotidase in synaptosome populations characterized by the presence of various neurotransmitters (Richardson 1983).

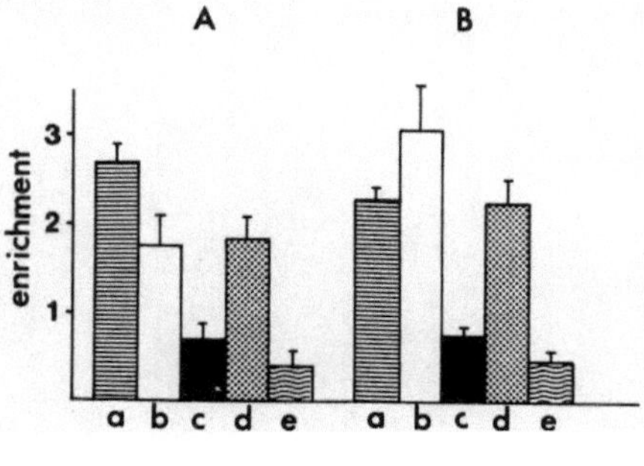

Fig. 9 A, B. Subcellular distribution of 5′-nucleotidase in rat cerebral cortex (*A*) and cerebellum (*B*). The data are expressed as a ratio of the activities of 5′-nucleotidase per mg protein in a subcellular fraction and in the original homogenate (i.e., homogenate = 1). Fractions measured were: **a** myelin; **b** synaptosomal; **c** mitochondrial; **d** microsomal; **e** soluble

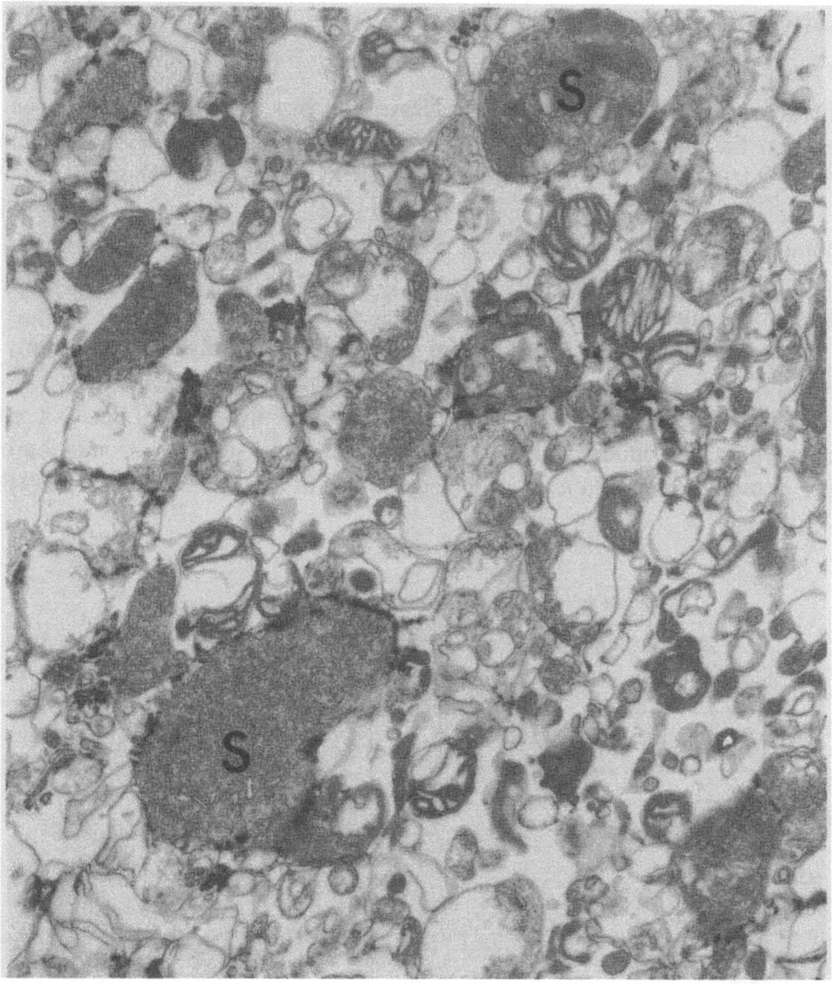

Fig. 10. Synaptosomal fraction from rat brain showing reaction product associated with several synaptosomes (*S*) in places where other possibly adhering membranes can be see. Magnification 1:24 000

It would, therefore, appear that most of the 5′-nucleotidase of rat brain subcellular fractions is associated with nonsynaptic structures, consistent with the localizations found in situ using EM-cytochemical techniques.

4 Biochemical Characterization of Membrane-Bound Nucleotidase Activity in Rat Brain

In the above discussion of AMPase localization in situ and in vitro it has been assumed that the hydrolysis of AMP is catalyzed only by 5′-nucleotidase. In principle, however, AMP could be broken down by high K_m nonspecific phosphatases. We have, therefore, examined the membrane-bound phosphatase activity of cerebral membranes in order to characterize the activity responsible for the hydrolysis of nucleotides. In particular we have focused on substrate specificity and the use of inhibitors of 5′-nucleotidase and nonspecific phosphatase.

At neutral pH, rat brain membranes were able to hydrolyze a wide variety of nucleotides. The best substrates were the 5′ribonucleotides, in particular 5′UMP, and 5′CMP (Fig. 11). 5′Deoxyribonucleotides were also hydrolyzed, but to a much lesser extent. Other modifications to the ribose moeity, such as introduction of a further phosphate as in 3′5′AMP or replacement with arabinose as in adenine-9-β-D-

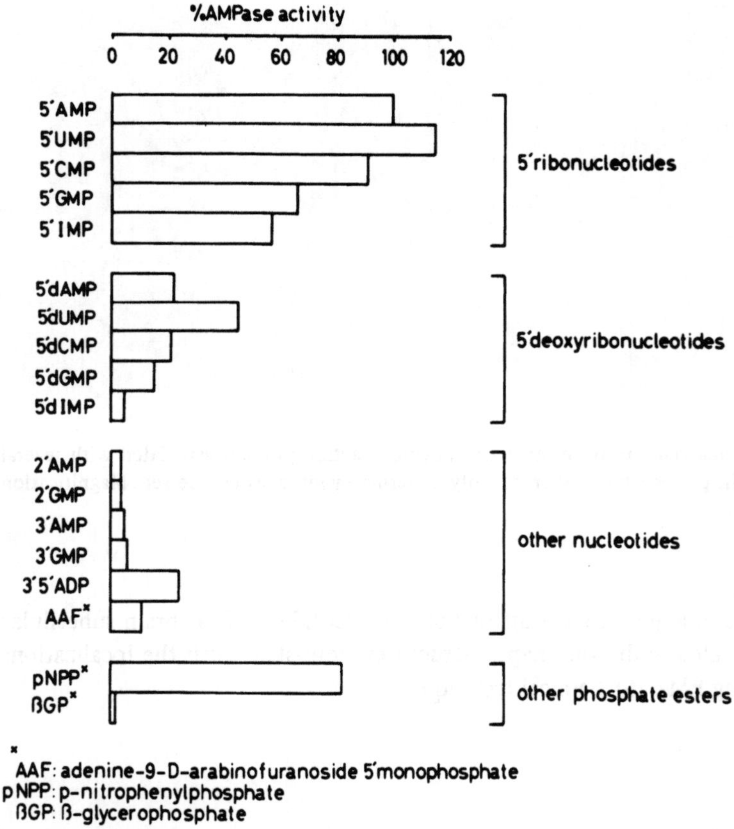

Fig. 11. Hydrolysis of 5′-nucleotidase and other phosphate esters by a crude membrane fraction from rat cerebral cortex. The rate of hydrolysis of 2 m*M* substrate is shown as a percentage of AMP hydrolysis. *AAF* adenine-9-β-D-arabinofuranoside 5′monophosphate; *pNPP* p-nitrophenylphosphate; βGP β-glycerophosphate

arabinofuranoside-5′monosphate also resulted in loss of ability to act as substrates. Nucleotides phosphorylated in the 2′ or 3′positions were inactive as substrates, as was the substrate for nonspecific phosphatases, β-glycerophosphate. On the other hand, such membrane fractions were able to hydrolyze p-nitrophenylphosphate (PNPP), a well-known substrate for nonspecific phosphatases, such as alkaline phosphatase. However, AMP hydrolysis was unaffected by addition of PNPP, indicating that the AMPase and pNPPase activities represent the actions of distinct enzymes. From the profile of nucleotide hydrolysis it can be deduced that a true 5′-nucleotidase is responsible for the hydrolysis of 5′AMP in cerebral membranes. This was confirmed by the sensitivity of 5′AMP hydrolysis to well characterized inhibitors of 5′-nucleotidase, such as α,β-methylene ADP (AOPCP) and concanavalin A (ConA), both in membrane fractions and in histochemical sections. In addition, 5′-AMP hydrolysis over a wide pH range is insensitive to the alkaline phosphatase inhibitor, bromotetramisole. Taken together, these data indicate that the hydrolysis of 5′AMP in brain tissue is catalyzed primarily by a true 5′-nucleotidase rather than by a nonspecific phosphatase. It may, therefore, be concluded that the localization data obtained by histochemical or EM-cytochemical methods using 5′-AMP as a substrate represent the distribution of 5′-nucleotidase activity.

The sensitivity of 5′-nucleotidase to inhibition by the lectin, ConA suggests that the cerebral enzyme, like that from other sources, is a glycoprotein. Indeed, it has recently been shown that the enzyme from bovine caudate nucleus presents N-glycosylation of the complex oligolactosaminic type with a fucose residue on the internal N-acetyl-D-glucosamine on the pentasaccharide core (Meflah et al. 1984).

Table 1. Inhibition of 5′-nucleotidase activity by lectins [a]

Lectin	Carbohydrate specificity	5′-Nucleotidase activity (% control) Membranes[b]	Solubilized enzyme
Concanavalin A	Glucosyl-, mannosyl-	24 ± 5*	28
Lentil lectin	Glucosyl-, mannosyl-	18 ± 3*	28
Wheat germ agglutinin	β-N-acetylglucosaminyl-, sialic acid	23 ± 5*	79
Soybean lectin	N-acetylgalactosaminyl-	98 ± 7	100
Helix pomatia lectin	N-acetylgalactosaminyl-	105 ± 10	101
Ulex europeaus agglutinin I	L-fucosyl-	102 ± 7	105
Peanut agglutinin	β-D-galactosyl-, β-D-gal-(1,3)-D-galNAc	97 ± 9	103
Bandireaea simplicifolia lectin I	α-D-galactosyl-	95 ± 9	103

[a] Lectins were screened for inhibition of 5′-nucleotidase activity at aconcentration of 0.1 mg/ml. Results are means ± SEM from five membrane preparations or the means of duplicate determinations from a single preparation of partially purified 5′-nucleotidase. [b] * = $P < 0.001$

The effects of various lectins on 5′-nucleotidase in rat brain membranes and a detergent-solubilized enzyme preparation is shown in Table 1. The inhibitory effects of ConA, lentil lectin, and wheat germ agglutinin suggest the presence of glucosyl or mannosyl, together with β-N-acetylglucosamyl or sialic acid residues. Preincubation of either membranes or solubilized 5′-nucleotidase with glycosidases also resulted in a significant inhibition of AMP hydrolysis (Table 2). Taken together with the carbohydrate specificities of lectins inhibiting 5′-nucleotidase, these data indicate that the hydrolysis of α-mannosyl or N-acetylglucosaminyl bonds results in loss of enzyme activity. Clearly, the integrity of cell membranes is not necessary for this effect since both lectins and glycosidases affected a solubilized 5′-nucleotidase preparation from cerebral membranes. It, therefore, seems likely that the observed effects on 5′-nucleotidase activity arise from direct effects on the 5′-nucleotidase itself. Qualitatively similar results have been obtained with a highly purified 5′-nucleotidase from rat liver plasma membranes (see Luzio et al., this volume). It, therefore, seems likely that the carbohydrate moeity of 5′-nucleotidase plays a role in allowing full expression of its enzyme activity.

Table 2. Effect of glycosidases on 5′-nucleotidase activity [a]

	5′-Nucleotidase activity (% control)[b]	
	Membranes	Solubilized enzyme
α-Mannosidase (Jack bean; 1.25 units/ml)	66 ± 10*	41
β-Glucosidase (almond; 2 units/ml)	99 ± 10	100
β-N-acetylglucosaminidase (Jack bean; 0.01 units/ml)	72 ± 5*	82
α-L-fucosidase (bovine epididimis; 0.2 units/ml)	101 ± 8	n.d.
α-Galactosidase (Asp. niger; 0.25 units/ml)	99 ± 7	n.d.
β-Galactosidase (*E. coli*; 0.75 units/ml)	101 ± 10	n.d.
Neuroaminidase (*Cl. perfringens;* 0.05 units/ml)	132 ± 33	118

[a] Preincubations with glycosidases were carried out for 30 min at 37°C before assaying 5′-nucleotidase activity. The results are means ± SEM obtained from eight membrane preparations or the means of duplicate determinations from a single preparation of partially purified 5′-nucleotidase [b] n.d. = not determined; * = $P<0.001$

5 Functional Aspects of 5'-Nucleotidase in Brain Tissue

The functional significance of ecto-5'-nucleotidase activity in the brain can be considered from two points of view. First, nucleotides appearing in the extracellular space may be utilized to form nucleosides like adenosine, which can act on extracellular receptors producing physiological effects, such as neuromodulation or changes in local cerebral blood flow. These humoral actions are then terminated by removing adenosine from the extracellular compartment via a high affinity uptake system. Secondly, 5'-nucleotidase might form part of a retrieval mechanism for purine or pyrimidine nucleotides. These are converted into a membrane-permeable form and can, therefore, be removed from the extracellular space. Evidently, in terms of both these functions, the relationship between AMPase activity and the nucleoside uptake system is a principle factor determining the extracellular concentration of nucleosides.

Two schemes depicting the possible relationships between nucleotidase and the nucleoside uptake mechanism (in this scheme represented by AMPase and the adenosine uptake system) are presented in Fig. 12. In the first scheme, the AMPase and transporter components are quite distinct. Adenosine produced from 5'AMP is released freely into the extracellular space before being taken up. This is clearly the scheme necessary where adenosine interacts with extracellular receptors before being inactivated via the uptake mechanism. On the other hand, if nucleotide hydrolysis was functioning primarily to clear the extracellular space of purines and pyrimidines, a second possible mechanism must be considered whereby AMPase and nucleoside transporter are tightly coupled to form a translocase complex (Fig. 12, Scheme II). In this case free nucleoside would not appear in the extracellular space. Such a coupling of nucleotide hydrolysis to uptake has been proposed in lymphocytes (Dornand et al. 1979; also Ch. 2.6 this volume) and in guinea pig heart (Frick and Loewenstein 1978).

In view of its obvious importance in regulating extracellular adenosine levels we have studied the relationship between AMPase activity and adenosine uptake in slices of rat hippocampus, a preparation which has proved to be of considerable importance in studies of neuronal function in the central nervous system (Lynch

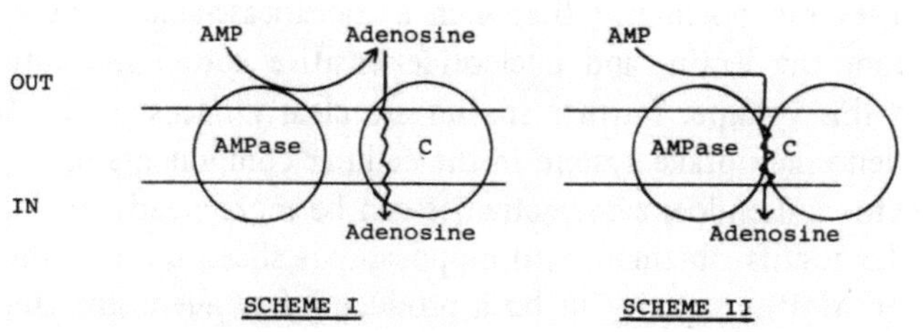

Fig. 12. Possible relationships between AMP hydrolysis and adenosine uptake into cells. *Scheme I:* AMPase and the adenosine carrier (*C*) are independent such that adenosine is produced in the extracellular space and subsequently taken up. *Scheme II:* AMPase and the adenosine carrier are tightly coupled such that hydrolysis of AMP occurs during uptake and no adenosine can accumulate in the extracellular space

Table 3. Effects of inhibitors of 5′-nucleotidase and of adenosine uptake on the hydrolysis of radioactively labeled 5′AMP and on the uptake of label in rat hippocampal slices[a]

	AMP hydrolysis	Uptake from AMP	Adenosine uptake
Control	100	100	100
Dipyridamole	84	42*	48***
AOPCP	27*	74**	95
Con A	23*	88**	97

[a] Results are from five experiments, each with six measurements in each experimental group. Incubations were for 15 min with 10 μM [^{14}C]-AMP or [^{3}H]-adenosine. Significance levels for the differences from control values were: *$P<0.01$; **$P<0.05$; ***$P<0.005$. All data are expressed as % control values

and Schubert 1980). Incubation of hippocampus slices with radioactively labeled adenosine results in uptake of the nucleoside in a time- and concentration-dependent manner (Reddington and Pusch 1983). This uptake was inhibited by dipyridamol, a well-known blocker of adenosine transport (Table 3). Likewise, the uptake of radioactive material from labeled 5′AMP was also inhibited to the same extent by dipyridamol, indicating the participation of the adenosine uptake mechanism in this process. Inclusion of the 5′-nucleotidase inhibitors, AOPCP or ConA inhibited AMP hydrolysis, as judged by accumulation of adenosine in the incubation medium, by 70–80%. However, under these conditions a reduction in uptake of labeled material of only ca. 20% was obtained (Table 3). The 5′-nucleotidase inhibitors had no effect on adenosine uptake. Taken together, these data suggest that although most of the adenosine produced from exogenous AMP by hippocampal slices accumulates in the medium, with some subsequent uptake (Fig. 12, Scheme I), there is an uptake component which is not influenced by inhibiting AMPase activity. This could indicate the presence in brain tissue of an AMPase directly coupled to the adenosine transporter to form a translocase complex (Fig. 12, Scheme II) as has been suggested to occur in lymphocytes (Dornand et al. 1979) and heart tissue (Frick and Loewenstein 1978). In the case of hippocampus slices this component of the AMPase activity accounts for 5–10% of the total AMP hydrolysis. However, the insensitivity of this component to inhibition by ConA and AOPCP raises the possibility that such a translocase-associated AMPase would be distinct from the lectin- and nucleotide-sensitive ecto-5′-nucleotidase discussed elsewhere in this volume. Further studies are clearly necessary on the AMPase activities and adenosine uptake system in the cellular components of CNS tissue in culture where ecto- and endoenzyme activities can be more clearly distinguished (see Appendix 1). The results obtained with hippocampus slices are nevertheless consistent with roles for AMPase activity in both producing free adenosine and in retrieving nucleotides from the extracellular space.

6 Summary

In the investigations reported here we have attempted to delineate the morphological localization and biochemical characteristics of 5′-nucleotidase in nervous tissue in situ. EM-cytochemistry shows the most common localization of this enzyme to be on glial cells in an ecto-position, but activity on the surface of neurons is also observed. Some intracellular reaction product can also be seen in certain specialized locations. These data are supported by EM-cytochemical examination of subcellular fractions from brain tissue. The enzyme under study is a true 5′-nucleotidase; non-specific phosphatases, such as alkaline phosphatase, do not contribute significantly to the total activity of tissue sections or membrane fractions. It is possible, however, that a small phosphatase component other than 5′-nucleotidase is associated with a translocase complex. 5′-Nucleotidase activity is attenuated by treatment of membranes with glycosidases suggesting that glycosylation might be necessary for maintenance of full enzyme activity. The functional role of 5′-nucleotidase activity is probably to regulate the concentrations of purines and pyrimidines in the extracellular space. This could be of importance both in processes of intercellular communication and in the recycling of purine and pyrimidine moeities. Other functions of the enzyme molecule independent of its enzymatic activity must, however, also be considered (see Ch. 3.3 this volume).

References

Bernstein HG, Weiss J, Luppa H (1978) Cytochemical investigation on the rat hippocampus with special reference to synaptic regions. Histochemistry 55:261–267

Blinzinger K, Kreutzberg GW (1968) Displacement of synaptic terminals from regenerating motoneurons by microglial cells. Z Zellforschung Mikrosk Anat 85:145–157

Bourre JM, Chanez C, Dumont O, Flexor MA (1982) Alteration of 5′-nucleotidase and Na^+, K^+-ATPase in central and peripheral nervous tissue from dysmyelinating mutants (jimpy, quaking, trembler, shiverer, and mld). Comparison with CNPase in the developing sciatic nerve from trembler. J Neurochem 38:643–649

Cammer W, Zimmermann TR (1981) Rat brain 5′-nucleotidase: developmental changes in myelin and activities in subcellular fractions and myelin subfractions. Dev Brain Res 1:381–389

Cammer W, Sirota SR, Zimmermann TR Jr, Norton WT (1980) 5′Nucleotidase in rat brain myelin. J Neurochem 35:367–373

Cammer W, Sacchi R, Kahn S (1985) Immunocytochemical localization of 5′-nucleotidase in oligodendroglia and myelinated fibers in the central nervous system of adult and young rats. Dev Brain Res 20:89–96

Cragg B (1979) Overcoming the failure of electromicroscopy to preserve the brain's extracellular space. Trends Neurosci 2:159–161

Dornand J, Bonnafous JS, Gavach C, Mani JC (1979) 5′Nucleotidase-facilitated transport by mouse lymphocytes. Biochimie (Paris) 61:973–978

Frick GP, Lowenstein JM (1978) Vectorial production of adenosine by 5′-nucleotidase in the perfused rat heart. J Biol Chem 253:1240–1244

Harwood JL, Hawthorne JN (1969) Metabolism of the phosphoinositide in guinea pig brain synaptosomes. J Neurochem 16:1377–1387

Heymann D, Reddington M, Kreutzberg GW (1984) Subcellular localization of 5′-nucleotidase in rat brain. J Neurochem 43:971–978

Israel M, Frachon-Mastour P (1970) Fractionnement du cortex cerebral du rat, distribution subcellulaire de la 5′-nucleotidase et de cholinesterases. Arch Anat Microsc Morphol Exp 59:383–392

Kai M, White GL, Hawthorne JN (1966) The phosphatidylinositol kinase of rat brain. Biochem J 101:328–337

Kreutzberg GW, Barron KD (1978) 5′-Nucleotidase of microglial cells in the facial nucleus during axonal reaction. J Neurocytol 7:601–610

Kreutzberg GW, Emmert H (1980) Glucose utilization of motor nuclei during regeneration: A [^{14}C]2-deoxyglucose study. Exp Neurol 70:712–716

Kreutzberg GW, Hussain ST (1982) Cytochemical heterogeneity of the glial plasma membrane: 5′-nucleotidase in retinal Mueller cells. J Neurocytol 11:53–64

Kreutzberg GW, Hussain ST (1984) Cytochemical localization of 5′-nuleotidase activity in retinal photoreceptor cells. Neuroscience 11:857–866

Kreutzberg GW, Barron KD, Schubert P (1978) Cytochemical localization of 5′-nucleotidase in glial plasma membranes. Brain Res 158:247–257

Lynch G, Schubert P (1980) The use of in vitro brain slices for multidisciplinary studies of synaptic function. Annu Rev Neurosci 3:1–22

Marani E (1977) The subcellular distribution of 5′-nucleotidase activity in mouse cerebellum. Exp Neurol 57:1042–1048

Meflah K, Hart J, Duflos Y, Bernard J (1984) 5′-Nucleotidase from bovine caudate nucleus synaptic plasma membranes: specificity for substrates and cations; study of carbohydrate moiety by glucosidases. J Neurochem 42:1107–1115

Nagata H, Mimori Y, Nakamura S, Kameyama M (1984) Regional and subcellular distribution in mammalian brain of the enzymes producing adenosine. J Neurochem 42:1001–1007

Phillips E, Newsholme EA (1979) Maximum activities, properties and distribution of 5′-nucleotidase, adenosine kinase and adenosine deaminase in rat and human brain. J Neurochem 33:553–558

Pilcher CW, Jones DG (1970) The distribution of 5′-nucleotidase in subcellular fractions of mouse cerebellum. Brain Res 24:143–147

Reddington M, Pusch R (1983) Adenosine metabolism in a rat hippocampal slice preparation: incorporation into S-adenosylhomocysteine. J Neurochem 40:285–290

Richardson PJ (1983) Presynaptic distribution of the cholinergic-specific antigen chol-1 and 5′-nucleotidase in rat brain, as determined by complement-mediated release of neurotransmitters. J Neurochem 41:640–648

Schubert P, Komp W, Kreutzberg GW (1979) Correlation of 5′-nucleotidase activity and selective trans-neuronal transfer of adenosine in the hippocampus. Brain Res 168:419–424

Schubert P, Lee KS, Reddington M, Kreutzberg GW (1983) Synaptic modulation by adenosine: electrophysiological and biochemical characteristics. In: Berne RM, Rall TW, Rubio R (eds) Regulatory functions of adenosine. Nijoff, The Hayge, pp 439–454

Suran AA (1974) 5′-Nucleotidase and an acid phosphatase of spinal cord. Comparative histochemistry and specificity of the enzymes in mouse and cat spinal cords. Cytological localization in mouse substantia gelatinosa. J Histochem Cytochem 22:802–811

Scott TG (1967) The distribution of 5′-nucletidase in the brain of the mouse. J Comp Neurol 129:97–114

Williams M (1984) Mammalian central adenosine receptors. Handb Neurochem 6:1–26

Withy RM, Mayer RJ, Strange PG (1982) The distribution of radioligand binding activity and 5′-nucleotidase activity in bovine caudate nucleus subcellular fractions. J Neurochem 38:1348–1355

The Role of 5′-Nucleotidases in the Heart

A. C. Newby[1] and Y. Worku[2]

1 Introduction

The demand for substrates and oxygen by the heart is closely matched by their supply through coronary circulation. Thus, an increase in force or rate of contraction is powered not by increasing the percentage extraction of substrates and oxygen from the blood but by increasing coronary blood flow (Eckenhoff et al. 1947). Conversely, a decrease in perfusion due to lowered systemic blood pressure or pathologically due to spasm or stenosis of the coronary arteries rapidly leads to an oxygen debt which if not compensated by vasodilation can be detected biochemically through an increase in lactate output, electrically in the E.C.G. and symptomactically through the experience of pain!

The autoregulation of coronary blood flow can be observed in isolated and hence denervated preparations implying the existence of a regulatory factor intrinsic to the heart. In 1964, after summarising earlier work, Berne propounded the hypothesis that adenosine is the endogenous regulator. He proposed that adenosine is formed in the myocardial cells when demand for ATP exceeds supply. Since adenosine is rapidly permeable through the plasma membrane it can diffuse to the vascular smooth muscle cells and mediate dilation. A substantial body of evidence supports this hypothesis (Berne 1980) although other endogenous and exogenous regulators certainly contribute to regulation of coronary blood flow in the intact animal. Interestingly, adenosine also antagonises the effect of β-adrenergic agonists on the myocardial cells (Schrader et al. 1977) and reduces the release of noradrenaline from endogenous nerve endings (Kahn and Malik 1980). These effects would reduce energy demands on the myocardium and may represent a "retaliatory" response of the myocardial cell to stimulation (Newby 1984). A similar function for adenosine may be deduced in the brain, kidney adipose tissue and in secretory cells in general (Newby 1984).

The adenosine hypothesis so stated, presupposes a biochemical link between ATP breakdown and adenosine formation; a plausible pathway is shown in Fig. 1. ATP breakdown primarily due to the demands for mechanical activity but also due to other metabolic and electrochemical demands will be matched by ATP formation

[1] Department of Cardiology, University of Wales College of Medicine, Heath Park, Cardiff CF4 4XN, UK

[2] Department of Clinical Biochemistry, University of Cambridge, Addenbrooke's Hospital, Hills Road, Cambridge CB2 2QR, UK

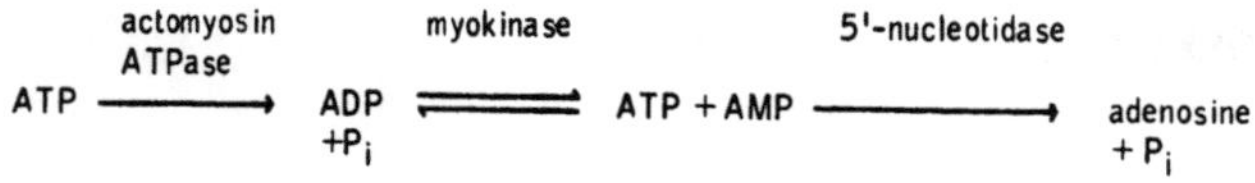

Fig. 1. A hypothetical pathway of adenosine formation

from glycolysis, oxidative phosphorylation and through the creatine kinase equilibrium. An excess of ATP degradation over ATP formation will lead to an increase in the free concentration of ADP which will stimulate the formation of AMP through the myokinase equilibrium. Owing to the high activity of myokinase in mammalian cells the enzyme will be close to equilibrium and since much larger amounts of ATP than AMP are present under basal conditions a large percentage change in AMP concentration may result from a small percentage fall in ATP concentration (Newsholme and Start 1973; Nishiki et al. 1978). The increased AMP concentration may then be sensed by a 5′-nucleotidase enzyme leading to the formation of adenosine and Pi. The adenosine so formed can either leave the cell via the symmetric nucleoside permease (Harley et al. 1979; Plagemann and Wohlheuter 1982), be deaminated by adenosine deaminase or be rephosphorylated to AMP by adenosine kinase (Arch and Newsholme 1978). The studies reported here support the hypothesis that a 5′-nucleotidase is involved in adenosine formation from cells undergoing ATP catabolism. The regulatory properties of the enzyme may contribute to understanding the regulation of adenosine formation in response to alterations in energy supply and demand.

2 The Ecto-5′-Nucleotidase of Rat Adipocytes

An AMP-hydrolysing activity from collagenase isolated rat adipocytes (Rodbell 1964) was shown to have the following properties (Newby et al. 1975). The enzyme was specific for 5′-mononucleotides, AMP being the preferred substrate (K_m = 43 μM). The enzyme required Mg^{2+} and was potently and competitively inhibited by nucleotide di- and triphosphates (Ki for ATP = 9 μM). The enzyme was located in the plasma membrane (Avruch and Wallach 1971). The AMP hydrolysing activity of intact adipocytes was equal to or greater than that of the homogenate despite the demonstrated impermeability of AMP into the fatcell. The 5′-nucleotidase of intact adipocytes was inhibited by 50 μMATP added to the extracellular medium but was unaffected by the ATP present in the cytoplasm at approx. 5 mM. The 5′-nucleotidase could be inhibited by exogenously added antibodies. Antibody-inhibited washed adipocytes failed to reveal any 5′-nucleotidase activity on homogenisation. We concluded that the 5′-nucleotidase was an ectoenzyme and that our experiments would have detected an intracellular 5′-nucleotidase activity if this contributed more than 20% of the ectoenzyme activity. In fact an intracellular, latent pool of 5′-nucleotidase was subsequently demonstrated (Stanley et al. 1980) which accounted for 20–30% of the total homogenate activity. The indentification of this pool relied on the demonstration of activity in the homogenate revealed

Table 1. Subcellular distribution of ecto-5′-nucleotidase in rat polymorphonuclear leucocytes[a]

Condition	Activity (% of activity of cells + Lubrol)	Activity + Lubrol / Activity − Lubrol (%)
Washed cells	77–84	
Washed cells, sonicated	66–79	
Washed cells + Lubrol	100	130–119
+ antiserum then washed	12–5	
+ antiserum, washed sonicated	16–11	
+ antiserum, washed, sonicated + Lubrol	45–46	280–420

[a] Polymorphonuclear leucocytes werde treated for 30 min at 4°C with 1/100 dilution of rabbit anti-(rat liver 5′-nucleotidase) serum. They were then washed five times at 4°C and resuspended in the original volume of Krebs-Ringer bicarbonate buffer. A portion was then sonicated according to Newby (1980). Fractions were assayed under conditions optimal for the ecto-5′-nucleotidase (Avruch and Wallach 1971; Newby et al. 1975) either in the presence or absence of 0.1% (wt/vol) of Lubrol PX. Values are those from two replicate experiments

only when detergents were added. It is thus crucial to rule out that this increase in homogenate activity was the result of detergent activation of the enzyme; an interpretation given to similar results by other workers (De Pierre and Karnovsky 1974). The evidence against detergent activation of the enzymes is as follows. (1) The fold activation of enzyme by a fixed concentration of detergent varies greatly for the same source of enzyme in different subcellular fractions (see for example Table 1). (2) Preparations of open membrane figures show low activation whereas preparations rich in membrane vesicles show higher activation. Indeed it is possible by vigorous homogenisation and hypotonic solutions to obtain preparations which are not activated at all by detergents (Stanley et al. 1980). (3) Detergents of widely different monomer and micellar structure yield the same total enzyme activity (Bailyes et al. 1982). (4) The latency of vesicles containing 5′-nucleotidase is similar to that of subcellular organelles such as lysosomes (Stanley et al. 1980). Thus increased activity of 5′-nucleotidase in homogenates of subcellular fractions is due to the exposure of latent activity.

3 Is Ecto-5′-Nucleotidase Involved in Nucleotide Catabolism in Cells Undergoing ATP Breakdown?

In the view of some workers the location of 5′-nucleotidase precluded its participation in intracellular nucleotide breakdown (Van den Berghe et al. 1977). However, the high activity of the enzyme and its strategic location at the cell margin (particularly in the heart) (Rubio et al. 1973) have led many workers to propose its participation in adenosine formation (Nakatsu and Drummond 1972; Berne 1980; Naito and Lowenstein 1982). Three hypotheses were proposed to account for adenosine formation (Newby 1975). (1) A separate 5′-nucleotidase either located in the cytosolic fraction or at the inner membrane face. (2) An intracellular pool of the ectoenzyme. (3) The ectoenzyme is a transmembrane protein whose active site is accessible both to extracellular and intracellular AMP. To test these hypotheses, we investigated the formation of adenosine, inosine and hypoxanthine in intact cells during ATP catabolism induced by cyanide or the non-metabolisable hexose, 2-deoxyglucose. Adipocytes were unsuitable for such studies since their fragility resulted in an unacceptable background of nucleotide degradation from broken cells (Grahn et al. 1981). Rat polymorphonuclear leukocytes were obtained at high purity, in sufficient numbers and with less than 2% broken cells under all experimental circumstances (Newby 1980). The ecto-5′-nucleotidase and ecto-β-glycerophosphatase activites were inhibited greater than 98% with a combination of 10 m*M* β-glycerophosphate and an antibody raised against purified ratliver 5′-nucleotidase (Newby 1980). Such inhibition did not diminish IMP hydrolysis to inosine and hypoxanthine in cyanide-poisoned polymorphonuclear leucocytes and a cytosolic 5′-nucleotidase sufficiently active to account for hypoxanthine formation was detected (Newby 1980). To measure adenosine formation, it was necessary to prevent its degradation by adenosine deaminase and its rephosphory-

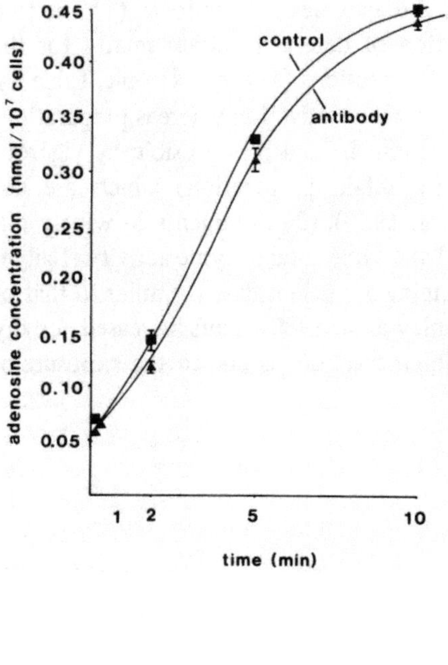

Fig. 2. Effect of inhibition of ecto-5′-nucleotidase on adenosine formation by polymorphonuclear leucocytes. Polymorphonuclear leucoytes were preincubated with 2′-deoxycoformycin and 5-iodotubercidin (■, ▲) and with an antiserum raised to ratliver plasma membrane 5′-ncleotidase (▲). They were then poisoned with 5 m*M* 2-deoxyglucose for the time intervals shown before trichloracetic acid extraction and determination of adenosine concentration by hplc. (Data are from Newby and Holmquist 1981)

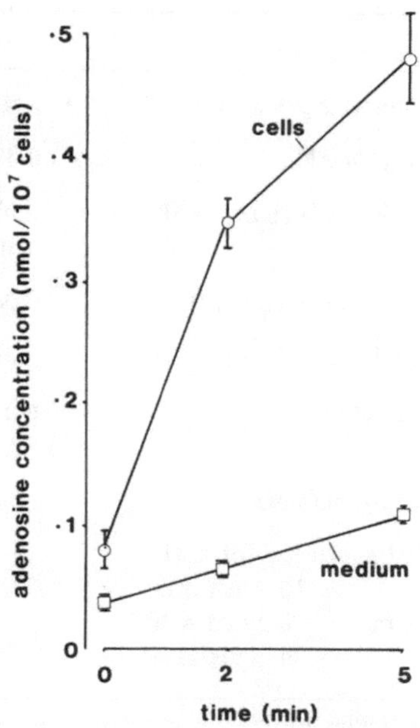

Fig. 3. Location of adenosine produced in polymorphonuclear leucocytes. Polymorphonuclear leucocytes were preincubated with 2′-deoxycoformycin and 5-iodotubercidin and then poisoned with 2-deoxyglucose. After the time intervals shown replicate cell suspensions were either extracted with trichloracetic acid (*cells*) or centrifuged for 15 s over a layer of silicone oil of density 1.03. The supernatant buffer (*medium*) was extracted with trichloracetic acid and adenosine concentration in both medium and cells plus medium determined. (Data are from Newby and Holmquist 1981)

lation by adenosine kinase. Selective inhibitors of both enzymes were shown to be effective in intact polymorphonuclear leukocytes (Newby 1981). Adenosine formation during the rapid catabolism of ATP induced in polymorphonuclear leucocytes by 2-deoxyglucose was not prevented by inhibition of ecto-phosphatases (Fig. 2, Newby and Holmquist 1981). Moreover, when cells were separated from the incubation medium adenosine was found inside the cells (Fig. 3) suggesting that it was formed from a cytoplasmic route (Newby and Holmquist 1981).

4 Does the Intracellular Pool of 5′-Nucleotidase Contribute to Breakdown of Cytosolic AMP and IMP?

An intracellular, membrane-bound pool which is kinetically and immunologically identical with the ectoenzyme has been described in several celltypes (Stanley et al. 1980; Widnell et al. 1982). Stanley et al. (1980) showed that in the liver this pool resides on the intraluminal face of vesicles which have a similar latency to lysosomes. Some similar data from polymorphonuclear leucocytes is shown in Table 1. Sonication of cells caused a small reduction in measured enzyme activity. Cells incubated with antiserum and then washed showed a greatly reduced surface activity. When such cells were sonicated a small activity (4–6% of total 5′-nucleotidase activity) was revealed. Detergent treatment revealed a large pool of intravesicular activity. All the enzyme could be inhibited by addition of antiserum at this stage (not shown in Table 1). These data indicated that only a small fraction of the total ecto-5′-nucleotidase of polymorphonuclear leucocytes could be available at the cytosolic face of the membrane. It also showed that inhibition of the ectoenzyme

Table 2. Comparison of cytosolic and ecto-5′-nucleotidase[a]

	Ecto-5′-nucleotidase	Cytosolic 5′-nucleotidase
Substrate preference	AMP ⟩ GMP ⟩ IMP	IMP ⟩ GMP ⟫ AMP
K_m for AMP	0.020	2.6
Effect of ATP and ADP	Inhibits competitively	Activates $n_H^{ATP} = 1.8 - 3.3$
Ka/Ki for ATP (m*M*)	0.004	0.34
Ka/Ki for ADP (m*M*)	0.0003	2.7
Effect of Pi	none	Inhibits, $n_H = 1.2 - 1.7$[b]
Ki for Pi (m*M*)	–	5.8
Estimated K_m for AMP with { 10 m*M* ATP, 0.2 m*M* ADP, 4 m*M* Pi	33.4	2.7

[a] Values of kinetic constants for rat heart ecto-5′-nucleotidase are taken from Sullivan and Alpers (1971). Values for kinetic constants for rat liver cytosolic 5′-nucleotidase were obtained from appropriate analysis of data obtained using assay methods previously described (Itoh 1981a; Worku and Newby 1983). K_m for AMP was estimated for cytosolic 5′-nucleotidase from a symmetric model assuming n_H for ATP = 4 and n_H for Pi = 1

[b] n_H = Hill coefficient

with antibody did not allow the cytoplasmic side of the protein to continue hydrolysing AMP. If one estimates the patency of the intracellular 5′-nucleotidase as that revealed by sonication divided by the revealed by sonication plus detergent one obtains a value of 12–17% (Table 1). This figure is similar to that obtained for lysosomes from the same cells (K. Docherty and A.C. Newby, unpublished). Thus there may be no ecto-5′-nucleotidase facing the cytosol in the living cell. In any event, the activity of the enzyme would be greatly suppressed by the cytosolic concentrations of ATP and ADP (see Table 2).

5 Which Enzyme is Responsible for Adenosine Formation in Polymorphonuclear Leukocytes?

The IMP hydrolysing activity detected in polymorphonuclear leucocytes (Newby 1980) was further characterised (Worku and Newby 1983). It was found to have the same K_m for IMP (0.2 m*M*) and the same electrophoretic mobility at pH 9.6 as the similar activity found in the cytosolic fraction of rat liver. The enzyme from either source also has the interesting property that it will catalyse the incorporation of a radiolabeled nucleoside moiety from labeled inosine into unlabeled IMP. This property is not shared by alkaline phosphatase or ecto-5′-nucleotidase (Worku and

Table 3. Comparison of the observed rates of adenosine formation in rat polymorphonuclear leucocytes and rat heart with the maximal potential calculated flux through the cytosolic 5′-nucleotidase[a]

	Polymophonuclear leucocytes + 5 mM 2-deoxyglucose	Rat heart perfused with buffer equilibrated with 95% N_2 : 5% CO_2
Activity of cytosolic 5′-nucleotidase	1160[b]	440[b]
Concentration of AMP (mM)	1.06[c]	1.19[d]
Observed rate of adenosine formation	28[c]	20[d]
Calculated from the Michaelis-Menten equation (K_m = 2.6 mM)	336	138

[a] The calculation assumes that all the AMP which accumulated was free and was distributed uniformly through the intracellular water space. Enzyme activity and adenosine formation rates are expressed in μmolmin^{-1} g^{-1} wet wt

[b] Data from Worku and Newby (1983)

[c] Data from Newby and Holmquist (1981)

[d] Data from Newby et al. (1983)

Newby 1982). Since more enzyme was available from the liver, it was purified from this source (Itoh 1981a). The major physical and kinetic properties of the cytoplasmic and ecto-5′-nucleotidase are compared in Table 2. The salient features are the relatively high K_m for AMP (2.6 mM) for the cytosolic enzyme and the fact that it is activitated rather than inhibited by ATP. Pi is an inhibitor of the cytosolic enzyme. Knowledge of the kinetic properties of the enzyme allowed its maximal activity to be measured in homogenates (Worku and Newby 1983). From this and the concentration of AMP inside polymorphonuclear leucocytes we were able to conclude that there was more than enough enzyme activity to explain the maximal observed rate of adenosine formation (Table 3). An attempt was made to model the process of adenosine formation by incubating purified ratliver 5′-nucleotidase with the concentrations of AMP, IMP, ATP Mg and Pi found inside polymorphonuclear leucocytes undergoing ATP breakdown. The adenosine formation rate showed a biphasic dependence on AMP concentration being low at time 0 when AMP concentration was low (Fig. 4) and also at ′0 min when AMP concentration was high. Hence it was necessary to express activity as a function of energy charge (Atkinson 1968; Itoh 1981b) to obtain a smooth function. When the data were plotted in this fashion (Fig. 5) it was clear that the cell-free model containing only cytosolic 5′-nucleotidase accounted for both the shape and magnitude of the adenosine formation rate vs energy charge relationship. We therefore concluded that the cytosolic 5′-nucleotidase was responsible for adenosine formation in this cell (Worku and Newby 1983).

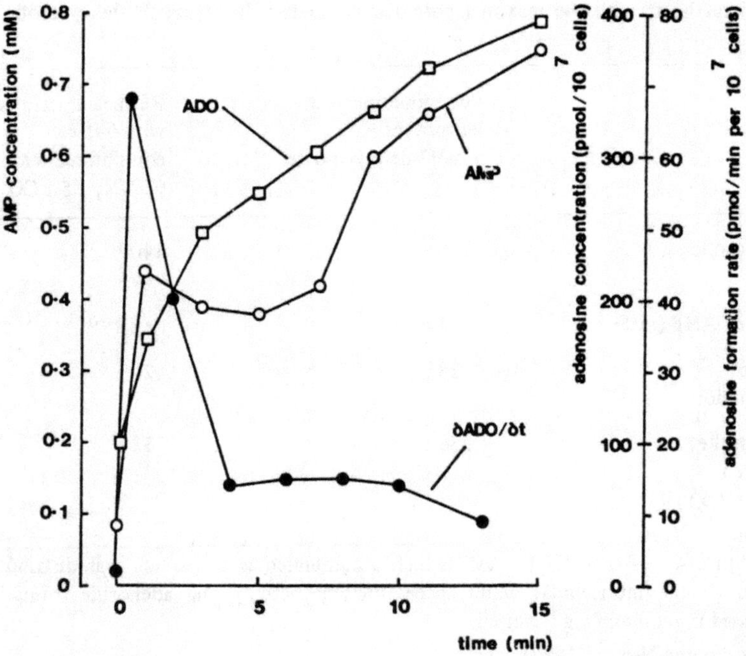

Fig. 4. The time dependence of AMP concentration and adenosine formation in 2-deoxyglucose poisoned polymorphonuclear leucocytes. AMP and adenosine concentration was measured 30 min before and at the time intervals shown after poisoning polymorphonuclear leucocytes with 5 m*M* 2-deoxyglucose. The rate of adenosine formation was estimated from the adenosine concentration at adjacent time points. (Data are from Worku and Newby 1983)

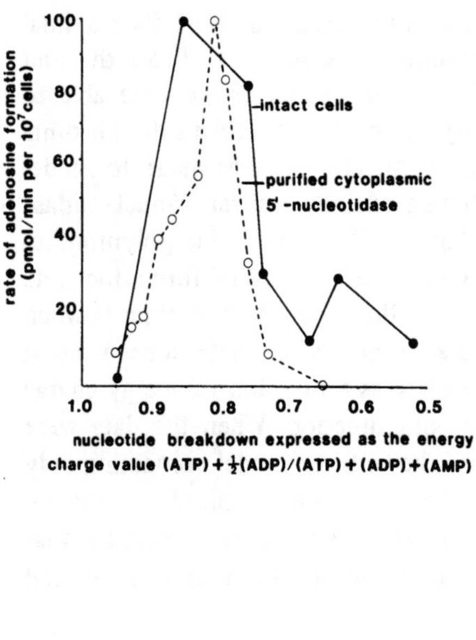

Fig. 5. Comparison of adenosine formation rates in polymorphonuclear leucocytes and in a cell-free system employing purified rat liver cytosolic 5′-nucleotidase. Using the data from Fig. 4 and the concentrations of ATP and ADP measured over the same time intervals, a plot of energy charge vs adenosine formation rate was constructed (●). Using the same concentrations of adenine nucleotides, the concentration of IMP and estimates of the intracellular free concentrations of Mg^{2+} (1 m*M*) and Pi (1.3 m*M*) adenosine formation from [^{3}H]AMP was determined using purified ratliver cytosolic 5′-nucleotidase (○). The activity was normalised to correspond the amount of enzyme present in 10^7 polymorphonuclear leucocytes.
(Data are from Worku and Newby 1983)

6 Does Cytosolic 5′-Nucleotidase Account for Adenosine Formation in the Heart?

Adenosine formation in cultured neonatal heart cells takes place by a cytoplasmic route which does not involved the ecto-5′-nucleotidase (P. Meghji, C.A. Holmquist and A.C. Newby, unpublished). Experiments with perfused hearts are complicated by the multiple compartments and by the uncertainly that exogenously added inhibitors of 5′-nucleotidase penetrate to their target. None-the-less, two groups have reported that inhibition of ecto-5′-nucleotidase does not prevent adenosine formation from the heart (Frick and Lowenstein 1976; Schütz et al. 1981). Schrader et al. (1981) elegantly demonstrated that nascent adenosine can be trapped as S-adenosyl homocysteine inside the tissue of the heart by simultaneous infusion of L-homocysteine. Thus adenosine formation in the heart takes place in the cytoplasmic compartment.

The activity of the cytosolic 5′-nucleotidase in rat heart is shown in Table 3. This activity may be enough to account for the rate of adenosine formation reported in a recent study of adenosine formation from rat heart (Table 3). However, the free concentration of AMP in the rat heart is not known precisely under these conditions and more data would be needed such as that obtained by Nishiki et al. (1978) to obtain an accurate estimate. Moreover, the calculated flux through the 5′-nucleotidase would only be sufficient to account for the observed rate of adenosine formation if (1) IMP and AMP did not compete for activity of the enzyme and (2) all the adenosine formed were transported from the cells by the nucleoside transporter. The catalytic efficiency (V_{max}/K_m) of adenosine kinase and adenosine deaminase are both much greater than of the nucleoside transporter (Fig. 6) implying that very little newly formed adenosine would leave the cell compared to that metabolised. Hence these indirect arguments suggest but cannot prove that the cytosolic 5′-nucleotidase accounts for adenosine formation in the heart. Such proof would establish the biochemical link between ATP breakdown and adenosine formation and confirm the role of adenosine as a "retaliatory metabolite" (Newby 1984).

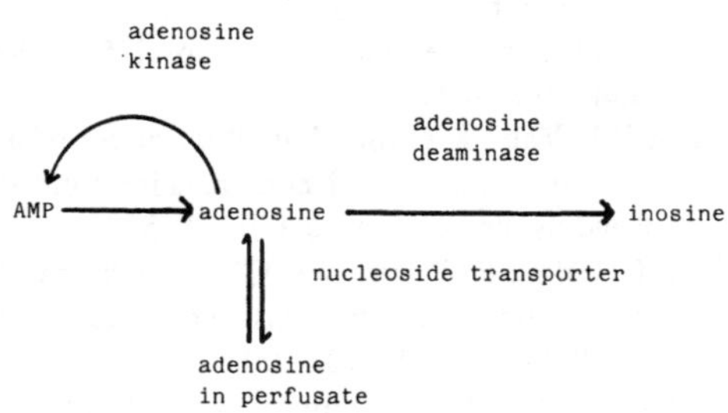

Process	Km	Vmax	Vmax/Km
adenosine kinase	0.4	60	156
adenosine deaminase	25	1000	40
adenosine transport	15	30	2.0

Fig. 6. Pathways of adenosine metabolism and transport. Values of K_m and V_{max} are taken from Arch and Newsholme 1978 and from Newby et al. 1983

7 Summary

A plasma membrane ecto-5′-nucleotidase is well-characterized in many cell types but its catalytic site dose not have access to intracellular AMP. Adenosine formation during cellular ATP catabolism takes places by an intracellular pathway which probably involves a distinct cytoplasmic 5′-nucleotidase.

References

Arch JRS, Newsholme EA (1978) The control of the metabolism and the hormonal role of adenosine. Essays Biochem 14:82–123

Atkinson DE (1968) The energy charge of the adenylate pool as a regulatory parameter. Interaction with feedback modifiers. Biochemistry 7:4030–4034

Avruch J, Wallach DFH (1971) Preparation and properties of plasma membrane and endoplasmic reticulum fragments from isolated rat fat cells. Biochim Biophys Acta 233:334–337

Bailyes EM, Hewby AC, Siddle K, Luzio JP (1982) Solubilization and purification of rat liver 5′-nucleotidase by use of a zwitterionic detergent and a monoclonal-antibody immunoadorbent. Biochem J 203:245–251

Berne RM (1964) Regulation of coronary blood flow. Physiol Rev 44:1–29

Berne RM (1980) The role of adenosine in the regulation of coronary blood flow. Circ Res 47:807–813

De Pierre JW, Karnovsky ML (1974) Ecto-enzymes of the guinea pig polymorphonuclear leukocyte. J Biol Chem 249:7121–7129

Eckenhoff JE, Hafkenshiel JH, Landmesser CM, Harmel M (1947) Cardiac oxygen metabolism and control of the coronary circulation. Am J Physiol 149:634–649

Frick GP, Lowenstein JM (1976) Studies of 5′-nucleotidase of the rat heart. J Biol Chem 250: 6372–6378

Grahn MF, Souness JE, Davies IJ (1981) The effect of incubation conditions on the release of fatty acids, adenine nucleotides and marker enzymes from rat epididymal fat-cells. Mol Physiol 1:119–132

Harley ER, Paterson ARP, Cass CE (1979) Initial rate kinetics of the transport of adenosine and 4-amino-7-(B-D-ribofuranosyl) pyrrolo [2,3-d] pyrimidine (Tubercidin) in cultured cells. Cancer Res 42:1289–1295

Itoh R (1981a) Purification and some properties of cytosol 5′-nucleotidase from rat liver. Biochim Biophys Acta 657:402–410

Itoh R (1981b) Regulation of cytosol 5′-nucleotidase by adenylate energy charge. Biochim Biophys Acta 659:31–37

Khan MT, Malik K (1980) Inhibitory effect of adenosine and adenine nucleotides on potassium-evoked efflux of [^{3}H] noradrenaline from the isolated heart: lack of relationship to prostaglandins. Br J Pharmacol 68:551–561

Lum CT, Marz R, Plagemann PGW, Wohlheuter RM (1979) Adenosine transport and metabolism im mouse leukaemic cells and in canine thymocytes and peripheral blood leukocytes. J Cell Physiol 101:173–200

Newby AC (1975) The properties and location of the rat fatcell 5′-nucleotidase. Ph D. Thesis, University of Wales

Newby AC (1980) Role of adenosine deaminase, ecto(5′-nucleotidase) and ecto-(nonspecific phosphate) in cyanideinduced AMP catabolism in rat polymorphonuclear leucocytes. Biochem J 186:907–918

Newby AC (1981) The interaction of inhibitors with adenosine metabolizing enzymes in intact isolated cells. Biochem Pharmacol 30:2611–2615

Newby AC, Holmquist CA (1981) Adenosine production inside rat polymorphonuclear leucocytes. Biochem J 200:309–403

Newby AC, Luzio JP, Hales CN (1975) The properties and extracellular location of 5′-nucleotidase of the rat fat-cell plasma membrane. Biochem J 146:625–633

Newsholme EA, Start C (1973) Regulation in metabolism, Wiley, London, pp 112–113

Nishiki K, Erecinska M, Wilson DF (1978) Energy relationships between cytosolic metabolism and mitochondrial respiration in rat heart. Am J Physiol 234:C72–C81

Rodbell M (1964) Metabolism of isolated fat-cells. J Biol Chem 239:375–380

Schrader J, Baumann G, Gerlach E (1977) Adenosine as an inhibitor of he myocardial effect of catecholamines. Pfluegers Arch Eur J Physiol 372:29–35

Schrader J, Schutz W, Bardenheuer H (1981) Role of S-adenosyl-homocysteine hydrolase in adenosine metabolism in mammalian heart. Biochem J 196:65–70

Schütz W, Schrader J, Gerlach E (1981) Different sites of adenosine formation in the heart. Am J Physiol 240:H963–H970

Stanley KK, Edwards MR, Luzio JP (1980) Subcellular distribution and movement of 5′-nucleotidase in rat cells. Biochem J 186:59–69

Sullivan JM, Alpers JB (1971) In vitro regulation of rat heart 5′-nucleotidase by adenine nunucleotides and magnesium. J Biol Chem 246:3057–3063

Van den Berghe G, Van Pottlesberghe C, Hers HG (1977) A kinetic study of the soluble 5′-nucleotidase of rat liver. Biochem J 162:611–616

Widnell CC, Schneider YJ, Pierre B, Baudhuin P, Trouet A (1982) Evidence for continual exchange of 5′-nucleotidase in cultured rat fibroblasts. Cell 28:61–70

Worku Y, Newby AC (1982) Nucleoside exchange catalysed by the cytoplasmic 5′-nucleotidase. Biochem J 205:503–510

Worku Y, Newby AC (1983) The mechanism of adenosine production in rat polymorphonuclear leucocytes. Biochem J 214:325–330

Ecto-Protein Kinase

Detection and Biochemistry of Cell Surface Protein Kinase

V. Kinzel, D. Kübler, E. Burow, and W. Pyerin [1]

1 Introduction

Protein phosphorylation represents an important physiological tool for modification and concomitant control of the biological function of proteins. The reaction is catalyzed by protein kinases (PK; EC 2.7.1.37) which in the presence of divalent metal ions transfer the terminal (the "γ") phosphoryl group from a nucleoside-triphosphate (NTP) to hydroxyl-carrying side chains of proteins:

$$\text{Protein} + \text{NTP} \xrightarrow[\text{Me}^{2+}]{\text{Kinase}} \text{Protein-PO}_4 + \text{NDP}.$$

A number of different PKs are known named either according to their preferred, but not necessarily physiological substrates, according to the mode of their regulation or to the amino acid residue phosphorylated (for review refs. Krebs and Beavo 1979; Cohen 1980; Flockhardt and Corbin 1982). Protein kinases have been recognized as key catalysts in metabolic cascades which are involved in extra-intracellular signal transduction (Nishizuka 1984) as well as in the control of metabolic pathways (Krebs and Beavo 1979; Cohen 1980).

In view of the important role of these intracellular protein phosphorylations, the detection of a cell surface located protein kinase activity (SPK; Ågren and Ronquist 1974; Mastro and Rozengurt 1976; Kübler et al. 1982a) capable of phosphorylating extracellular proteins as well as proteins located at the outer surface of the cell appears of particular interest (Fig. 1). Recent data seem to suggest that

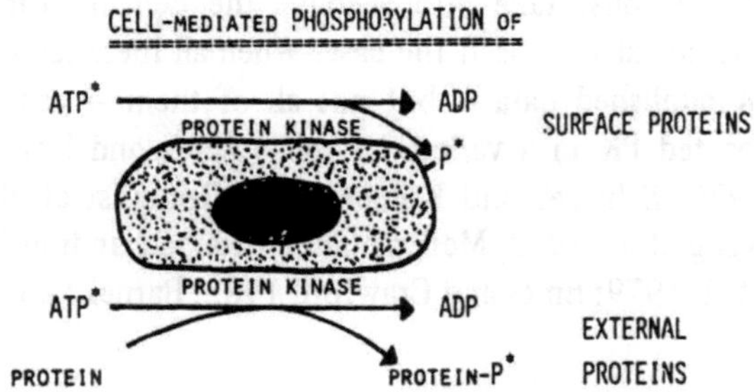

Fig. 1. Schematic drawing

[1] German Cancer Research Center, Institute of Experimente Pathology, Im Neuenheimer Feld 280, 6900 Heidelberg, FRG

SPK enables cells to interact specifically with their environment (Kübler et al. 1982b). Considering the high degree of specificity required in cell-cell recognition and signaling, it has been postulated that enzyme-substrate interactions may contribute to such mechanisms.

The only rational way to gain insight into the physiological significance of SPK and its substrates consists in the first instance in the characterization of the components of this enzyme-substrate system with respect to biochemistry and cell biology, and the analysis of its distribution among cells of different species and tissues. In this chapter, we report on a detailed study of this enzyme system in HeLa cells – a human carcinoma cell line. However, SPK has meanwhile been detected on the surface of all cells analyzed by us so far as reported in the next chapter by Kübler et al. in this volume.

2 Detection of Surface-Located Protein Kinase

2.1 General Considerations: Pitfalls

The dectection of SPK is impeded by the difficulty of demonstrating a potentially minor PK activity on the high "background" of the various intracellular PKs. In order to exclude their participation in the detectable reaction, an intact membrane barrier for the radioactive cosubstrate (usually) ATP had to be employed. Even though the intact membrane may not represent an absolute hindrance, this form of assay nevertheless makes sense: The question is, namely, whether conditions can be chosen which restrict the detectable reaction to the surface. The enzymology has to be studied under conditions which guarantee the cell's intactness during the experiment.

It was anticipated that surface-located substrates should be the physiological ones of SPK. For their phosphorylation, the cells had to be incubated under appropriate conditions with [γ-^{32}P]ATP as depicted schematically in Fig. 1. A failure of the detection of products may not necessarily indicate the absence of substrate or SPK since a predominant degradation of ATP (or even the phosphoprotein) may prevent the SPK reaction. The search for nonpenetrable extracellular PK substrates must take into account the demand for the integrity of the cells, which has not always been the case. When all these factors have been considered, a number of published data – but not all of them – can be taken as evidence for surface-located PK in a variety of cells (Ågren and Ronquist 1974; Mastro and Rozengurt 1976; Schläger and Köhler 1976; Ronquist et al. 1977; Remold-O'Donnell 1978; Kang et al. 1978; McPershon and Ramachandran 1980; Somarrin et al. 1981; Chiang et al. 1979; Emes and Crawford 1982; Barnekow et al. 1982; Amano et al. 1984).

2.2 Experimental Conditions

Details regarding materials and methods have been published elsewhere (Kübler et al. 1982a). The phosphorylation reaction with HeLa monolayer cultures at 37°C was carried out under isoosmotic conditions (NaCl 70 m*M*, Tris 30 m*M*, EDTA

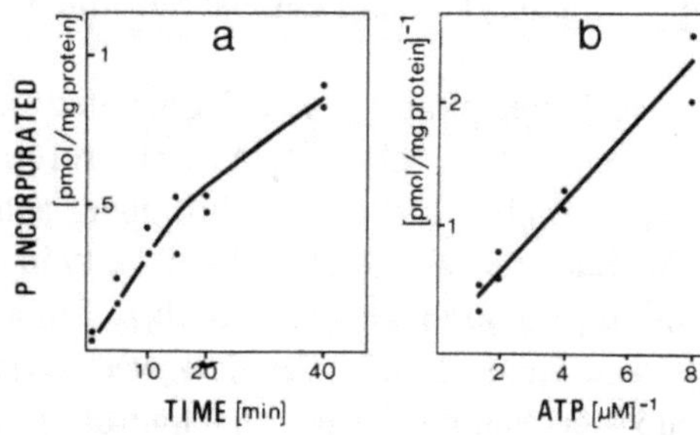

Fig. 2a, b. Incorporation of [^{32}P] into acid-precipitable material on incubation of cells with [γ-^{32}P]ATP. **a** Time dependence. 10^6 HeLa cells/dish were phosphorylated at 37°C with [γ-^{32}P] ATP (0.5 *μM*). At the times indicated, the reaction was terminated by removal of the supernatant and addition of cold 5% trichloroacetic acid. Radioactivity was determined after extraction of remaining [γ-^{32}P]ATP. **b** Cosubstrate dependence (LineweaverBurk plot). 1.2x10^6 cells/dish were incubated for 15 min as in **a**, with different ATP concentrations

0.5 m*M*, bovine serum albumin 0.1% (wt/vol), glucose 75 m*M*; pH 7.2, 290±10 mOsm); the mixture contained in addition 5 m*M* Mg acetate for complexing ATP and 5 m*M* potassium phosphate to trap $^{32}P_i$ released from [γ-^{32}P]ATP by ATPase. During incubation under these conditions the capacity of the cells to exclude trypan blue (TB) – a parameter for cell viability – was not altered as followed up to 2 h. Inclusion of [γ-^{32}P]ATP led to the generation of acid precipitable radioactivity, the amount of which was dependent on incubation time, concentration of ATP (Fig. 2), and cell number.

2.3 Detectable Utilization of Extracellular [γ-^{32}P]ATP

Linear transformation of the data was possible only at low concentrations of [γ-^{32}P]ATP, resulting in an apparent K_m of the order of 2–5 *μM*. Usually the concentration of ATP was 0.5 *μM* in order to exclude pharmacological side effects (for review see following chapter). This concentration is in contrast to intracellular ATP in the millimolar range. Any labeled ATP penetrating the cell would therefore instandly be diluted at least a thousandfold. Carring out such a "dillution" experimentally, i.e., mixing labeld ATP with unlabeled ATP to 1 m*M* extracellular final concentration, abolished any detectable reaction. The omission of Mg^{++} from the incubation mixture also prevented a detectable phosphorylation reaction, although intracellularly occurring protein phosphorylation could be considered independent of extracellular Mg^{++}, at least within the time frame of the experiment. The generation of $^{32}P_i$ and $^{32}PP_i$ from extracellular [γ-^{32}P]ATP by HeLa cells (Kübler et al. 1980) raised the possibility that intracellularly labeled ATP generated from these might contribute to the visible reaction. However, replacement of [γ-^{32}P] ATP by equal amounts of either $^{32}P_i$ or $^{32}PP_i$ did not yield a detectable reaction, thus, indicating that 5 m*M* unlabeled phosphate in the incubation mixture was a sufficient trap. Therefore, under the conditions employed, a detectable utilization of extracellular [γ-^{32}P]ATP obviously does not occur within HeLa cells, but takes place only at their external surface.

2.4 Phosphorylation of Cellular Substrates

The acid precipitable material generated during the reaction did not lose ^{32}P on treatment with DNAse, RNAse, chloroform-methanol (2:1), and hydroxylamine. It was unstable, however, in hot alkali and hot trichloroacetic acid (Kinzel et al. 1976; Kübler et al. 1982a). The stability in hydroxylamine ruled out the presence of acyl-linked phosphate (as described for an intermediate in an ATPase reaction; Post and Kume 1973). The data were consistent with the generation of phosphomonoester bonds. Partial acid hydrolysis and high voltage paper electrophoresis revealed phosphoserine and phosphothreonine as products.

On separation of optimal cell cultures on polyacrylamide gels in the presence of SDS after phosphorylation, a reproducible labeling pattern was obtained exhibiting two predominantly radioactive peptides (see following chapter). On electronic cell separation of intact from damaged cells after phosphorylation (see Sec. 2.6) it was found that the radioactive pattern of the intact fraction resembled that obtained with optimal cultures. In contrast, the composition of radioactive peptides obtained from the damaged fraction differed and was not reproducible, probably due to the difficulty of generating a homogenous as well as reproducible "damaged cell". The radioactivity disappeared from phosphorylated cells after a short treatment (4 min) with 0.05% trypsin, thus, indicating that in the case of intact cells it was removed from their surface. (The disappearance from the damaged fraction at the same time, however, demonstrates that the argument "trypsin-sensitive" for "cell-surface located" is not sufficient per se.)

In order to exclude the possibility that some of the phosphorylated proteins may have originated from damaged cells present in each culture (usually TB values were below 1%), i.e., released and subsequently attached to intact cells, the following experiment was conducted. Sonicates of HeLa cells were phosphorylated with [γ-^{32}P]ATP and subsequently incubated with optimal cultures. None of the radioactive proteins present in the sonicate were firmly bound to unlabeled cells.

2.5 Localization of Protein Kinase Activity

The evidence for the utilization of extracellular [γ-^{32}P]ATP in a protein kinase reaction as well as the fact that radioactive products could be removed from intact cells by short treatment with trypsin indicated indirectly that the catalytic activity was exposed at the cell surface as well. Proof of that was obtained – in accordance with criteria for ecto- and exoenzymes (DePierre and Karnovsky 1972) – by the ability of intact cells to metabolize nonpenetrable substrates, namely, phosvitin and casein. These proteins do not disturb the intactness of cells even at high concentrations (5 mgml^{-1}). A concentration-dependent release of the cytosolic enzyme lactic dehydrogenase (LDH) or of radioactive proteins from cells prelabeled with radioactive amino acids was not observed (Kübler et al. 1982c). Multiplication of HeLa cells was not affected by the presence of phosvitin (1 mgml^{-1}) during 2 days. (The degree of phosphorylation of phosvitin by HeLa cells is related to time, cell number, amount of phosvitin, and temperature; the label is attached to serine and threonine residues; see Kübler et al. 1982a; 1983.) In all cells studied so far, removal

Table 1. Effect of diazotized sulfanilic acid on the phosphorylation of phosvitin[a]

Diazotized sulfanilic acid (m*M*)	Inhibition (%)
0	0
0.3	50
3.0	69.3

[a] HeLa cells were incubated under standard conditions for 15 min at approx. 5°C with the diazonium salt of sulfanilic acid at concentrations as indicated. For the phosphorylation reaction phosvitin (1 mgml^{-1}) and [γ-^{32}P]ATP were added. The incubate was kept for another 15 min at 37°C. The supernatants were separated by SDSgel electrophoresis. The degree of phosvitin phosphorylation was determined by photometric scans of the autoradiograph

of phosvitin by a simple washing procedure after phosphorylation did not result in traces of radioactive phosvitin in the cells, indicating the inability of this protein to adhere to or penetrate intact cells.

Based on the criteria developed in the classical paper by DePierre and Karnovsky (1974), the surface-localization of protein kinase activity was further proven by the application of the nonpenetrating diazonium salt of sulfanilic acid. Treatment of HeLa cells with diazotized sulfanilic acid at concentrations up to 3 m*M* did not influence the intracellular LDHactivity. In the same buffer system a labeling of HeLa cell surface proteins was observed upon treatment with diazotized iodosulfanilic acid [^{125}I], thus, indicating that the action of sulfanilic acid was restricted to the cell surface. The phosphorylation of surface proteins as well as of phosvitin (Table 1) was inhibited by sulfanilic acid in a concentration-dependent fashion. However, the inhibition was only achieved in the presence of diazotized sulfanilic acid, whereas after its removal phosphorylation was not significantly altered. The reason for this behavior is not clear. It may be explained by a unique architecture of the SPK association (see following chapter, Sec.2.3) or by a reversible inactivation of the phosphorylation reaction.

Further evidence for the surface location of the PK activity is contributed by a unique feature of the enzyme in this location. Three washes with the substrate phosvitin (1 mgml^{-1}; in the absence of cosubstrate ATP) erase the capability of HeLa cells to phosphorylate phosvitin without, however, impairing the cells viability or even their ability to multiply (see following chapter). The enzyme activity can be recovered in the supernatant. This enzyme release induced by a nonpenetrable substrate argues also for its surface location.

With respect to the origin of the protein kinase activity, the possibility had to be considered that the enzyme was taken up from the medium after its previous release from damaged cells. To test this possibility, HeLa cells were incubated with different amounts of the 4000 x g supernatant of HeLa sonicates in amounts equivalent to up to 7% damage; subsequently, the cells were washed as usual and analyzed for surface phosphorylation. In no case did the values exceed the control data, thus, making an origin from damaged cells unlikely. Moreover, the pattern of radioactive peptides was also not altered by the pretreatment with sonicates.

2.6 Interference of Damaged Cells

In view of the large number of PKs described intracellularly, the degree of participation of membrane-damaged cells had to be evaluated. In such cells added [γ-^{32}P] ATP can reach intracellular PKs and substrates. Further, such cells have probably lost their intracellular ATP and thereby the capacity to diminish the specific radioactivity of added labeled ATP.

For this purpose the phosphorylation of cellular proteins in cultures with different degrees of cell damage (analyzed by TB uptake) was determined (Table 2, for details see Kübler et al. 1982a). Cultures with a large percentage of TB positive cells exhibited a high degree of phosphorylation. However, the relation of the degree of damage and the specific radioactivity was nonlinear. On extrapolation to zero TB positive cells, the phosphorylation of the intact portion of cells became evident. Furthermore, intact cells were separated from damaged cells by electronic cell sorting on the basis of differential stain uptake (using a computerized cell sorter FACS II from Becton Dickinson) (Fig. 3). For this purpose phosphorylated cells were incubated with DAPI (4′, 6-diamidino-2-phenylindol), a stain which is taken up immediately by damaged cells but much more slowly by intact cells, i.e., within hours. This compound produces fluorescence on interaction (predominantly) with double-stranded DNA. Analysis of separated cells showed that damaged cells were 10 to 20 times more intensely phosphorylated than the intact portion. This result permits the calculation that in optimal cultures with 1% damaged cells approximately 80 to 90% of the radioactivity is associated with the intact fraction.

With respect to the capacity of damaged cells to phosphorylate exogenous substrates, experiments with normal cultures as well as those with a purposely high degree of damage have been carried out. On increasing the percentage of damaged

Table 2. Phosphorylation of cultures with different proportions of damaged cells under standard conditions

Trypan blue positive cells % Total	^{32}P incorporated pmol/10^6 cells[a]	pmol/damaged cell x 10^3
0.5	0.215 ± 0.038	0.043
5.7	0.269 ± 0.018	0.0047
68.5	3.011 ± 0.489	0.0043

[a] Results of three cultures/group, ± SD

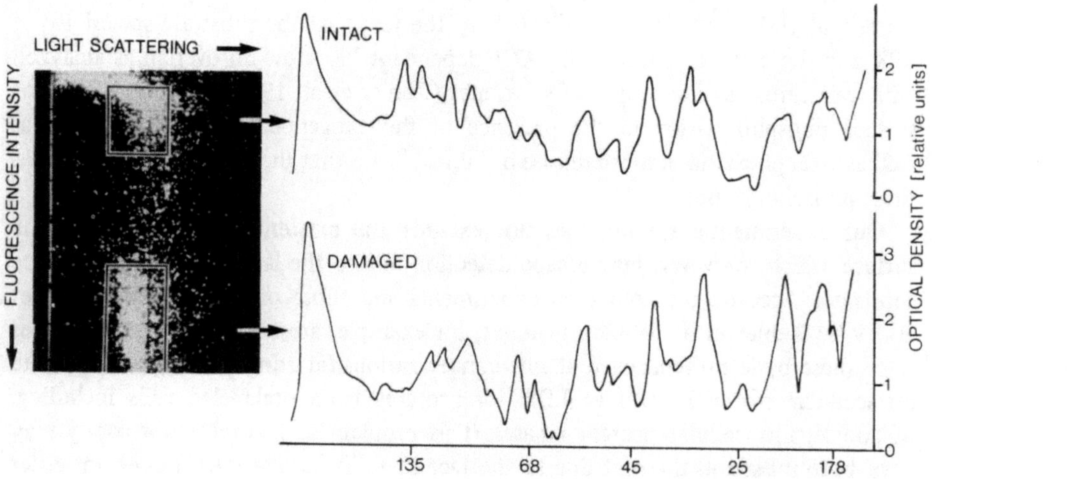

Fig. 3. Profile of [^{32}P]-labeled proteins of phosphorylated cells sorted electronically into an intact and a damaged portion. Generation of cultures with at purposely high rate of damaged cells, phosphorylation of cultures for 15 min and subsequent separation into intact and damaged portions by cell sorting was performed. Since the sorting procedure took about 4 h, cell phosphorylation was performed in two batches with a 2 h interval in accordance with the stability of the phosphorylation pattern. *Left side,* image on storage oscillograph as defined by fluorescence intensity and light scattering with the electronically preselected areas of cells to be collected. The other cells were discarded. Sorted cells were fixed in trichloroacetic acid (5% final concentration). Comparable amounts of protein-bound radioactivity were separated by SDSgel electrophoresis (7.5 to 15% gradient). Shown are the photometric scans of the autoradiograph. Molecular weight markers are given in kD

cells, the phosphorylation of phosvitin is not increased, and the same holds for the phosvitin-induced release of SPK. Both the phosphorylation of phosvitin and the induced release of SPK, therefore, appear to be functions of intact cells – a further indication of the surface location of the responsible enzyme.

3 Biochemistry of Cell Surface Protein Kinase

3.1 Characterization of Surface Protein Kinase

As outlined above, biochemical characterization of SPK is limited to conditions compatible with an intact membrane barrier. This was, therefore, carefully controlled. For the phosphorylation of the same spectrum of cell surface proteins, analyzed by SDS gel electrophoresis, either ATP or GTP (although the latter with less efficiency) was utilized by the enzyme. The surface phosphorylation was influenced neither by cyclic AMP nor by the heat- and acid-stable inhibitor specific for the catalytic subunit of cAMP-dependent PK (in absence or presence of cAMP). These data, in addition to the utilization of GTP, show that a cAMP-dependent PK was not responsible for the observed phosphorylation. The absence of this enzyme at the surface of HeLa cells is also indicated by the fact that the substrate specificity of SPK and the catalytic subunit of cAMP-dependent PK show no overlap as analyzed with calf serum as the source of substrates (Kübler et al. 1982b). The reduction of surface phosphorylation in the presence of the exogenous substrate phosvitin as well as after phosvitin-induced release of SPK indicate that the same enzyme catalyzes these phosphorylations.

Our experimental set up does not exclude the existence of other PKs at the surface which, however, may escape detection due to the lack of a suitable surface substrate. According to our own experiments and those of others (Rahmsdorf et al. 1979; Kübler et al. 1982c), histones, for example, are not suitable as substrates since these basic proteins even at low concentrations (starting at 10 $\mu g ml^{-1}$) liberate intracellular material, such as LDH, or proteins from prelabeled cells including, of course, intracellular protein kinases. This problem of histone cytotoxicity may have been missed in the past due to the fact that TB (unlike DAPI) does not enter histone-treated cells for a certain period of time even though they are damaged (Kübler et al. 1982c). The lack of TB uptake in the presence of histones may, thus, have been accepted as a sign for their intact state (e.g., Chiang et al. 1979). Such released intracellular enzyme appears to be responsible for the cAMP-dependent phosphorylation of histones by cells (Kübler et al. 1982c).

Evidence has been compiled for the existence of a surface-derived tyrosine kinase which differs from the SPK described here (Barnekow et al. 1982).

3.2 Characterization of Released Surface Protein Kinase

SPK was released from intact HeLa cells as described (see following chapter). The enzyme was found by sedimentation analysis (5–20% sucrose) to possess a mol. wt. of ca. 150 000, to accept phosvitin as well as casein as substrate (although with preference to the former; ratio of phosphorylation approx. 1.7), and to utilize GTP as cosubstrate reaching approximately half the efficiency observed with ATP. The amino acid side chains phosphorylated in phosvitin and casein were serine and threonine with comparable amounts of the respective amino acid phosphate appearing in hydrolysates of both phosvitin and casein (somewhat more phosphothreonine in β-casein). SPK was inhibited by heparin, a sulfated mucopolysaccharide, at low concentrations, i.e., 50% inhibition (I_{50}) was reached below 1 $\mu g ml^{-1}$.

3.3 Isolation and Characterization of a SPK-Related Phosvitin Kinase in HeLa Cell Extracts

A full understanding of SPK demands a more detailed knowledge of its enzymology and biochemistry. Since SPK is difficult to isolate in reasonable amounts even by inducible release, we decided first to purify and characterize related enzyme(s) from whole cells. For this purpose HeLa cells were extracted in the presence of Nonidet P-40 (0.5%) in order to favor the extraction of kinases located towards or within the plasma membrane (Borun et al. 1967).

This extract contained a number of different protein kinase activities including phosvitin phosphorylating forms (Fig. 4). While all of them phosphorylated casein as well, only the phosvitin kinase activity eluting at ionic strength above 250 m*M* NaCl (range III; enzyme named accordingly "kinase P III") resembled SPK in utilizing GTP as cosubstrate, and this was therefore purified.

For purification, the majority of histone – and other phosvitin kinases as well as of cellular proteins was removed by washing extract-loaded DEAE-columns with 250 m*M* NaCl. The subsequently applied gradient dissociated P III from the exchanger at 320–350 m*M* NaCl. A coeluting histone kinase was removed by phospho-cellulose chromatography, and a final purification was accomplished by affinity chromatography employing phosvitin-sepharose resulting in a total purification of $> 10^4$-fold.

P III catalyzes the phosphorylation of phosvitin and casein at apparent K_m values of approx. 0.5 μM and 4 μM, respectively. V_{max} is approx. three fold higher for phosvitin. For ATP and GTP, the apparent K_m values are 3–4 μM and 4–5 μM, respectively. GTP, however, lowered V_{max} by a factor of about 3. ITP as well as dATP and dGTP can also serve as cosubstrate, while pyrimidine-triphosphates cannot. The kinase is not influenced by cAMP, the calmodulin/Ca^{2+}system or the

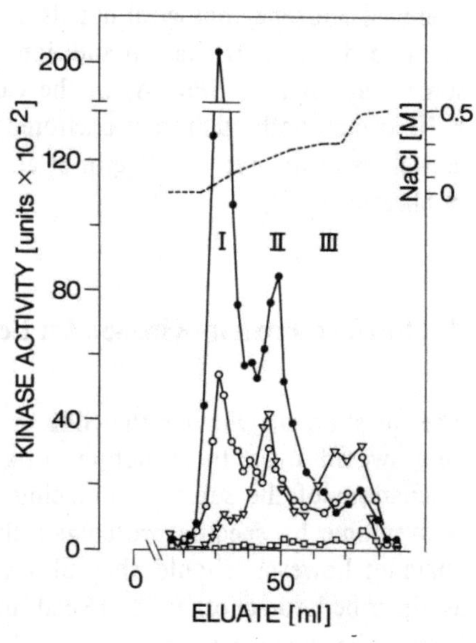

Fig. 4. Pattern of protein kinases of HeLa cells. HeLa cells were extracted by PBS/0.5 % NP-40, centrifuged at 105 000 x g_{av}, and the extract chromatographed on DEAE-cellulose as described in Pyerin et al. 1981. Symbols: NaCl gradient (- - - -); protein kinase activity without added protein (□) and with addition of phosvitin (▽) or of histone in the presence (•) or in the absence of 5 μM cAMP (○)

heat- and acid-stable kinase inhibitor. P III is inhibited by heparin at low concentrations, i.e., I_{50} below 1 $\mu g ml^{-1}$, and it is stimulated by Na^+ and K^+. In phosvitin and α-casein, phosphorylation occurs at serine and threonine residues, in β-casein exclusively at threonine. The mol. wt. of kinase P III is approx. 120,000 ± 11,000 as determined by gel filtration (Ultrogel ACA-34) and density gradient centrifugation (5–20% sucrose; sedimentation coefficient 6.1–6.15). Enzyme preparations contain a phosphorylatable protein of M_r 26–28,000 which does not seem to dissociate from the enzyme during gradient centrifugation and, therefore, could represent a functional part of P III. Similar results have been described for the casein kinase type II (Hathaway and Traugh 1983).

In the light of their obvious similarity, a direct comparison was made between SPK and P III in their ability to phosphorylate substrates in calf serum and plasma. Out of a large number of proteins present, only a few were phosphorylated by kinase P III, including a protein with M_r 135,000 in serum and a protein (possibly two) of M_r above 300,000 in plasma. Cell-bound SPK phosphorylated preferentially these proteins. Identical results were obtained if SPK was released without any protein supplement or in the presence of serum. If phosvitin was employed to release SPK, it obviously out-competed the proteins in serum and plasma.

The existence of an intracellular counterpart of SPK stresses the importance of taking all those precautions described above to come to an unambiguous demonstration of the existence of SPK. On the other hand, it implies that further biochemical and enzymological details of SPK may be gained by studying kinase P III in more detail. P III has, in addition, an intracellular distribution which reflects in some way that of SPK. As in the case of the bound and released form of SPK, P III occurs both membrane-anchored, i.e., bound to, e.g., microsomal membranes, as well as in the cytosol (Pyerin et al. 1981; Pyerin et al., manuscript submitted for publication).

4 Surface Protein Kinase: An Ecto- or Exoenzyme?

The question of whether the SPK in HeLa cells is an ecto- or exoenzyme cannot be answered unless the structure in its plasma membrane-bound form as well as the mechanism of the substrate inducible release are known. If for instance anchoring is overcome by specific proteolysis clipping off a certain part it might be an ectoenzyme; however, should the substrate induce a dramatic conformational change as described for another PK (Reed and Kinzel 1984) which causes the discharge, it might be an exoenzyme.

With respect to the function and the biological potential of protein phosphorylation the "exo-ecto-question" appears to be less important, particularly in view of the unique feature of being specifically released on interaction with substrate thereby becoming an exoenzyme. It is this property which gives the system, in addition, the character of a potentially remote message.

5 Summary

Evidence for cell surface located protein kinase in HeLa cells has been reported. SPK was shown to be capable of phosphorylating cell surface proteins as well as suitable exogenous substrates: detection was carried out with complete cells utilizing their membrane barrier to prevent penetration of cosubstrates, substrates, and macromolecular probes as well as by a choice of conditions limiting the detectable phosphorylation reaction to the cell surface. Cellular substrates become phosphorylated at serine and threonine residues. In combination with evidence for the extracellular utilization of [γ-^{32}P]ATP and the phosphorylation of extracellular substrates, such as phosvitin and casein, it is concluded that the PK activity is located at the cell surface. Neither substrates nor the enzyme are derived from damaged cells. The substrate-effected release of SPK from intact cells gives additional proof of the surface location of the enzyme and facilitates a more detailed biochemical analysis and a comparison of SPK with PKs isolated from complete cells. The SPK is independent of cyclic nucleotides, it utilizes ATP and GTP in a ratio of 2.1:1 with a K_m for ATP of 2–5 μM, it phosphorylates phosvitin and casein in a ratio of 1.7:1. In its released form the SPK has a M_r of 150,000, it is inhibited by heparin (I_{50} at < 1 μgml^{-1}). By these criteria, SPK resembles P III, an enzyme which has recently been isolated and characterized from HeLa extracts. Physiological aspects of SPK are discussed in the next chapter by Kübler et al.

References

Ågren G, Ronquist G (1974) (^{32}P) Phosphoryl transfer by endogenous protein kinase at the glia and glioma cell surface in culture into extrinsic acceptor proteins. Acta Physiol Scand 92:430–432

Amano F, Kitagawa T, Akamatsu Y (1984) Protein kinase activity on the cell surface of a macrophage-like cell line, J774.1 cells. Biochim Biophys Acta 803:163–173

Barnekow A, Boschek CB, Ziemiecki A, Friis RR, Bauer H (1982) Demonstration of the Rous sarcoma virus pp60^{src} and its associated protein kinase on the surface of intact cells. In: Winnacker E, Schoene HH (eds) Genes and tumor genes. Raven, New York, p 65

Borun TW, Scharff MD, Robbins E (1967) Preparation of mammalian polyribosomes with the detergent Nonidet P-40. Biochim Biophys Acta 149:302–304

Chiang TM, Kang ES, Kang AH (1979) Ecto-protein kinase activity of fibroblasts. Arch Biochem Biophys 195:518–525

Cohen P (1980) In: Cohen P (ed) Molecular aspects of cellular regulation,vol 1. Elsevier/North-Holland Biomedical, Amsterdam

DePierre JW, Karnovsky ML (1972) Ectoenzymes, sialic acid, and the internalization of cell membrane during phagocytosis. In: Ward PA, Lepow TH (eds) Inflammation: mechanisms and control. Academic, New York, p 55

DePierre JW, Karnovsky ML (1974) Ecto-enzymes of the guinea pig polymorphonuclear leucocyte. J Biol Chem 249:7111–7120

Emes CH, Crawford N (1982) Ecto-protein kinase activity in rabbit peritoneal polymorphonuclear leucoytes. Biochim Biophys Acta 717:98–104

Flockhart DA, Corbin JD (1982) Regulatory mechanisms in the control of protein kinases. CRC Crit Rev Biochem 12:133–186

Hathaway GM, Traugh JA (1983) Casein kinase II. In: Corbin JD, Hardman JG (eds) Methods in enzymology, vol 99. Academic, New York, p 317

Kang ES, Gates RE, Farmer DM (1978) Localization of the catalytic subunit of a cyclic-AMP-dependent protein kinase(s) and acceptor proteins on the external surface of the fat cell membrane. Biochem Biophys Res Commun 83:1561–1569

Kinzel V, Kübler D, Mastro AM, Rozengurt E (1976) Phosphoproteins of the cell surface as generated by endogenous or exogenous protein kinse: stability of the ^{32}P-labeled product. Biochim Biophys Acta 434:281–285

Krebs EG, Beavo JA (1979) Phosphorylation-dephosphorylation of enzymes. Annu Rev Biochem 48:923–959

Kübler D, Pyerin W, Kinzel V (1980) Generation of pyrophosphate from extracellular ATP at the surface of HeLa cells. Eur J Cell Biol 21:231–233

Kübler D, Pyerin W, Kinzel V (1982a) Protein kinase activity and substrates at the surface of intact HeLa cells. J Biol Chem 257:322–329

Kübler D, Pyerin W, Burow E, Garcon T, Kinzel V (1982b) Interaction of cells with their proximal environment via surface protein kinase. In: Akoyunoglou J, Evangelopoulos J, Georgatsos J, Palaiologos G, Trakatellis A, Tsiganos CP (eds) Cell function and differentiation, part A. Proc Spec FEBS Meeting. Liss, New York, p 499

Kübler D, Pyerin W, Kinzel V (1982c) Assays of cell surface protein kinase: importance of selecting cytophilic substrates. Eur J Cell Biol 26:306–309

Kübler D, Pyerin W, Burow E, Kinzel V (1983) Substrate effected release of surface-located protein kinase from intact cells. Proc Natl Acad Sci USA 80:4021–4025

Mastro AM, Rozengurt E (1976) Endogenous protein kinase in outer plasma membrane of cultured 3T3 cells. J Biol Chem 251:7899–7906

McPershon MA, Ramachandran J (1980) Corticotropin stimulates cyclic nucleotide independent protein kinase activity of intact adrenocortical cells. Biochem Biophys Res Commun 94: 1057–1065

Nishizuka Y (1984) The role of protein kinase C in cell surface signal transduction and tumor promotion. Nature (Lond) 308:693–698

Post RL, Kume S (1973) Evidence for an aspartyl phosphate residue at the active site of sodium and potassium ion transport adenosine triphosphatase. J Biol Chem 248:6993–7000

Pyerin W, Balbach N, Kübler D, Kinzel V (1981) Protein kinases in HeLa cells and in human cervix carcinoma. Z Naturforsch Sect C Biosic 36:552–561

Rahmsdorf HJ, Malchow D, Gerisch G (1979) Cell surface protein kinases in *dictyostelium:* are they artifacts. Cell Biol Int Rep 3:237–245

Reed J, Kinzel V (1984) Ligand-binding site interaction in adenosine cyclic 3′, 5′-monophosphate dependent protein kinase catalytic subunit: circular dichroic evidence for intramolecular transmission of conformation change. Biochemistry 23:968–973

Remold-O'Donnell E (1978) Protein kinase activity associated with the surface of guinea pig macrophages. J Exp Med 148:1099–1104

Ronquist G, Ågren G, Eklund S, Wernstedt C (1977) Cyclic 3′-5′-GMP independent protein kinase at the outer surface of intact Ehrlich cells. Ups J Med Sci 82:1–5

Schläger EJ, Köhler G (1976) External cyclic AMP-dependent protein kinase activity in rat C-6 glioma cells. Nature (Lond) 260:705–707

Sommarin M, Henriksson T, Jergil B (1981) Cyclic AMP-dependent protein phosphorylation on the surface of rat hepatocytes. FEBS Lett 127:285–289

Physiological Aspects of Surface Protein Kinase Activity

D. KÜBLER, W. PYERIN, M. FEHST, and V. KINZEL [1]

1 Introduction

A detailed report on the detection and biochemistry of cell surface protein kinase (SPK) has been given in Ch. 4.1 by Kinzel et al. These data have been worked out with HeLa cells, a human carcinoma line. It could be assessed that the SPK activity, in principle, has two target directions towards substrates located at the cell surface as well as extracellular substrates (Kübler et al. 1982).

This chapter is concerned with physiological aspects of the SPK system. These are discussed with respect to (1) the molecular behaviour of SPK as a cell surface constituent; (2) the widespread phenomenon of SPK; and (3) endogenous and exogenous substrates.

Apart from the functional property of the enzyme as a catalyst of protein phosphorylation, SPK exhibits a unique physiological feature by the fact that it can be selectively released from intact cells into the extracellular fluid through interaction with substrate (Kübler et al. 1983). Although the mechanism of such an inducible release process is unknown, the possibility of SPK movement away from the "mother" cell may indicate cell-environmental interaction via PK activity also at some distance.

In order to address the question of whether or not SPK and extracellular protein phosphorylation represent a general principle, different cell types were analyzed under appropriate conditions. With respect to the relevance of SPK release for the organism, body fluids were also analyzed for such kinase activity.

The physiological significance of the SPK is intimately linked with that of physiological substrates. Since the activity range of SPK is directed towards cell surface proteins and extracellular substrates, our investigations followed both lines. Whereas presumably natural substrates of the SPK are located at the cell surface, natural extracellular substrates are unknown. However, a substrate, at least present in the environment of cells in vitro, has been found in calf and other sera. This particular serum protein is recognized by the SPK in a highly selective manner.

The data given in this chapter, although far from yielding a complete picture, delineate a first contour of a physiological system which might be denominated as *extracellular protein phosphorylation.*

[1] German Cancer Research Center, Institute of Experimental Pathology, Im Neuenheimer Feld 280, 6900 Heidelberg, FRG

2 Release of Surface Protein Kinase from HeLa Cells

2.1 Criteria for Liberation

Extracellular substrates, such as phosvitin (a protein from egg yolk) or casein (a protein from milk), can be quite effectively phosphorylated by cells (Kübler et al. 1983). Since none of the extracellular products was detectably associated, with the cells, but stayed in the cell supernatant we analyzed the reverse, i.e., the strength of the association of SPK with the cell surface. For this purpose monolayer or supension cells were incubated without cosubstrate ATP in physiological buffer solution as used for the detection of surface protein kinase (SPK; see previous chapter by Kinzel et al.) in the absence or presence of exogenous protein (as indicated in the particular experiment). After certain periods of time the supernatant fluid was taken off, clarified by centrifugation (2500 x g for 10 min), and, unless not present during incubation, completed with phosvitin (usually 1 $mg ml^{-1}$) for the detection of kinase activity. The supernatant was further incubated with [γ-^{32}P] ATP (0.5 μM final conc.) for 15 min. Incorporation of [^{32}P] from [γ-^{32}P]ATP was determined by scintillation counting of the material precipitated with trichloroacetic acid (20%, vol/vol) on Whatman ET 31 paper (Kübler et al. 1983) or autoradiography of the material separated by SDS gel electrophoresis.

The result of such an experiment is shown in Fig. 1. Supernatants of cells incubated in the absence of protein display no significant amounts of kinase activity as indicated by the negligible phosvitin phosphorylation; serum albumin, when present during cell incubation, resulted in very small phosvitin kinase activity. If phosvitin or casein, however, were included during cell incubation, significant protein kinase activity was detectable in the supernatant after removal from cells.

These results strongly suggest that the presence of substrates was responsible for the appearance of phosvitin kinase in the supernatant. Biochemical characteriza-

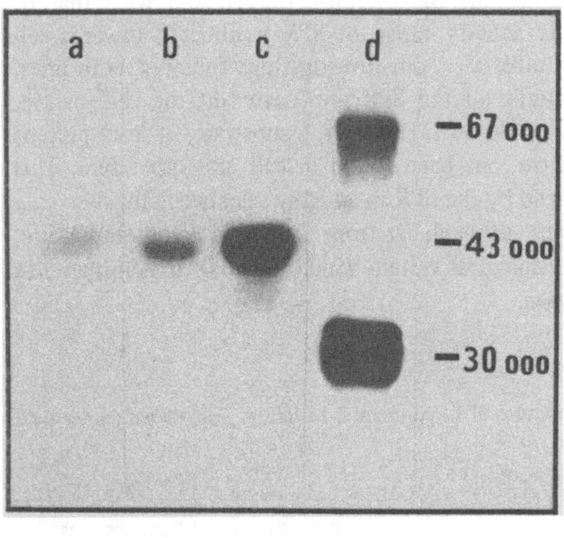

Fig. 1 a–d. Release of surface protein kinase from intact HeLa cells in the presence of (**a**) no protein, (**b**) bovine serum albumin (1 $mg ml^{-1}$), (**c**) phosvitin (1 $mg ml^{-1}$), (**d**) casein (1 $mg ml^{-1}$); for details see text; aliquots from incubates were subjected to electrophoresis on SDS gels and autoradiographed; M_r of standard proteins given on the *right*

tion of released enzyme with cell-bound SPK (see previous chapter) and the data presented in Sec. 2.2 indicate that it is SPK which is released from the cells by the action of substrate.

2.2 Properties of Surface Protein Kinase Release

The appearance of SPK in the cell supernatant in the presence of substrate (phosvitin) was followed with respect to incubation time, concentration of phosvitin, and temperature. For this purpose separate sets of cultures were incubated in the presence of 1 mgml^{-1} phosvitin. The time course of SPK release shows a plateau after 10 min. Different phosvitin concentrations during cell incubation resulted in an optimum SPK yield at 1 mgml^{-1}, preferably at 37°C (rather than at 24°C or 4°C).

In order to correlate the amount of released SPK with the cell number employed, cultures were allowed to grow in plates of different diameter (3 cm, 5 cm, and 10 cm) to differend cell densities and were then analyzed. On comparison of the amount of protein kinase activity released per cell with cell densitiy, it became evident that the enzyme release decreases with increasing density (Fig. 2). Hence, cell density is an important parameter, which determines the degree of release. Therefore, cell density has to be taken into account if a comparison of SPK content of different cell types (vide inferior) is being approached on the basis of the induced kinase release.

The substrate dependence of SPK release was further evidenced by experiments where the same set of cells was iteratively treated with buffer containing either phosvitin (1 mgml^{-1}) or no protein. Only in the presence of substrate was a significant amount of PK activity recovered in the supernatant, whereby it became evident that the release is not complete with the first or second wash even though the time course of release comes to a plateau with a single wash. The reason for this par-

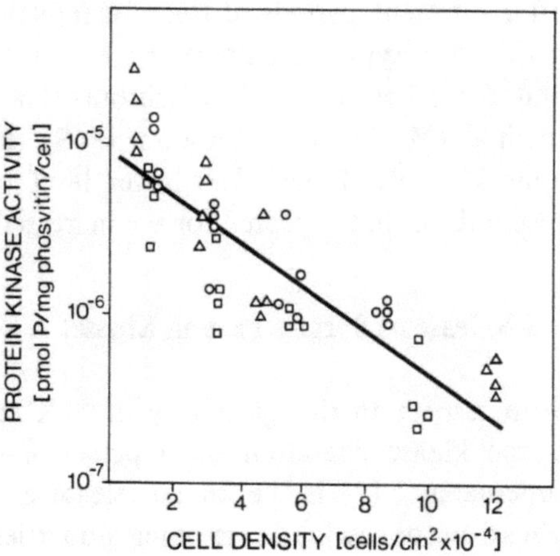

Fig. 2. Dependence of protein kinase release on cell number and cell density. HeLa cells grown in different sized plates to cell densities as indicated were incubated for 10 min with phosvitin at 1 mgml^{-1} in various amounts of incubation fluid: ○, 0.7 ml in 3 cm plates; ∇, 2.0 ml in 5 cm plates; □, 7.8 ml in 10 cm plates. Protein kinase activity in cell-free supernatants was determined for 5 min. *Symbols* represent values for single plates (three independent experiments); the *straight line* was determined by linear regression from all values

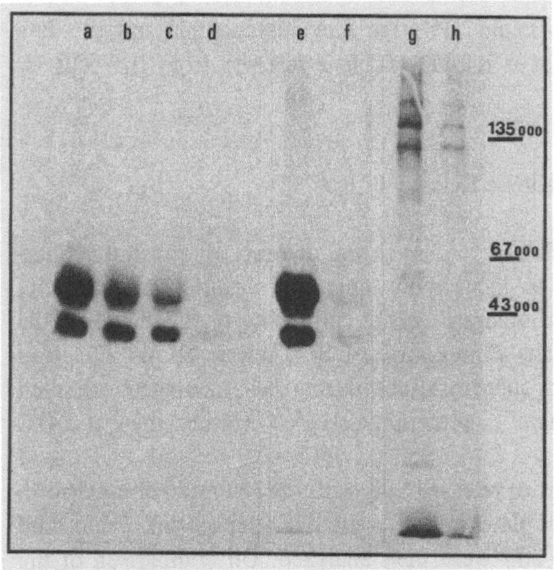

Fig. 3 a–h. Depletion of surface protein kinase by iterative incubation with substrate. HeLa cells were incubated four times with fresh phosvitin solution and protein kinase activity determined with **(a–d)** cell-free supernatants or **(e)** in the presence of cells before surface kinase release or **(f)** after treatment; aliquots from incubates were subjected to eletrophoresis on SDS gel and autoradiographed. [^{32}P]-phosphoprotein spectra of intact cells were obtained from **(g)** fresh culture or **(h)** from surface enzyme-depleted culture after the fourth wash with phosvitin

ticular behavior of SPK is presently unkown. It may reflect new elements in cell surface architecture so far not discovered.

The subsequent release of phosvitin kinase from cells treated successively four times for 5 min each with fresh phosvitin solution provides conclusive evidence for the identity of this enzyme with the SPK. As indicated by phosvitin phosphorylation, the kinase activity in these supernatants decreased with almost no activity in the fourth wash (Fig. 3). Thereby the cells lost completely their capacity to phosphorylate phosvitin directly; simultaneously the phosphorylation of surface proteins was drastically reduced. Thus, the appearance of phosvitin kinase in the cell supernatant was accompanied by concomitant diminution of cellular SPK.

The question arose whether the cells are able to restore the original level of SPK. To this end a series of cells was treated twice with fresh phosvitin solution under sterile conditions in order to remove most of their SPK. The depleted cells were further incubated under sterile conditions with complete medium, and assayed after different periods of time for inducible release of PK activity. For this investigation two types of cultures were employed: one with exponentially growing cells and the other with cells synchronized at the G_1/S border with the amethopterin method (Mueller and Kajiwara 1969). In both cases restoration occurred within approximately 1 day. The lower level of recoverable SPK after that period corresponds to that expected for the increased cell density.

2.3 Release of Surface Protein Kinase: A Selective Process?

With respect to the specificity of SPK release, two questions have been asked: (1) is the kinase liberation accompanied by the appearance of other enzymes in the supernatant? (2) Is the kinase release a consequence of the substrate character of phosvitin (or casein) or are other properties of these compounds responsible?

In order to answer the *question of the selectivity of the kinase release* cell-free supernatants were assayed according to published methods for subcellular marker enzymes, such as 5′-nucleotidase (Ipata 1967), Na^+/K^+-ATPase (Blostein 1968), lactic dehydrogenase (Wroblewski and LaDue 1955), NADH/cytochrome c reductase (Sottocasa et al. 1967), and succinate/cytochrome c reductase (Avruch and Hoelzl Wallach 1971). The liberation of proteins, in general, was assayed with cells prelabeled with radioactive amino acids. In these experiments the inclusion of phosvitin in the incubation fluid did not cause the appearance of enzymes (besides the kinase) and proteins which were not seen in the control, thus, strengthening the view that release of SPK occurred selectively. The cell supernatant obtained in the presence of phosvitin was unable to diminish radioactivity or size of [^{32}P]-prelabeled phosvitin as determined by SDS gel electrophoresis. This result excludes the liberation of phosphoprotein phosphatase or phosvitin-digesting protease activities.

Although unlikely from the above data, we tested whether the SPK leaves the cell surface in soluble form or as particulate matter (vesiculated) by sedimentation velocity experiments. Phosvitin-released material together with molecular weight markers (which by themselves did not alter phosvitin kinase activities) were loaded on top of a sucrose gradient (5–20%, vol/vol) and centrifuged at 105,000 x g_{av} for 17 h. Kinase activity was recovered in the molecular weight in the order of 150,000 suggesting that the SPK had been liberated in soluble form.

Is the substrate character of phosvitin responsible for the kinase release? An answer to this question would immediately shed light on the mechanism of release of SPK. To address this problem, a number of compounds were compared with phosvitin for their capacity to induce SPK release. The selection of these substances was based (1) on their ability to mimic phosvitin in one or another property, and, or course, (2) in their inability to interfere with the phosvitin kinase assay. By these criteria the materials given in Table 1 were chosen. However, polyphosphates (chain length 3, 4, and 6–8) had to be excluded from the test since they inhibited the enzyme assay. For the actual analysis supernatants raised in the presence of the particular compound (15 min incubation at 37°C) were withdrawn, centrifuged (2500 x g for 10 min), and assayed after inclusion of 1 $mg\,ml^{-1}$ phosvitin and 0.5 μM [γ-^{32}P]ATP. Controls were performed in the absence of phosvitin.

A small, but reproducible, SPK release occurred when calf serum was present during incubation. Whether this results from the particular substrate shown to be present in serum is not deducible at present. Several other proteins such as albumin, fibronectin, epidermal, or fibroblast growth factor were ineffecitve. Also, polyglutamic acid, a peptide of similar size and charge as phosvitin, or the heptapeptide Kemptide, a substrate for the catalytic subunit of cAMP-dependent PK (Kemp 1979) did not cause removal of SPK; neither did DNA (as polyphosphate), cyclic AMP or ATP in different concentrations.

If the efficiency of phosvitin, a polyphosphorylated protein, as effector of SPK release were dependent on the degree of its phosphorylation, then a partial dephosphorylation sould result in an altered kinase release response. However, such an approach failed since it turned out that enzymatic dephosphorylation of phosvitin caused the protein to precipitate.

Table 1. Significant release of surface protein kinase[a]

Substance (conc.)	PK activity in supernatant
Phosvitin (0.1–5 $mg ml^{-1}$)	+
Casein (0.1–2 $mg ml^{-1}$)	+
Calf serum (1%)	(+)
Serum albumin (1%)	(–)
Fibronectin (0.05–0.2 $mg ml^{-1}$)	–
Polyglutamic acid (0.2 $mg ml^{-1}$)	–
Kemptide (200 μM)	–
DNA (0.002–0.2 $mg ml^{-1}$)	–
ATP (0.25–2.5 μM)	–
cAMP (5 μM)	–
Epidermal growth factor (0.1 $mg ml^{-1}$)	–
Fibroblast growth factor (0.2 $mg ml^{-1}$)	–

[a] 10^6 HeLa cells per 5 cm dish were incubated for 15 min at 37°C in a total volume of 2 ml of incubation buffer in the presence of the compounds indicated. Protein kinase activity in cell-free supernatants, after completion to 1 $mg ml^{-1}$ phosvitin, was assayed for 15 min; aliquots from incubates were subjected to electrophoresis on SDS gels and autoradiographed

As far as the mechanism is concerned, the enzyme release detectable in presence of serum, although small, seems to argue against a discharge due to cleavage by common proteases since serum is known to contain a number of protease inhibitors. It may be that substrates induce a conformational change of the enzyme that leads to its discharge. Conformational changes on binding of substrate have been observed, for example, in cAMP-dependent PK (Reed and Kinzel 1984, a, b).

The contribution of damaged cells which have lost their membrane barrier to the phosvitin-induced release of PK appears to be negligible for the following reasons: (1) on increase of the percentage of damaged cells in a culture the phosphorylation of external phosvitin is diminished rather than increased (see previous chapter by Kinzel et al.); (2) intracellular marker enzymes, such as LDH, have not been detected in the cell free supernatants of optimal cultures (Sec. 2.3); and (3) dense cultures exhibited a lower SPK release per cell than sparse cultures (Fig. 2), although one might expect more damaged cells in crowded cultures. These data support the conclusion that SPK release is a function of intact cells.

The phosvitin-induced kinase liberation differs basically from exocytosis (Poste and Allison 1973) by the fact that the kinase is already exposed at the cell surface. The discharge of SPK appears to be a result of the direct interaction of substrate with the enzyme. At present, however, there is no explanation why SPK release occurs stepwise, but not completely during the first wash with phosvitin (Sec. 2.2), although sufficient substrate should be available. Also, there is no hint why the continued presence of calf serum in the culture medium does not deplete the surface kinase. One is tempted to speculate that the accessibility of SPK at the cell surface is somehow restricted by particular spatial properties of the surface architecture. The relatively fast kinetics of SPK release (Sec. 2.2) is dissimilar from spontaneous shedding (Black 1980) as this process represents a comparably slow and continuous release of cellular material.

3 Prevalence of Surface Protein Kinase

3.1 Surface Protein Kinase in Other Cell Types

If the observations obtained with HeLa cells are to be of any physiological significance, two questions have to be addressed (1) do other cells exhibit SPK as well as the phenomenon of substrate-induced liberation of SPK? (2) Are the expression of SPK and its release a result of the more or less artificial tissue culture situation of those cells?

With regards to the first question, we investigated cells from various species and tissues, including those from human, rat, mouse, rabbit, and chicken, and as well as those of epidermal or mesenchymal type (Table 2). The following three criteria for SPK were applied: (1) phosphorylation of cell surface proteins and analysis of their labeling pattern; (2) phosphorylation of exogenous substrates in the presence of cells; and (3) phosvitin-induced kinase release from optimal cultures, including the analysis of LDH as cytosolic marker. It became evident that not only HeLa cells, but all cells investigated were positive by these criteria.

Since this study did not exclude that the cell culture situation per se is responsible for the expression of SPK we went one step further and analyzed cells taken directly from the organism: red blood cells from man, rat, mouse, rabbit, chicken, and frog could be shown to release protein kinase on treatment with phosvitin without substantial loss of intracellular material. In these cells, however, phosphorylation of surface proteins, if there was any, was not demonstrable, probably due to rapid ATP hydrolysis expressed by these cells.

There are some reports in the literature which indicate the existence of SPK activities on a variety of other cells (see previous chapter).

Table 2. Prevalence of cell surface protein kinase[a]

Human	HeLa (m)	Cervix carcinoma	Cell line
	HeLa (s)		
	A 431 (m)	Vulva carcinoma	
	Radji (s)	Lymphoblastoid	
	Molt 3 (s)		
	BJAB (s)		
	HEL (m)	Lung fibroblast	
Rat	hepatocytes (m)	Liver	
Rabbit	RK 13	Kidney	
Mouse	3T3 (m)	Fibroblast	
	SV 3T3 (m)		
Chicken	CEC (m)	Embryo fibroblast	Secondary culture
	RSV CEC (m)		
Mouse	Cerebellum (m)	Cerebellum	Primary culture

[a] Monolayer (m) cultures grown to about 10^6 cells/5 cm plate or suspension (s) cultures with $5-8 \times 10^5$ cells ml^{-1} were investigated as given in the text. For surface protein phosphorylation cells were incubated for 15 min at 37°C with 0.5 μM [γ-^{32}P]ATP. 1 $mgml^{-1}$ phosvitin or 1% serum were taken for phosphorylation of exogenous substrates for 15 min. SPK release was assessed by phosvitin in the absence of cosubstrate ATP and PK assay with cell-free supernatant was performed for 15 min. Incorporation of [^{32}P] in cell surface protein or exogenous substrate was analyzed by SDS gel electrophoresis and autoradiography

3.2 Phosvitin Kinase Activity in Body Fluids

The general occurrence of SPK detectable even in cells taken directly from the organism, allows us to ask whether possibly released phosvitin-kinase is present in body fluids. Included in the study were human serum, plasma, cerebrospinal fluid (CSF) and ascites, calf serum and plasma, tumor ascites from rat and mouse, and the content of a hamster liver cyst.

The freshly taken fluids were clarified of particulate matter by centrifugation at 2500 x g for 10 min and incubated with [γ-^{32}P]ATP in the presence or absence of 1 $mgml^{-1}$ phosvitin for 10 min at 37°C. As analyzed by SDS gel electrophoresis and autoradiography, among those fluids tested cerebrospinal fluid, tumor ascites from rat and mouse, as well as the fluid from the hamster liver cyst exhibited phosvitin kinase activity. Neither in plasma nor in serum of different sources was such kinase activity detectable. The latter result appears to be intriguing in view of the fact that erythrocytes can be forced to release their SPK by interacting with substrates, but may again reflect a particular anchoring of SPK in the cell surface. Whether the lack of protein kinase detection in blood fluids is due to absence of protein kinase or to presence of inhibitors remains to be established. The source of phosvitin kinase found in some of the body fluids, however, cannot be defined unless criteria are available to identify the different families of phosvitin kinase isoenzymes, in particular SPK (Pyerin et al. manuscript submitted).

4 Substrates of Surface Protein Kinase

4.1 Substrates at the Cell Surface

The question of the physiological significance of a given PK is ultimately linked to the physiological function of its substrates and the role of their phosphorylation and dephosphorylation. The labeling pattern obtained with intact cells represents probably natural products of the SPK reaction. As reported for HeLa cells the autophosphorylation of optimal cultures using exogenous [γ-^{32}P]ATP led to a reproducible labeling pattern including two predominantly radioactive peptides with M_r about 116,000 as revealed by SDS gel electrophoresis and autoradiography. The autophosphorylation protocol applied to other cell types (Sec. 3.1) resulted in a characteristic labeling pattern for each cell (Fig. 4). Yet surprisingly often, there

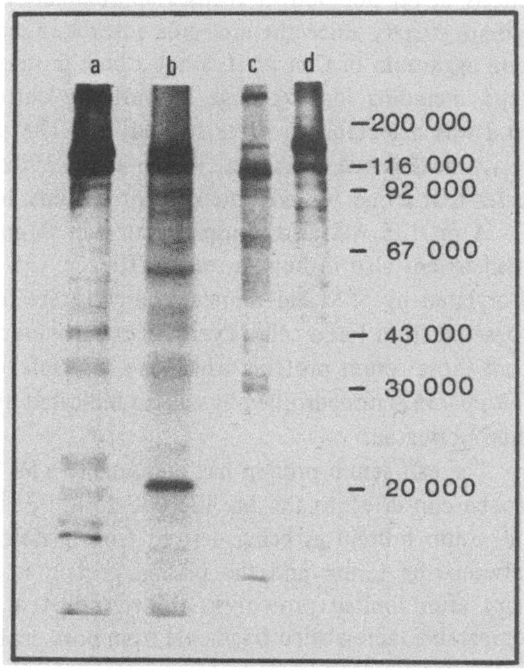

Fig. 4 a–d. Phosphorylation of cell surface located proteins. Intact cultures of (**a**) BJAB (human lymphoblastoid line), (**b**) mouse cerebellum cells, (**c**) chicken embryo cells, and (**d**) HeLa cells were incubated in presence of 0.5 μM [γ-^{32}P]ATP for 15 min; cell lysates were subjected to electrophoresis on SDS gel and autoradiographed

is a clear similarity among all cells with positive surface phosphorylation for the two above mentioned high molecular weight polypeptides.

The [^{32}P]-phosphoprotein pattern generated in HeLa cells within 15 min proved to be stable for at least 2 h as indicated by reincubation of phosphorylated cells under growth conditions. This result suggests a comparably slow turnover of the phosphorylated surface proteins.

The extraction of the two predominant phosphoproteins from membranes requires the presence of nonionic detergents (Nonidet P40, octyl-L-β-D-glucopyranoside), a fact which might emphasize that these proteins are inserted within the plasma membrane.

As mentioned above, intact erythrocytes failed to exhibit protein phosphorylation on incubation with [γ-^{32}P]ATP, although SPK activity can be liberated from intact cells. This result is probably due to rapid ATP hydrolysis by these cells as detected by thin-layer chromatography. It does not necessarily mean that there are not potential substrates present at their surface. In view of the large ATPase activity, however, SPK of erythrocytes may have another physiological "purpose" which may reside strictly in extracellular function(s).

4.2 Selective Phosphorylation of a Protein in Serum

Upon searching for substrates in the appropriate physiological environment of cells, a particular substrate was found in calf serum. Among the large number of proteins therein a single protein can be selectively phosphorylated by intact cells (Fig. 5). According to preliminary characterization, the serum protein migrates in the presence of SDS at about M_r 135,000 (pp 135), under nondenaturing conditions at M_r 500,000; it focuses approximately at pH 4.5 and contains a carbohydrate moiety since the molecule adsorbs to lectin columns carrying either wheat germ agglutinin or Con A. It is not a bulk protein since even after some purification steps, including ion exchange and affinity chromatography, only a faint protein band was detectable in silver stained gels. The pp 135 is phosphorylated by SPK in a time-dependent fashion (Kübler et al. 1982). Phosphoamino acid analysis revealed that serine, but not threonine or tyrosine, became phosphorylated.

A pp 135 was also phosphorylated in serum of rat, rabbit, chicken, and to a small extent also in human serum. The pp 135 in calf serum is specifically phosphorylated by SPK and represents a substrate for the phosvitin kinase isoenzymes extracted from HeLa cells (Pyerin et al., manuscript submitted); it differs, however, from those serum proteins which are substrates of cAMP-dependent PK (Fig. 5b). The pp 135 is not adsorbed by cells as indicated by incubation with serum previously phosphorylated.

The calf serum protein has presumably a M_r 300,000 precursor in plasma which can be converted to the M_r 135,000 form by clotting. This excludes the origin of the serum protein as being derived from platelet decomposition. The relationship between the serum and the plasma protein is also indicated by phosphopeptide maps after limited proteolysis (Cleveland et al. 1977) which yield a number of comparable radiolabeled fragments from both proteins.

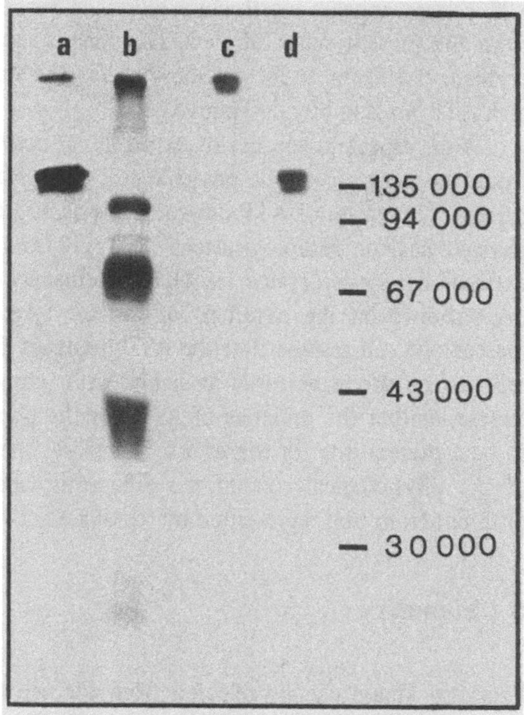

Fig. 5 a–d. Phosphorylation (15 min) of (**a**) calf serum (10% in incubation fluid) by intact HeLa cells, (**b**) calf serum by catalytic subunit of cAMP-dependent protein kinase from rat muscle (for details see Kübler et al. 1979), (**c**) calf plasma (10% in incubation fluid) by intact HeLa cells, (**d**) calf plasma phosphorylated by cells and coclotted with human blood (see text); samples were separated on SDS gels and autoradiographed

5 Cosubstrates of Surface Protein Kinase

Extracellular ATP effects certain alterations in intact cells, including changes in membrane permeability with respect to ions (Hempling et al. 1969; Perdue 1971; Parker and Snow 1972; Trams 1974; Aiton and Lamb 1975; Rozengurt et al. 1977; Yatani et al. 1982), nucleotides (Cockcroft and Gomperts 1973; Makan 1978), sugars (Chang et al. 1974; Gould 1979); and proteins (Dahlquist et al. 1974) as well as changes in the uptake of p-nitrophenylphosphate (Rozengurt and Heppel 1975), in the translocation of hyaluronic acid (Sato et al. 1977) and the biosynthesis of macromolecules (Kitagawa 1980). ATP is also believed to represent the cotransmitting factor (Kolb and Wakelam 1983) which gave rise to the concept of purinergic neurotransmission (Burnstock 1972). It may well be that some of these ATP-dependent effects are related to protein phosphorylation. As long as the observed phenomena, however, remain confirmed to ATP, but are not effected by GTP, phosphorylation via SPK seems to be excluded as the trigger since SPK will accept GTP as cosubstrate. Moreover, some of these effects usually require rather large ATP concentrations, whereas the cell surface directed phosphorylation is detectable already by the use of micromolar ATP concentrations.

Therefore, it might be possible that cellular responses to high ATP concentrations are somehow related ATP hydrolysis, whereas those caused by low ATP concentrations may be linked to protein phosphorylation. Thus, considering the rapid turnover of ATP by erythrocytes, in their case only the substrate-induced

SPK release appears worthwhile following for its physiological appearance, at least from the present point of view. This idea is supported by data according to which erythrocytes seem to be responsible for the down regulation on the occasion of high ATP levels in blood (Trams 1980).

With respect to sources of extracellular cosubstrate ATP in the organism which would be necessary for a physiological SPK reaction, three can be considered at present: (1) release of ATP during cytolysis, i.e., extracellular ATP originating during physiological or pathological cell decay; (2) translocation of cytoplasmic ATP into extracellular space (Trams 1974); (3) formation of extracellular ATP by corresponding enzymes on the external cell surface (Ågren et al. 1971). Considering these sources one can assume that the ATP necessary for phosphorylation reactions at the cell surface shoud be easily available, with respect to the physiological role of SPK release, neither the presence of ATP nor the phosphorylation reaction per se appear to be a prerequisite for this event. Therefore, one may have to question whether part of the physiological role of the SPK entity may reside in a function or message different from that represented by its kinase activity.

6 Summary

Phosvitin kinase can be released from the cell surface of intact HeLa cells by substrate (in the absence of cosubstrate); cytosolic marker enzymes were not detectable in the supernatant. The degree of discharge depends on substrate concentration, incubation time, and temperature; it is inversely related to cell density. The appearance of released protein kinase is accompanied by a reduction in the capability of cells to phosphorylate cell surface or exogenous proteins, thus, indicating that it is the surface protein kinase (SPK) which had been released. This unique physiological feature of SPK differs basically from spontaneous shedding or induced exocytosis. Four subsequent incubations with the substrate phosvitin were sufficient to deplete cells of most of their SPK; within approx. 24 h HeLa cells restore their SPK. The mechanism of release is unknown. SPK was shown to be present in all cells studied so far, including those from human, rat, mouse, chicken, and frog, by the following criteria: (1) phosphorylation of cell surface proteins and incubation of cells with [γ-^{32}P]ATP and /or (2) phosvitin-effected release of SPK from intact cells. In some instances the first criterion was not demonstrable probably due to rapid ATP hydrolysis. Positive surface phosphorylation revealed a reproducible phosphoprotein pattern particularly with respect to two intensely labeled proteins with M_r in the range of 116,000. The relative stability of the surface label suggests that SPK is probably not involved in fast regulation of these proteins. The wide occurrence of SPK as well as the similarity of surface substrates among different cells may indicate an essential biological function of SPK. Released SPK may serve as a "message" dispersed on contact with substrate effectors. In this respect the detection of a "physiological" exogenous substrate in serum ($M_r \sim 135{,}000$) with a possibly larger precursor in blood plasma may be a clue.

References

Ågren G, Ponten J, Ronquist G, Westermark B (1971) Formation of extracellular adenosine triphosphate by normal and neoplastic human cells in culture. J Cell Physiol 77:331–336

Aiton JF, Lamb JF (1975) The transient effect of ATP on K movements in HeLa cells. J Physiol (Lond) 248:14–15P

Avruch J, Hoelzl Wallach DF (1971) Preparation and properties of plasma membrane and endoplasmic reticulum fragments from isolated rat fat cells. Biochim Biophys Acta 233:334–347

Black PH (1980) Shedding from the cell surface of normal and cancer cells. Adv Cancer Res 32:75–199

Blostein R (1968) Relationships between erythrocyte membrane phosphorylation and adenosine triphosphate hydrolysis. J Biol Chem 243:1957–1965

Burnstock G (1972) Evolution of the autonomic innervation of visceral and cardiovascular systems in vertebrates. Pharmacol Rev 21:247–324

Chang KJ, Marcus NA, Cuatrecasas P (1974) Cyclic adenosine monophosphate-dependent phosphorylation of specific fat cell membrane proteins by an endogenous membrane–bound protein kinase. J Biol Chem 249:6854–6865

Cleveland DW, Fischer SG, Kirschner MW, Laemmli UK (1977) Peptide mapping by limited proteolysis in sodium dodecyl sulfate and analysis by gel electrophoresis. J Biol Chem 252:1102–1106

Cockcroft S, Gomperts BD (1979) ATP induces nucleotide permeability in rat mast cells. Nature (Lond) 279:541–542

Dahlquist R, Diamant B, Krüger PG (1974) Increased permeability of the rat mast cell membrane to sodium and potassium caused by extracellular ATP and its relationship to histamine release. Int Arch Allergy Appl Immunol 46:655–675

Gould MK (1979) Multiple roles of ATP in the regulation of muscle sugar transport. Trends Biochem Sci 4:10–13

Hempling HG, Stewart CC, Gasic G (1969) The effect of exogenous ATP on the electrolyte content of TA_3 tumor cells. J Cell Physiol 73:133–140

Ipata PL (1967) A coupled optical enzyme assay for 5′-nucleotidase. Anal Biochem 20:30–36

Kemp BE (1979) Phosphorylation of synthetic peptide analogs of rabbit cardiac troponin inhibitory subunit by the cyclic cAMP-dependent protein kinase. J Biol Chem 254:2638–2642

Kitagawa T (1980) Selective inhibition of macromolecular synthesis in transformed mouse cells by means of ATP-treatment. Biochem Biophys Res Commun 94:167–173

Kolb HA, Wakelam MJ (1983) Transmitter-like action of ATP on patched membranes of cultured myoblasts and myotubes. Nature (Lond) 303:621–623

Kübler D, Gagelmann M, Pyerin W, Kinzel V (1979) Isolation of the catalytic subunits of cAMP-dependent protein kinases from different mammalian tissues on the basis of charge differences of their subunits. Hoppe-Seyler's Z Physiol Chem 360:1421–1431

Kübler D, Pyerin W, Kinzel V (1982) Protein kinase activity and substrates at the surface of intact HeLa cells. J Biol Chem 257:322–329

Kübler D, Pyerin W, Burow E, Kinzel V (1983) Substrate-effected release of surface-located protein kinase from intact cells. Proc Natl Acad Sci USA 80:4021–4025

Makan NR (1978) Induction of permeability change and restoration of membrane permeability barrier in transformed cell cultures. Exp Cell Res 114:417–427

Mueller GC, Kajiwara K (1969) Synchronization of cells for DNA synthesis: In: Habel K, Salzman NP (eds) Fundamental techniques in virology. Academic, New York, p 21

Parker JC, Snow RL (1972) Influence of external ATP on permeability and metabolism of dog red blood cells. Am J Physiol 223:888–893

Perdue JF (1971) The isolation and characterization of plasma membranes from cultured cells. J Biol Chem 246:6750–6759

Poste G, Allison AC (1973) Membrane fusion. Biochim Biophys Acta 300:421–465

Reed J, Kinzel V (1984a) Ligand/binding site interaction in adenosine cyclic 3′, 5′-monophosphate dependent protein kinase catalytic subunit: circular dichroic evidence for intramolecular transmission of conformational change. Biochemistry 23:968–973

Reed J, Kinzel V (1984b) Near- and far-ultraviolet circular dichroism of the catalytic subunit of adenosine cyclic 5′-monophosphate dependent protein kinase. Biochemistry 23:1357–1362

Rozengurt E, Heppel LA (1975) A specific effect of external ATP on the permeability of transformed 3T3 cells. Biochem Biophys Res Commun 67:1581–1588

Rozengurt E, Heppel LA, Friedberg I (1977) Effect of exogenous ATP on the permeability properties of transformed cultures of mouse cell lines. J Biol Chem 252:4584–4590

Sato C, Kojima K, Nishizawa K (1977) Translocation of hyaluronic acid in cell surface of cultured mammalian cells after X-irradiation and its recovery by added adenosine triphosphate. Biochim Biophys Acta 470:446–452

Sottocasa GL, Kuylenstierna B, Ernster L, Bergstrand H (1967) An electron-transport system associated with the outer membrane of liver mitochondria. J Cell Biol 32:415–438

Trams EG (1974) Evidence for ATP action on the cell surface. Nature (Lond) 252:480–481

Trams EG (1980) A proposal for the role of ecto-enzymes and adenylates in traumatic shock. J Theor Biol 87:609–621

Yatani A, Tsuda Y, Akaike N, Brown AM (1982) Nanomolar concentrations of extracellular ATP activate membrane Ca channels in snail neurones. Nature (Lond) 296:169–171

Ectoglycosylation

Ectoglycosyltransferase Activities During Neuronal Differentiation in Cell Cultures

P. Mandel, Y. Matsui, D. Lombard, and H. Dreyfus [1]

1 Introduction

It is generally admitted that glycosyltransferases and the synthesis machinery of glycoconjugates are localized in the intracellular membrane structures such as the endoplasmic reticulum (Cook et al. 1965), Golgi apparatus (Schachter et al. 1970; Fleischer et al. 1969) and mitochondrial membranes (Bosman 1971; Broquet et al. 1975). Recently several reports concluded that glycosyltransferases also exist at the cell curface with their active sites exposed to the extracellular space (see for review Shur and Roth 1975; Cacan et al. 1976). The methods applied for detection of these ectoenzymes were direct enzymatic assays of intact cells or membranes as well as immunocytochemistry using antisera against glycosyltransferases and autoradiography (for review see Shur and Roth 1975; Shur 1982). Most frequently the addition of a radiolabeled sugar from a sugar nucleotide to the plasma membrane of intact cells was used to test ectoglycosyltransferase activities. According to the results obtained, the presence of ectoglycosyltransferase activities was reported at the cell surface of lymphocytes (Cacan et al. 1976; Hoflack et al. 1979), leukemic cells (Bernacki and Porter 1978), cultured nerve cells from chick embryo retinas (Roth et al. 1971), embryonic liver (Arnold et al. 1973), sperm (Shur and Bennet 1979), teratocarcinoma cells (Shur and Roth 1975; Shur 1982) and migrating mesenchymal cells (Hakomori 1981; Johnson 1977). The results obtained suggest an involvement of surface glycosyltransferases in several developmental processes: e.g., embryogenesis, cell migration, and tumor genesis (see for review Hakomori 1981; Turley and Roth 1979) as well as in plasma membrane surface properties such as electrochemical potentials (Forrester et al. 1964), membrane permeability (Rahmann et al. 1976), cell-cell recognition, adhesion, and interactions (Roseman 1970), contact inhibition, regulation and maintenance of growth factor receptor function (Hakomori 1973) and crypticity of immunogenic loci (Currie and Bagshawe 1969). However, glycosyltransferase activity towards glycoconjugate substrates does not define the molecular mechanism for the complex phenomena observed. Surface glycosyltransferases may only be involved in recognition and subsequent binding to extracellular substrates.

A great variety of sialoglycoproteins, and sialoglycolipids has been described in the central nervous system, particularly at synaptic junctions (for review see Ledeen and Yu 1976; Mahler 1979). One type of sialoglycolipids, the gangliosides, which are main-

1 Centre de Neurochimie du CNRS, 5, rue Blaise Pascal, 67084 Strasbourg Cedex, France

ly located in the cell plasma membrane, are asymmetrically disposed on the outer membrane leaflet; they are very abundant in synaptic membranes. Although their role and mechanism of action is still poorly understood, they are suspected to play, together with glycoproteins, an important role in neurotransmission (Svennerholm 1980; Yamakawa and Nagai 1978).

Upon oncogenic transformation in vitro and in tumors in vivo, gangliosides of cell surface membranes alter their chemical composition. Blocked synthesis as well as neosynthesis of some gangliosides could be observed (Hakomori 1981; Robert et al. 1977). The blocked synthesis may be due to suppressed or repressed glycosyltransferase activities and the neosynthesis may be caused by activation or derepression of enzymes with less restricted substrate specificity. Some of these changes may also be a consequence of transformation and gene activation (see for review Hakomori 1981).

Since it has been suggested that several enzymes involved in ganglioside and glycoprotein synthesis may exist at the external cell surface (Horowitz 1982) the question was raised whether and how they contribute to the multiple functions of different nerve cells. To answer this question, the first step was to establish clearly the existence, and characterization of the ectoenzymes able to synthesize the glycoconjugates.

Cell culture provides a model system for such an investigation. However, several parameters have to be controlled in order to obtain unequivocal results. The cell population should be homogenous. It should be demonstrated that the supposed ectoenzymatic reactions occur at the external surface of the membrane and are not a result of intracellular synthesis followed by a transfer to the outside of the cell. Moreover, the possibility of a reaction produced by enzymes released by the cells into the culture medium during measurements of ectoenzyme activities should be excluded.

In this paper we report the existence of galactosyl-, fucosyl- and sialyltransferases at the external surface of chick-embryo neurons in culture (Pettmann et al. 1979) and the increase of some ectoglycosyltransferase activities during neuronal maturation and synapse formation. The possible sources of error have been excluded (Matsui et al. 1983).

2 Methodological Procedures

Before the beginning of the test for the ectoenzyme activities the neurons cultured in monolayer according to Pettmann et al. (1979) were carefully washed three times with serum-free Dulbecco's modified Eagle medium (DMEM) and then incubated in a Petri dish at 37°C in fresh medium in an humidified atmosphere (95% air and 5% CO_2) containing the labeled substrates and an external acceptor which does not penetrate into the cell. This was bead-coupled ovomucoid or asialofetuin in most of the experiments.

After various periods of incubation, the medium containing the Sepharose beads was collected while the cell monolayer was washed with 0.147 M NaCl. The beads were washed ten times by centrifugation, using a solution of saturated NaCl, collected on glass filters and finally dried with absolute ethanol (Matsui et al. 1983).

For identification of endogenous (plasma membrane) surface acceptors of labeled sugars, the cultured neurons were detached from the Petri dishes, precipated with 10%

trichloroacetic acid (TCA) and washed of GF/B glass filters ten times using 10 ml of 5% TCA, ten times using 10 ml of water and finally ethanol. The filters containing the beads and those containing the cell precipitates were then dried under an infrared lamp, introduced into a vial containing 10 ml of Rotiszint 22 (Roth) and the radioactivity was determined. In some experiments neuronal cells were incubated without exogenous acceptors, and after incubation and washing submitted to subcellular fractionation according to Eichberg et al. (1964).

Sugar-nucleotides do not cross the plasma membrane. However, during the incubation the labeled sugar nucleotides undergo a partial hydrolysis. The labeled sugar or neuraminic acid may penetrate into the cell and after incorporation into newly synthesized glycolipids or glycoproteins a transfer to the plasma membrane could occur. An error in evaluation of the ectoglycosyltransferase activity on autoreceptors of the plasma membrane surface could thus occur. In order to avoid this error it is possible to inhibit sugar-nucleotide hydrolysis using mononucleotides GMP, AMP and UDP-gal, CMP-sialic acid and GDP-fucose, respectively. Under these conditions, the quantitiy of free sugars liberated from sugar-nucleotides was about 4%, that of sialic acid ca. 10%. However, the mononucleotides also inhibit the corresponding transferase activities. Therefore, it seemed more convenient to evaluate quantitatively the error linked to sugar-nucleotide hydrolysis; thus we measured the hydrolysis of the sugar-nucleotides and the potential penetration of sugars liberated during the incubation time. Moreover, cold galactose, fucose, and neuraminic acid (final concentration 1 mM) were added to the medium containing the labeled sugar-nucleotides in order to reduce considerably the specific activity of the eventual intracellular precursor as compared to the extracellular substrate.

After incubation as described above for the determination of ectoglycosyltransferase activities the sugar-nucleotide hydrolysis was measured; the remaining ^{14}C-sugar-nucleotides and the free ^{14}C-sugars obtained from the hydrolysis of ^{14}C-sugar-nucleotides were analyzed by paper chromatography (Matsui et al. 1983). The radioactivity contained in each compound was determined using a TLC scanner. Since cold sugars and sialic acid were present in the incubation medium at a final concentration of 1mM, it appeared clear that the fraction of labeled sugars liberated by hydrolysis during incubation was strongly diluted and only a very small fraction of labeled compounds, an eventual source of error, could penetrate into the cells.

3 Characterization of the Ectoglycosyltransferases

3.1 Effect of Divalent Cations

Mg^{2+} activates ectosialyltransferase and fucosyltransferase activity (+ 40% with 10 mM Mg^{2+}). In contrast, ectogalactosyltransferase activity is inhibited by 55% in the presence of 5 mM Mg^{2+}.

Mn^{2+} (5 mM) activates all three transferases tested (33 to 62%) with a striking increase in activity with higher concentrations (10 mM) for galactosyltransferase (+ 120%).

This differential effect of cations might be of physiological significance in the regulation of ectoglycosyltransferase activities. However, more evidence is necessary to support this suggestion.

4 Ectoglycosyltransferase Activities of Chick Embryo Neurons in Culture

As was shown in previous reports, these cells can be maintained in culture during 7 days in an excellent condition. During this time, they differentiate and develop synapses containing synaptic vesicles. The presence of dopamine (DA) and GABA (Louis et al. 1981) and the neuropeptides, somatostatin, enkephalins, and substance P could be demonstrated in these neurons (Louis et al. 1983). Similarly, the existence of a high affinity uptake for DA and GABA (Louis et al. 1981) and binding sites for GABA (De Feudis et al. 1981) and γ-OH butyrate (Maitre and Mandel 1984) which do not exist in glial cell cultures were also demonstrated.

As shown in Fig. 1 surface UDP-Gal: galactosyltransferase, GDP-Fuc: fucosyltransferase and CMP-NeuAe: sialyltransferase activities increase linearly for at least 90 min when measured towards endogenous and exogenous bead coupled acceptors (Fig. 1). The glycosylation of exogenous acceptors was strikingly higher than that observed for endogenous acceptors. Similarly, an increase of the glycosyltransferase activities measured as a function of neuronal protein concentrations from 0.5 to over 1 mg was observed (Fig. 2). In order to avoid a possible effect of cell density which might influence the glycosyltransferase activities, different sized dishes (d= 2.5, 6 and 10 cm) were used; thus approximately the same cell density could be prepared. Concerning

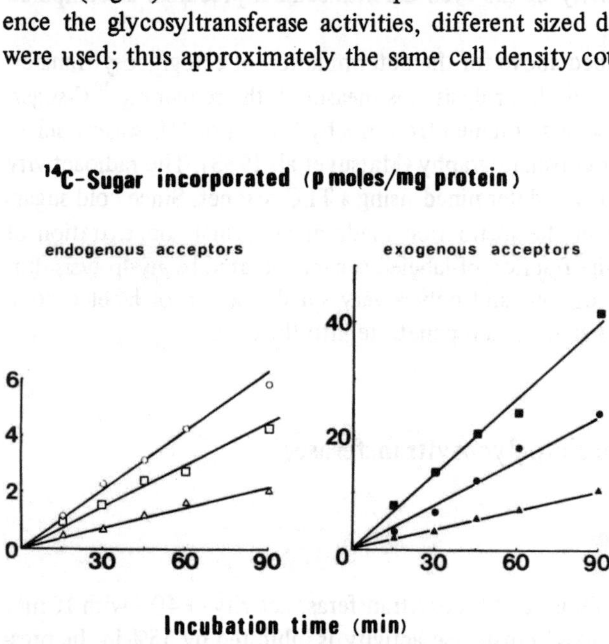

Fig. 1. Effect of incubation time on ectoglycosyltransferase activities determined with endogenous cellular acceptors or exogenous acceptors. The culture, 7 days old, was incubated with standard assay medium for various incubation times. ○● Galactosyltransferase; □■ fucosyltransferase; △▲ sialyltransferase. Each value is the mean of five experiments. (For details see Matsui et al. 1983)

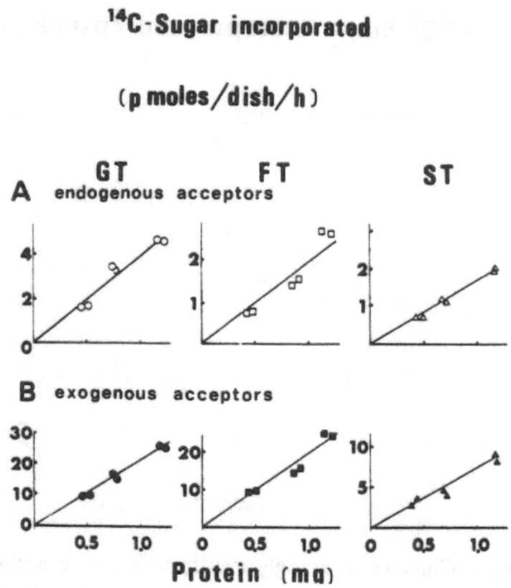

Fig. 2 A,B. Ectoglycosyltransferase activities as a function of the content of proteins per dish with endogenous acceptors (A) or exogenous acceptors (B). **A** the cells were cultured in different sized dishes (d = 2.5, 6 and 10 cm) with the same cell density. After 7 days, the cells were assayed for glycosyltransferase activities. **B** the cells of 7-day-old cultures were incubated with standard assay medium which contained various concentration of exogenous acceptors: ovomucoid (*GT*), and/or asialofetuin (*FT, ST*). *GT* ectogalactosyltransferase; *FT* ectofucosyltransferase; *ST* ectosialyltransferase. *Open symbols* activities for cellular acceptors; *closed symbols* activities for exogenous acceptors. Each value is the mean of three experiments

the exogenous acceptors, fucosyl and sialyltransferase activities reached maximal values at 5 and 7 mg of asialofetuin, respectively while galactosyltransferase activities were not apparently saturated with less than 15 mg of ovomucoid (Matsui et al. 1983). The transglycosylation to the bead coupled acceptors due to the adsorption of radioactive material or to enzymes possibly released from intact and/or damaged cells during 60 min of incubation were 20, 10 and 25% of the values recorded when intact cells were used for galactosyl-, fucosyl- and sialyltransferase, respectively. It is noteworthy that similar or very close values were obtained with the exogenous acceptors when the incubation was performed without cells or at times close to zero. This strongly suggests that the level of glycosylation, if any, due to the release of enzymes from the cells, was extremely low (Matsui et al. 1983).

5 Changes in Glycosyltransferase Activities During Neuronal Differentiation

The galactosyl-, fucosyl-, and sialyltransferase activities were measured in 1-, 3-, 5- and 7-day-old cultures. All three enzymatic activities were already present in neurons after 1 day in culture and were higher with exogenous acceptors as substrates (Fig. 3). The sialyltransferase activities per unit protein on endogenous and exogenous accep-

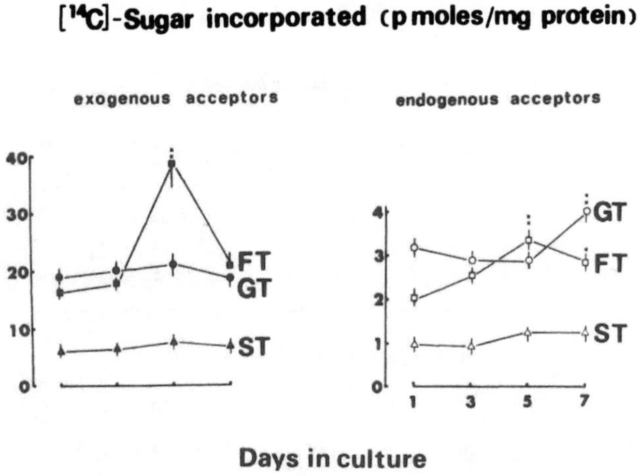

Fig. 3. Changes of ectoglycosyltransferase activities toward cellular acceptors and exogenous acceptors in 1- to 7-day-old cultures. The cells cultured in dishes of 10 cm diameter were assayed at various times after culture with standard assay medium. Each value is the mean ± S.D. of 5 experiments. $**p < 0.01$, $*p < 0.05$ (Student's t-test). Significance values refer to a comparison with the activities of 1-day-old cultures

tors did not change during the culture time, suggesting an adapted high capacity of this enzyme largely involved in glycoprotein and glycolipid synthesis. The galactosyltransferase activity on the endogenous acceptors increased in 7-day-old cultures while that toward exogenous acceptors did not change. The fucosyltransferase activity toward the endogenous acceptors increased gradually from 3 to 5 days in culture, which is the period of synapse formation and then decreased at day 7. The fucosyltransferase activity toward exogenous acceptors also increased during synapse formation (3-5 days), then decreased at 7 days; the activity returned to values of 3-day-old cultures. This decrease of the enzymatic activity may also be related to the selective disappearance of some surface glycoproteins observed during maturation of the nervous system (Littauer et al. 1980).

In the 7-day-old cultures which already possess multiple synapses, the ectoglycosyltransferase activities of galactosyltransferase and sialyltransferase with exogenous acceptors could reach five fold higher levels and in the case of fucosyltransferase seven fold higher values than those measured with endogenous acceptors.

6 Discussion

Ultrastructural and biochemical evidence have suggested the existence of ectoglycosyltransferase activities at the cell surface in various cell types (Cacan et al. 1976; Hoflack et al. 1979; Bernacki and Porter 1978).

In the central nervous system these enzymes may be involved in the metabolism of membrane glycoconjugates, glycoproteins, and glycolipids which are rather abundant in nerve cells. Moreover, some of these glycoconjugates namely glycoproteins and gangliosides may play in important role in cell recognition, migration, cell attachment and interaction (neuron-glia) as well as in synaptic structure and function. In this respect extrusion of neurotransmitters from synaptic vesicles followed by a repair of the plasma membrane should be considered.

The data we reported demonstrate the existence at the neuronal surface of UDP-galactose: galactosyltransferase, GDP-fucose: fucosyltransferase and CMP-NeuAc: sialyltransferase activities which are able to transfer galactose, fucose and sialic acid units either on the endogenous cell membrane acceptors or possibly on acceptors of neighboring cells. The surface galactosyl-, fucosyl-, and sialyltransferase activities were shown to be time-dependent, substrate (exogenous acceptors) dependent and cell-number (protein) dependent. They are activated by divalent cations and can be inhibited by an excess of one of the products (monophosphate nucleosides). Artifacts or misleading errors were considered and excluded by our methodology. Maximal error which may result from intracellular synthesis after sugar-nucleotide hydrolysis and sugar uptake was found not to exceed 12.7%, 4% and 8.3% of the ectogalactosyl-, ectofucosyl- and ectosialyltransferase activities, respectively. Enzymatic activities possibly released from the cells were negligible (close to blank values). This observation, in addition to trypan blue exclusion observed in 99.5% of cells, excludes an effect of enzymes extruded from altered cells.

In our experimental conditions (excess of free sugars in the medium), the demonstration that the active sites of galactosyl-, fucosyl- and sialyltransferase located on the external surface are turned toward the extracellular milieu, are strongly supported by the transfer of substrate to exogenous substrates, ovomucoid and asialofetuin whether coupled or not to resin beads. The activities of ectogalactosyl-, fucosyl- and sialyltransferase were also determined with soluble ovomucoid and asialofetuin as exogenous acceptors, since the use of bead coupled acceptors limits the accessibility of the enzymes to the substrates. Actually in these conditions an increase of the activities of ectogalactosyl-, fucosyl- and sialyltransferases dependent on time, protein, and exogenous acceptor concentration could be demonstrated.

The ectofucosyltransferase activity on exogenous acceptors increased markedly between 3 and 5 days in culture and during the same period there was also an increase of endogenous neuronal fucosylated acceptors. Both activities decreased in 7-day-old cultures when synaptic contacts had been established. These results suggest that the increase in fucosyltransferase activity and fucosylated acceptors are concomitant with the formation of synaptic contacts. Now that the existence and some properties of these neuronal ectoenzymes have been established, the question must be considered whether there are correlations between these enzymatic activities and processes in which neuron recognition is involved. In this respect, synaptogenesis seems to offer a useful tool. In our experimental conditions synaptic junction-like structures have been observed starting from the 3rd day in culture, in parallel with biochemical concomitants of synaptogenesis such as an increase in tyrosine hydroxylase and DOPA decarboxylase activities and the appearance of a high affinity uptake system. Moreover, an increase in the amount of endogenous fucosyl acceptors was observed in a synaptosomal-enriched fraction in sucrose density gradients. When fucosylated acceptor proteins sedimenting between 0.32 M and 0.8 M sucrose were applied to SDS polyacrylamide gel electrophoresis, at least 5 fucosylated protein peaks appeared. It is noteworthy that after 5 days in culture when synaptogenesis apparently slows down, ectofucosyltransferase activity decreases.

It is also interesting to report that cultured glial cells showed a different pattern of glycosylation when compared to neurons. Glycolipids are mostly fucosylated during maturation of glial cells, while glycoproteins are essentially fucosylated in neurons (in preparation). One may speculate that these events express specific cellular recognition in agreement with the specialization of the two cell types, neurons and glial cells.

In neurons, glycoproteins have been shown to play an important role in synapse formation. Many reports showed the important role of glycoproteins and especially fucosyl-glycoproteins in neural function (Littauer et al. 1980; Chen and Rinehart 1982; Zanetta et al. 1978). Recently, the existence of glycoproteins which disappear after the period of synaptogenesis in cerebellum was also reported (Dontenwill et al. 1985). Our present results suggest a role for ectoglycosyltransferase and especially ectofucosyltransferase in the synthesis of specific glycoproteins during neuronal cell maturation, and furthermore suggest for these glycoproteins a role in synaptic function.

References

Arnold D, Hommel E, Risse HJ (1973) Glycosyltransferase activities in intact liver cells of embryonic chick. Biochem Biophys Res Commun 54: 100-107

Bernacki RJ, Porter CW (1978) Biochemical and ultrastructural studies of ectoglycosyltransferase systems of murine L1210 leukemic cells. J Supramol Struct 8: 139-152

Bosmann HB (1971) Identification of products of autonomous mitochondrial protein and glycoprotein synthesis. Nat New Biol 234: 54-56

Broquet P, Morelis R, Louisot P (1975) Biosynthesis of cerebral glycoproteins–studies on mitochondrial mannosyl transferase. J Neurochem 24: 989-995

Cacan R, Verbert A, Montreuil J (1976) New evidence for cell surface galactosyltransferase. FEBS Lett 63: 102-106

Chen KY, Rinehart Jr CA (1982) Identification of exposed surface glycoproteins in undifferentiated and differentiated mouse N-18 neuroblastoma cells. Biochim Biophys Acta 685: 61-70

Cook GMW, Laico MT, Eylar EH (1965) Biosynthesis of glycoproteins of Ehrlich ascites carcinoma cell membranes. Proc Natl Acad Sci US 54: 247-252

Currie GA, Bagshawe KD (1969) Tumour-specific immunogenicity of methylcholanthrene-induced sarcoma cells after incubation in neuraminidase. Br J Cancer 23: 141-149

DeFeudis FV, Ossola L, Sarlieve LL, Schmitt G, Rebel G, Varga V, Mandel P (1981) GABA and muscimol binding processes in CNS tissue culture preparations. In: Amino acid neurotransmitters. DeFeudis FV, Mandel P (eds) Raven Press NY, pp 405-410

Dontenwill M, Roussel G, Zanetta JP (1985) Immunohistochemical localization of a lectin-like molecule, R1, during the postnatal development of the rat cerebellum. Develop Brain Res 17: 245-252

Eichberg J, Whittaker VP, Dawson RMC (1964) Distribution of lipids in subcellular particles of guinea pig brain. Biochem J 92: 91-100

Fleischer B, Fleischer S, Ozawa H (1969) Isolation and characterization of Golgi membranes from bovine liver. J Cell Biol 43: 59-79

Forrester JA, Ambrose EJ, Stoker MGP (1964) Microelectrophoresis of normal and transformed clones of hamster kidney fibroblasts. Nature (Lond.) 201:945-946

Hakomori SI (1973) Glycolipids–their chemical pattern, synthesis and degradation in normal and tumor cells. In: Tumor lipids. Biochemistry and metabolism, Wood R (ed) American Oil Chemist's Society Press, Champaign (Illinois, USA), pp 269-284

Hakomori SI (1981) Glycosphingolipids in cellular interaction, differentiation and ontogenesis. Ann Rev Biochem 50: 733-764

Hoflack B, Cacan R, Montreuil J, Verbert A (1979) Detection of ectosialyltransferase activity using whole cells–correction of misleading results due to the release of intracellular CMP-N-acetylneuraminic acid. Biochim Biophys Acta 568: 348-356

Horowitz MI (ed) The glycoconjugates, vol. III (1982) Academic Press New York, pp 369

Johnson KE (1977) Extracellular matrix synthesis in blastula and gastrula stages of normal and hybrid frog embryos: 3. Characterization of galactose-labeled and glucosamine-labeled materials. J Cell Sci 25: 335-354

Ledeen RW, Yu RK (1976) Gangliosides of nervous system. In: Glycolipid methodology, Witting LA (ed) American Oil Chemist's Society Press, Champaign (Illinois, USA), pp 187-214

Littauer UZ, Giovanni MY, Glick MC (1980) A glycoprotein from neurites of differentiated neuroblastoma cells. J Biol Chem 255: 5448-5453

Louis JC, Pettmann B, Courageot J, Rumigny JF, Mandel P, Sensenbrenner M (1981) Developmental changes in cultured neurons from chick embryo cerebral hemispheres–an ultrastructural and neurochemical study. Exp Brain Res 42: 63-72

Louis JC, Rougeot C, Bepoldin O, Vulliez B, Mandel P, Dray F (1983) Presence of somatostatin, enkephalins and substance P-like peptides in cultured neurons from embryonic chick cerebral hemispheres. J Neurochem 41: 930-938

Mahler HR (1979) Glycoproteins of the synapse. In: Complex carbohydrates of nervous tissue, Margolis RU, Margolis RK (eds), Plenum Press New York, pp 165-184

Maitre M, Mandel P (1984) Gamma-hydroxybutyrate, a putative neurotransmitter in the central nervous system. CR Acad Sci Paris 298: 341-345

Matsui Y, Lombard D, Hoflack B, Harth S, Massarelli R, Mandel P, Dreyfus H (1983) Ectoglycosyltransferase activities at the surface of cultured neurons. Biochem Biophys Res Commun 113: 446-453

Pettmann B, Louis JC, Sensenbrenner M (1979) Morphological and biochemical maturation of neurons cultured in the absence of glial cells. Nature (Lond.) 281: 378-380

Rahmann H, Rosner H, Breer H (1976) Functional model of sialo-glyco-macromolecules in synaptic transmission and memory formation. J Theor Biol 57: 231-237

Robert J, Rebel G, Mandel P (1977) Glycosphingolipids from cultured astroblasts. J Lipid Res 18: 517-522

Roseman S (1970) Synthesis of complex carbohydrates by multiglycosyltransferase systems and their potential function in intracellular adhesion. Chem Phys Lipids 5: 270-297

Roth S, McGuire EJ, Roseman S (1971) Evidence for cell surface glycosyltransferase–their potential role in cellular recognition. J Cell Biol 51: 536-547

Schachter H, Jabbai I, Hudgin R, Pinteric L, McGuire E, Roseman S (1970) Intacellular localization of liver sugar nucleotide glycoprotein glyosyltransferases in a golgi-rich fraction. J Biol Chem 245: 1090-1100

Shur BD, Roth S (1975) Cell-surface glycosyltransferases. Biochim Biophys Acta 415: 473-512

Shur BD, Bennett D (1979) Specific defect in galactosyltransferase regulation on sperm-bearing mutant alleles of the T-T locus. Dev Biol 71: 243-259

Shur BD (1982) Cell surface glycosyltransferase activities during fertilization and early embryogenesis. In: The glycoconjugates, vol. III, Horowitz MI (ed) Academic Press Inc., pp 146-185

Svennerholm L (1980) Gangliosides and synaptic transmission. In: Structure and function of gangliosides. Svennerholm L, Mandel P, Dreyfus H, Urban PF (eds) Plenum Press, pp 533-544

Turley EA, Roth S (1979) Spontaneous glycosylation of glycosaminoglycan substrates by adherent fibroblasts. Cell 17: 109-115

Yamakawa T, Nagai Y (1978) Glycolipids at the cell surface and their biological functions. TIBS 3: 128-131

Zanetta JP, Ghandour MS, Vincendon G, Gombos G (1978) Postnatal development of rat cerebellum–massive and transient accumulation of concanavalin A binding glycoproteins in parallel fibre axolemma. Brain Res 142: 301-319

Modulation Theory of Adhesion: Role of the Ectoenzymes Glycosidase and Glycosyltransferase

W. E. G. MÜLLER [1]

1 Introduction

It is amazing, but still plausible, that in all animal systems hitherto studied there is only one basic principle by which cell-cell recognition occurs. From sponges to higher eukaryotes there is only a small number of cell adhesion molecules which are involved in cell-cell adhesion. Specificities and histogenetic patterning are achieved by modulating processes acting on cell adhesion molecules. In sponges (secondary aggregation system) and in other eukaryotes, the activity and function of cell adhesion molecules are assumed to be modulated by enzymatic processes (glycosylation and deglycosylation). Strong evidence from experiments with sponges is available which indicates that modulation can also occur nonenzymatically by reversible interaction of cell adhesion molecules with lectins. The modulation theory requires only a few kinds of adhesion molecules in order to explain specific cell-cell interaction. In contrast, the chemoaffinity theory (Sperry 1963) requires a multiplicity of different adhesion molecules which are pairwise complementary. The data available also disfavor the assumption that cell adhesion takes place via weak forces (Curtis 1967).

The hypothesis that enzymatic processes modulate the binding properties of a few cell-cell adhesion molecules, implies that the chemical alterations of these molecules occur either intracellularly or extracellularly. The latter possibility has been experimentally proven in the sponge cell system, *Geodia cydonium*. It could be demonstrated that the plasma membrane of sponge cells contains an ectoenzyme, the β-glucuronidase, whose active site faces the external medium. This enzyme attacks the cell surface-localized aggregation receptor. Moreover, a second ectoenzyme (in a broader sense) is present in the extracellular milieu, the aggregation factor-associated glucuronosyltransferase which uses the deglucuronylated aggregation receptor as acceptor. The latter process proceeds again at the cell exterior.

1 Institut für Physiologische Chemie, Universität Mainz, Abteilung "Angewandte Molekularbiologie", Duesbergweg, 6500 Mainz, FRG

2 Cell Adhesion Molecules in Sponges

Since the pioneering work of Wilson (1907), marine sponges afford the classic model for studies of cell-cell interactions on a molecular level. Later, Galtsoff (1929), who demonstrated the existence of a soluble aggregation factor, and Moscona (1963) contributed basic ideas to the understanding of cellular recognition in sponges. The detailed work on the underlying molecular mechanisms was performed by Humphreys (1970), Burger (1977), and ourselves (Müller 1982).

Owing to their independent history of some 1000 m.y., the sponges have maintained some original features of early metazoans. Their cells are characterized by an unusually high reversible differentiation potency and by a comparably high motility. Because, firstly, the junctions between sponge cells show only little specialization; secondly, the distances between them are unusually high (between 300 and 1000 Å); and thirdly, the intercellular matrix into which they are embedded consists of a fluid phase (carbohydrate-rich compounds) and a solid phase (fibers, fibrils, and spicules). The sponge organism provides its cells with the structural prerequisites and the dynamic potentials by which they can arrange themselves in a highly flexible and dynamic way in tissue units. For the stabilization of such a multicellular system, highly developed cell-cell recognition mechanisms are present in sponges which enable their cells to carry out contact inhibition of movement, contact promotion, and contact guidance. Two distinct recognition systems have been elucidated; both of them are Ca^{++} dependent. In the following, we discuss primarily our own work, performed with the Tetractinomorphans (siliceous sponge) *Geodia cydonium* (Jam.), and *Cliona celata* (Grant).

2.1 Ca^{++}-Dependent Primary Aggregation System

Cells of marine sponges, belonging to the classes Hexactinellida (unpublished), Calcarea (Müller 1982), and Demospongiae (Müller 1982) are provided with the primary Ca^{++}-dependent aggregation system. Detailed work with *Geodia cydonium* (Müller and Zahn 1973; Müller et al. 1974) and *Cliona celata* (α-stage) (Müller et al. 1982a) revealed that at least 18 m*M* of Ca^{++} are required for formation of 80–500 μm large aggregates.

By an immunochemical approach it could be demonstrated that primary aggregation in *Geodia cydonium* is mediated by a specific ligand which was termed primary aggregation factor (pAF) (Müller et al. 1985). After purification almost to homogeneity, the factor turned out to be composed of three subunits with the following mol. wts.: 16,000, 15,000, and 13,500. The pAF is a glycoprotein (48% protein, 34% neutral carbohydrates, 13% hexuronic acid, 4% hexosamines) which is heat labile and rapidly destroyed by proteases. The purified pAF is still provided with the property to bind to surfaces of both homologous and heterologous urea-treated sponge cells. As known so far, pAF-mediated binding is second-order homophilic (pAF-pAF) (Fig. 1). No data on the molecular events proceeding during pAF-pAF interaction are available.

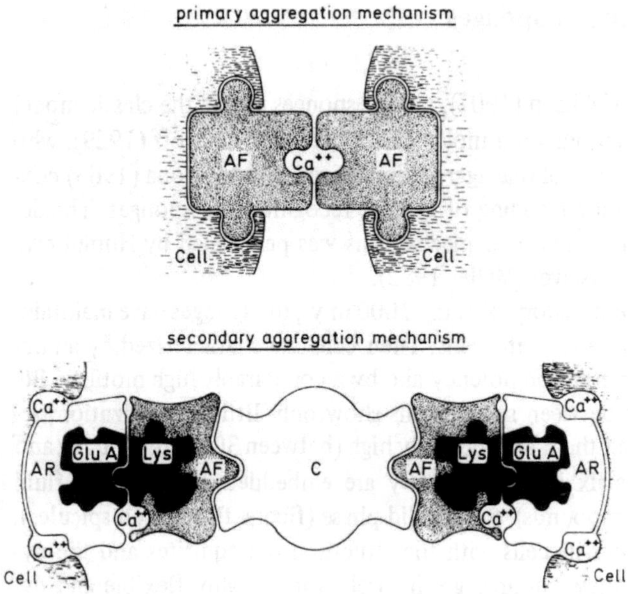

Fig. 1. The two molecular mechanisms proposed to explain primary and secondary aggregation mechanisms in the sponge *Geodia cydonium*. *AF* aggregation factor; *AR* aggregation receptor; Ca^{++} calcium ion; *C* core structure

2.2 Ca^{++}-Dependent Secondary Aggregation System

Most Demosponges (Müller 1982) and Hexactinellidians (*Aphrocallistes vastus*; Müller et al. 1984) are provided in addition to the primary cellular recognition system with a secondary adhesion apparatus. Basically, it functions in the same way in the different models used by Humphreys (Henkart et al. 1973), Burger (Weinbaum and Burger 1973) and ourselves (Müller and Zahn 1973; Müller et al. 1976). We worked primarily with *Geodia cydonium*, a siliceous sponge weighing up to 50 kg.

The secondary cell adhesion system is characterized by the presence of the secondary aggregation factor molecule (sAF), which is readily obtained in a soluble form after incubation of *Geodia* tissue in isotonic solution, containing 20 m*M* EDTA. Re-addition of the sAF to single cells results in the formation of large aggregates (5 mm and larger). After incubating them for some days a rearrangement of the cells in the aggregates to a functioning sponge occurs (Müller and Zahn 1973). The sAF is a complex macromolecule with a sedimentation coefficient of 90. In its "native" form the sAF appears as spheres (diameter: 1,000 Å) with a concave cup structure (Fig. 2A). The center is electron dense after platinum shadowing. After Nonidet treatment of the *Geodia* particle, a core structure is obtained which appears as "sunbursts" with a circular center (diameter: 1,000 Å) and 25 radiating arms (Fig. 2B).

The *Geodia* sAF is intracellularly assembled and becomes firmly associated with the cell membrane during the release into the extracellular space (Müller et al. 1982b). There, the sAF is embedded into the dense intercellular matrix and establishes cell-cell contact. The *Geodia* sAF is provided with the following three functional subunits

Fig. 2 A,B. Electron micrographs of aggregation factors. **A** *Geodia cydonium*: native aggregation factor, x 70,000. **B** *Geodia cydonium*: core structure, x 55,000

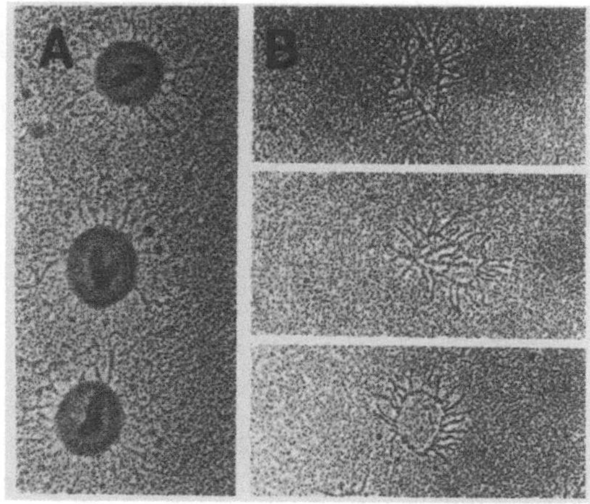

(Müller et al. 1978b); the glucuronosyltransferase (which is involved in the "sorting out" process), the galactosyltransferase (of unknown function) and that subunit which carries the binding site to the cell membrane. The latter subunit (the low molecular AF) is not covalently linked with the core structure (Müller et al. 1974) (Fig. 1).

The low molecular AF-subunit, which consists of 91% protein and 6% carbohydrate binds to a specific molecule on the cell surface, the aggregation receptor (AR) (Weinbaum and Burger 1973; Müller et al. 1976) (Fig. 1). The chemical groups which are involved in the recognition of the AR by the AF are L-lysine (in the AF molecule) (Müller et al. 1974) and glucuronic acid (in the AR unit) (Vaith et al. 1979). Common to all sponge models investigated (Burkart and Burger 1981; Müller 1982) is the fact, that adhesion between cells, which is based on the secondary aggregation system is third-order heterophilic (AR-sAF-AR); AR and sAF are different cell adhesion molecules [CAMs in the nomenclatur of Edelman (1983)] and sAF is present as binding ligand. The "activity" of these molecules is modulated by the following enzymic and nonenzymic processes.

2.2.1 Modulation by Enzymic Interactions

The *Geodia* system served as an appropriate model for the biochemical investigation of the reversible transition from aggregation-susceptible to aggregation-deficient cells, the basis for a controlled cell movement and positioning. The ability of the cells to form aggregates is controlled by the binding capacity of the AR (Müller et al. 1979c). As mentioned above, the AR carries a glucuronic acid moiety in a terminal position. This sugar unit can be reversibly removed from the AR.

β-Glucuronidase. The β-glucuronidase was solubilized and partially purified from *Geodia* cell membranes (Müller et al. 1979c). It has broad pH optimum between 6.5 and 4.0; its activity is independent of the presence of divalent cations, but requires 200 m*M* NaCl for full enzymatic activity. The temperature optimum was determined to be around 20°C. The mol. wt. of the enzyme was estimated to be approximately 80,000.

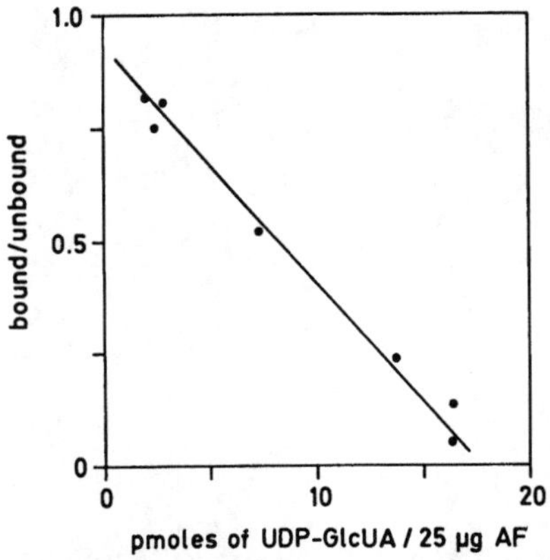

Fig. 3. Binding of UDP-D-[U-^{14}C] glucuronic acid (spec. activity 321 mCi mmol^{-1}; Amersham) to the secondary aggregation factor-glucuronosyltransferase particle. The binding experiments (volume of the assay: 1 ml) were performed in calcium and magnesium ion-free artificial seawater in the presence of 25 μg of purified aggregation factor (Müller and Zahn 1973). 3-500 n*M* of UDP-D-[U-^{14}C] glucuronic acid was added and incubation was performed for 20 min at 20°C. Subsequently, the aggregation factor was separated from free UDP-glucuronic acid by gel chromatography, using a P 300 (Bio-Rad) column (2 x 100 cm). Fractions of 3.3 ml each were collected. The UDP-glucuronic acid-aggregation factor complex appeared at a V_e-V_0 range between 1.0 and 1.18. The radioactivities in the fractions containing the bound as well as the unbound UDP-[^{14}C] glucuronic acid were determined. The data were plotted according to Scatchard (1949)

Glucuronosyltransferase. This transferase was found to be firmly associated with the sAF molecule (Müller et al. 1978b). The glucuronosyltransferase uses deglucuronylated AR as an acceptor. This enzyme is characterized by a Michaelis constant for the UDP-glucuronic acid substrate of 1.1 m*M* and for the acceptor of 216 n*M*. The enzyme has a pH optimum of 8.2. For full activity the presence of Mg^{2+} and 300 m*M* NaCl is required. The temperature optimum was determined to be 20°C.

The sAF-transferase particle has a high affinity for UDP-D-glucuronic acid (Fig. 3). The Scatchard analyses revealed an association constant of 5.3 x 10^7 M^{-1} and 95 UDP-GlcUA binding sites/molecule of sAF (M_r of the sAF: 1.3 x 10^8). Future studies will show whether (if at all) the transmembraneous transport of UDP-D-glucuronic acid is mediated by the sAF complex.

Interrelation Between Glucuronosyltransferase and Glucuronidase. From the experimental data given earlier (Müller et al. 1979c), we postulated that the glucuronic acid can, under defined conditions, be removed from the AR by the cell surface-bound β-glucuronidase, resulting in an inactivation of its binding function. The biological consequence of this enzymatic process is cell separation (Fig. 4). The deglucuronylation is a reversible event. Catalyzed by the sAF-associated glucuronosyltransferase the glucuronic acid unit can be reattached to the AR by which this molecule is reactivated. After this process the cells have regained their aggregation potency. This proposed mechanism of an enzymatically controlled cell adhesion has some striking similarities with the Roseman theory proposed in 1970 (Roseman 1970). At present, we have no experimental evidence on the mechanisms by which glucuronylation and deglucuronylation are controlled. One possibility could be an alteration in the cell sur-

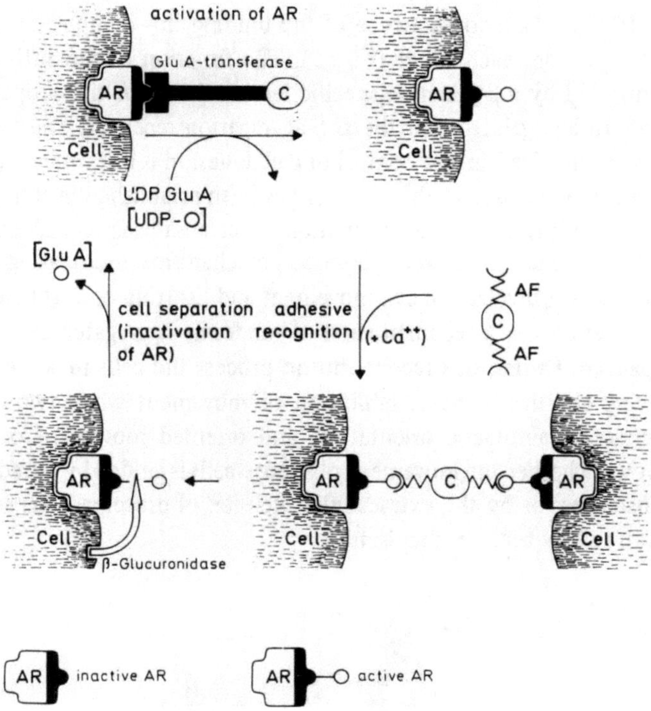

Fig. 4. Proposed molecular mechanism for cell aggregation and cell separation via activation or inactivation of the aggregation receptor (*AR*). *C* Core structure; *AF* secondary aggregation factor; *Glu A* and ○ glucuronic acid

face pH micromilieu. The glucuronosyltransferases, associated with the sAF, have pH optima of 8.2 (Müller et al. 1978b), while the cell membrane-bound glycosidases are active around pH 5.0 (Müller et al. 1979c). No data are available which give direct experimental evidence for a pH or ion change of the sponge cell membrane during the process of cell separation. However, from previous studies (Müller et al. 1978b), it is known that the sAF from *Geodia* alters the cell surface charge after its binding to homologous cells. Because alterations of the cell surface charge can also be caused by a pH change at the cell surface (Curtis 1967), we postulate that pH fluctuations govern the activity of both the glycosidases and glycosyltransferases.

A very similar modulation mechanism for cell adhesion molecules has recently been described for the N-CAM system from higher eukaryotes (Rothbard et al. 1982). The authors propose that the fluctuation of sialic acid content of N-CAM during development might be enzymatically controlled (Edelman 1983) by a balanced glycosylation-deglycosylation mechanism.

2.2.2 Modulation by Nonenzymatic Interactions

A second aggregation regulating system has been detected on the surface of sponge cells (Müller et al. 1979a). The key molecule in this system is the anti-aggregation receptor, which can bind with its terminal galactose to the sAF-AR complex (Müller et al. 1979b). As a consequence of the binding this complex is destroyed and the cells separate from each other (Fig. 5). The function of the anti-aggregation receptor is controlled by a β-galactose-specific lectin which has the property to mask reversibly the terminal galactose of the anti-aggregation receptor. The conclusion that the lectins are involved in the control of cell adhesion is not restricted to sponges. A definition role of lectins in this process has been established in slime molds (Barondes and Rosen 1976); a comparable function in vertebrates is very likely (Kitamura 1980).

We assume that the two described mechanisms, modulating secondary aggregation process, are involved in cell movement and "sorting out" (Müller 1982). These events are essential for a reconstitution of randomly aggregated cell clumps to a functional organism. During this reconstitution process the cells must be provided with the following abilities: contact inhibition of movement when two locomoting cells make contact, chemotactic orientation, and oriented movement on a substrate. It is not known whether the movement of sponge cells is guided by a gradient of morphogenetic substances, or by the extracellular skeleton of proteinaceous and glycoproteinaceous nature or by both mechanisms.

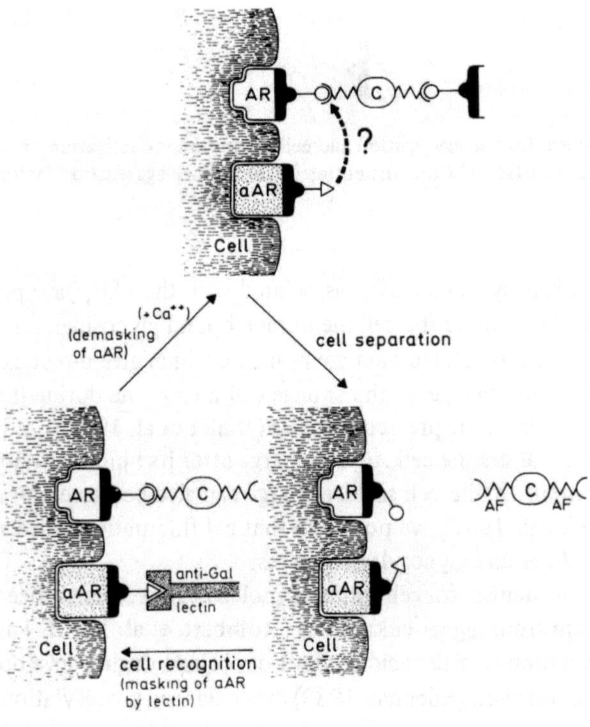

Fig. 5. A tentative model for the biological role of the *Geodia cydonium* lectin during the reaggregation of homologous cells. *AR* aggregation receptor; *C* core structure; *AF* secondary aggregation factor; *aAR* anti-aggregation receptor; ▷ galactose

3 Cell-Cell Adhesion in Vertebrates

It is conceivable that the elucidation of the mechanism(s) of cell-cell adhesion in vertebrates, with the aim of understanding their role in histogenesis especially in the development of the nervous system had at the beginning a higher priority than comparable studies with lower eukaryotes. However, due to the higher biological and biochemical complexity of the cell-cell recognition systems in higher eukaryotes and especially in vertebrates, the basic concept for the understanding of cell-cell recognition had been worked out in lower animals primarily using sponges. Today it is known that in vertebrates the following two distinct cell adhesion systems exist.

3.1 The Ca^{++}-Independent Adhesion System

Based on the pioneering works of Moscona (Hausman et al. 1976), Lilien (Balsamo and Lilien 1974), and Glaser (Merell et al. 1975), Edelman (1976, 1983) succeeded in the identification and purification of those molecules, which are involved in the Ca^{++}-independent system. They were termed cell adhesion molecules (CAM). So far, they have been isolated from retina and brain (N-CAM), from liver (L-CAM), and from astrocyte-like glial cells with neurons (GN-CAM). The N-CAM is a large sialoglycoprotein with a M_r of 120,000. This molecule (single polypeptide chain) is present on cell surfaces of neurons and striated muscle precursors as an integral membrane protein. The N-CAM system has been associated with the formation of neurite bundles, the sorting out of neurites and somata in cell aggregates, and the organization of cell layers in retinal tissue (Brackenbury et al. 1981; Edelman 1983). Isolated N-CAM shows self-aggregation and microheterogeneity. The latter characteristic is attributed to its carbohydrate portion, which is composed predominantly of sialic acid. Immunological studies revealed that N-CAM binding is second-order homophilic, which means that two identical CAMs form the linkage. Ca^{++} is not required for this process. The CAMs have been shown to function cell and tissue specifically, but not species specifically (Chuong et al. 1982). Sialic acid itself is not directly involved in the adhesion process, but modulates its strength. In this context it is important to note that embryonic N-CAM has a sialic acid content of up to 26%, while adult molecules contain only 9% of this sugar. During development, N-CAM undergoes in vivo a conversion from a sialic acid-rich to a sialic acid-low form. Concomitantly with this change, an alteration of the strength of the CAM-CAM interaction from low to high is assumed. This means that sialic acid might modulate the free energy of binding between CAM molecules. During the transition of N-CAM from the sialic acid-rich to the sialic acid-low state the free energy strongly increases. The CAM molecules on the cell surface are mobile in the plane of the membrane and are characterized by a diffusion constant similar to that of the majority of the other surface receptors (Gall and Edelman 1981). These data support the modulation theory of adhesion (Edelman 1976), which had been substantiated first in sponges (Weinbaum and Burger 1973; Müller et al. 1976).

3.2 The Ca^{++}-Dependent, Trypsin-Resistant Adhesion System

By application of a different tissue-dissociation technique (treatment with trypsin in the presence of Ca^{++} by which CAMs are effectively destroyed) trypsin-resistant cell adhesion molecules (CATs) have been detected and identified (Urushihara et al. 1979; Brackenbury et al. 1981). The following two distinct CAT systems can be distinguished: the Ca^{++}-dependent mechanism in neural cells and the different Ca^{++}-dependent mechanism in liver cells. These CAT systems and the CAM systems are antigenically distinct cell adhesion mechanisms (Brackenbury et al. 1981). The CATs diplay little or no organ specificity. Referring to the neural system, inhibition of the CAT system causes an increase in the thickness of neurite fasciles, an effect not observed with antibodies against N-CAM. This result may indicate that the CAM and CAT systems play different roles during development (Brackenbury et al. 1981). They seem to act simultaneoulsy in cells during initial binding events in neural retina cells (McClay and Marchase 1982) and perhaps successively during embryogenesis (Brackenbury et al. 1981). So far, CATs have not yet been solubilized from cell membranes and purified.

4 Summary

Today it is well established that in eukaryotes the nucleus is filled with the genetic information for the synthesis of all proteins which function either as structural, functional, or controlling elements both in nucleus, cytoplasm, and in the cell membrane. On the other hand, the cell membrane of unicellular and multicellular organisms is that structure which contains the receptors for extracellular signals. These "messengers" are predominantly involved in the control of the cell and tissue-specific readout of the genetic information. Moreover, the cell membranes have acquired an additional, essential function during the phylogenetic evolution from the unicellular to the multicellular organisms; they carry the cell, tissue, and species-specific recognition system(s) and they are provided with mechanisms to recognize and destroy foreign cell entities. In animals, the cells are bound together by adhesion molecules which constitute a relatively loose meshwork of large extracellular organic molecules. The affinities of the adhesion molecules are–at least in the sponge model– controlled by ectoenzymes. The enzymatic reactions proceed on the cell surface or in the extracellular space. In addition to the adhesion molecules, cells of most multicellular eukaryotes are held together by intercellular junctions, such as gap and tight junctions and desmosomes. Hence, such cell-cell interactions govern the positioning of cells within embryos during morphogenetic movements and establish a controlled cell-cell, cell-tissue, and tissue-organism interaction. These interactions are not static, but dynamic and, therefore, require at least the following two mechanisms; firstly, the formation of stable linkages between cell surface molecules of adjacent cells and secondly, destruction of cell-cell binding by molecules and/or enzymatic processes. The latter process is also known as "sorting out".

Besides the two separate kinds of cellular interaction (via cell-cell adhesion molecules and via intercellular junctions) a third mechanism has evolved to stabilize the position of a cell in a functional organism, the cell-substrate adhesion system. This involves the interaction of the cells with the extracellular matrix, such as fibronectin, collagen, or lamin (Hay 1981).

Acknowledgment. This work was supported by a grant of the Deutsche Forschungsgemeinschaft (Mu 348/8-1).

References

Balsamo J, Lilien J (1974) Functional identification of three components which mediate tissue-type specific adhesion. Nature (Lond) 251: 533-534

Barondes SH, Rosen SD (1976) Cell surface carbohydrate-binding proteins: Role in cell recognition. In: Barondes SH (ed) Neuronal recognition. Plenum, New York, pp 331-356

Brackenbury R, Rutishauser U, Edelman GM (1981) Distinct calcium-independent and calcium-dependent adhesion systems of chicken embryo cells. Proc Natl Acad Sci USA 78: 387-391

Burger MM (1977) Mechanisms of cell-cell recognition: Some comparisons between lower organisms and vertebrates. In: Karkinen-Jäöskeläinen M, Saxén L, Weiss L (eds) Cell interactions in differentiation. Academic, New York, pp 357-376

Burkart W, Burger MM (1981) The contribution of the calcium-dependent interaction of aggregation factor molecules to recognition. A system providing additional specificity forces. J Supramol Struct Cell Biochem 16: 179-192

Curtis ASG (1967) The cell surface: Its molecular role in morphogenesis. Academic, New York, p 70

Edelman GM (1976) Surface modulation in cell recognition and cell growth. Science (Wash DC) 192: 218-226

Edelman GM (1983) Cell adhesion molecules. Science (Wash DC) 219: 450-458

Gall WE, Edelman GM (1981) Lateral diffusion of surface molecules in animal cells and tissues. Science (Wash DC) 213: 903-905

Galtsoff PS (1929) Heteroagglutination of dissociated sponge cells. Biol Bull Mar biol Lab (Woods Hole) 57: 250-260

Hausman RE, Knapp LW, Moscona AA (1976) Preparation of tissue-specific cell aggregating factors from embryonic neural tissues. J Exp Zool 198: 417-422

Hay ED (1981) Cell biology of extracellular matrix. Plenum, New York

Henkart P, Humphreys S, Humphreys T (1973) Characterization of sponge aggregation factor. A unique proteoglycan complex. Biochemistry 12: 3045-3050

Humphreys T (1970) Biochemical analysis of sponge cell aggregation. Symp Zool Soc Lond 25: 325-334

Kitamura K (1980) The changes in lectin activity during the development of embryonic chick skin. J Embryol Exp Morphol 59: 59-69

McClay DR, Marchase RB (1982) Calcium-dependent and calcium-independent adhesive mechanisms are present during initial binding events of neural retina cells. J Cell Biochem 18: 469-478

Merrell R, Gottlieb DI, Glaser L (1975) Embryonal cell surface recognition. Extraction of an active plasma membrane component. J Biol Chem 250: 5655-5659

Moscona AA (1963) Studies on cell aggregation: Demonstration of materials with selective cell-binding activity. Proc Natl Acad Sci USA 49: 742-747

Müller WEG (1982) Cell membranes in sponges. Int Rev Cytol 77: 129-181

Müller WEG, Zahn RK (1973) Purification and characterization of a species-specific aggregation factor in sponges. Exp Cell Res 80: 95-104

Müller WEG, Müller I, Zahn RK (1974) Two different aggregation principles in reaggregation process of dissociated sponge cells (*Geodia cydonium*). Experientia (Basel) 30: 399-902

Müller WEG, Müller I, Zahn RK, Kurelec B (1976) Species-specific aggregation factor in sponges VI. Aggregation receptor from the cell surface. J Cell Sci 21: 227-241

Müller WEG, Zahn RK, Kurelec B, Müller I (1978a) Species-specific aggregation factor in sponges. Nature and alteration of the cell surface charge. Wilhelm Roux' Arch Dev Biol 184: 29-40

Müller WEG, Zahn RK, Kurelec B, Uhlenbruck G, Vaith P, Müller I (1978b) Aggregation of sponge cells. Glycosyltransferases associated with the aggregation factor. Hoppe-Seyler's Z Physiol Chem 359: 529-537

Müller WEG, Kurelec B, Zahn RK, Müller I, Vaith P, Uhlenbruck G (1979a) Aggregation of sponge cells. Function of a lectin in its homologous biological system. J Biol Chem 254: 7479-7481

Müller WEG, Zahn RK, Kurelec B, Müller I, Vaith P, Uhlenbruck G (1979b) Aggregation of sponge cells. Isolation and characterization of an inhibitor of aggregation receptor from the cell surface. Eur J Biochem 97: 585-591

Müller WEG, Zahn RK, Kurelec B, Müller I, Uhlenbruck G, Vaith P (1979c) Aggregation in sponge cells. A novel mechanism of controlled intercellular adhesion, basing on the interrelation between glycosyltransferase and glycosidase. J Biol Chem 254: 1280-1287

Müller WEG, Zahn RK, Conrad J, Kurelec B, Uhlenbruck G (1982a) Aggregation of sponge cells: Stage dependent, distinct adhesion mechanisms in *Cliona celata*. Eur J Cell Biol 28: 243-250

Müller WEG, Zahn RK, Bernd A, Dawes K, Müller I, Uhlenbruck G, Dorn A (1982b) Molecular basis of cell-cell recognition: Isolation, synthesis, and function of an aggregation complex from the sponge *Geodia cydonium*. In: Frazier WA, Glaser L, Gottlieb DJ (eds) Cellular recognition Liss, New York, pp 806-816

Müller WEG, Conrad J, Zahn RK, Steffen R, Uhlenbruck G, Müller I (1984) Cell adhesion molecule in the hexactinellid *Aphrocallistes vastus*. Differentiation 26: 30-35

Müller WEG, Conrad J, Zahn RK, Gramzow M, Kurelec B, Uhlenbruch G (1985) Identification and isolation of the primary aggregation factor from the cell membrane of the sponge *Geodia cydonium*. Molec Cell Biochem 67: 55-64

Roseman S (1970) The synthesis of complex carbohydrates by multiglycosyltransferase systems and their potential function in intercellular adhesion. Chem Phys Lipids 54: 217-221

Rothbard JB, Brackenbury R, Cunningham BA, Edelman GM (1982) Differences in the carbohydrate structures of neural cell-adhesion molecules from adult and embryonic chicken brains. J Biol Chem 257: 11064-11069

Scatchard G (1949) The attraction of proteins for small molecules and ions. Ann NY Acad Sci 51: 660-672

Sperry RW (1963) Chemoaffinity in the orderly growth of nerve fiber patterns and connections. Proc Natl Acad Sci USA 50: 703-710

Urushihara H, Ozaki HS, Takeichi M (1979) Immunological detection of cell surface components related with aggregation of chinese hamster and chick embryonic cells. Dev Biol 70: 206-216

Vaith P, Müller WEG, Uhlenbruck G (1979) On the role of D-glucuronic acid in the aggregation of cells from the marine sponge *Geodia cydonium*. Dev Comp Immunol 3: 259-275

Weinbaum G, Burger MM (1973) A two-component system for surface guided reassociation of animal cells. Nature (Lond) 244: 510-512

Wilson HV (1907) On some phenomena of coalescence and regeneration in sponges. J Exp Zool 5: 245-258

Ectoacetylcholinesterase

Molecular Forms of Acetylcholinesterase

M. Vigny[1], M. Lazar[2], and J. Koenig[3]

The mechanism of cholinergic transmission requires the rapid inactivation of the neurotransmitter, acetylcholine (Dale 1914). In their early investigations, Marnay and Nachmannsohn (1937) found that in muscles acetylcholinesterase (AChE) activity is highly concentrated in the regions of the neuromuscular junctions or end plates. The function of this enzyme at this level is well established and we can assume a similar role in all cholinergic synapses. In contrast, the presence of this enzyme in other tissues or cells, such as lymphocytes or erythrocytes, remains obscure.

The electric organs of the marine ray *Torpedo* and of the fresh water teleost *Electrophorus* (electric eel) correspond to modified muscles. They possess a high density of synapses and contain an exceptionally high level of AChE. For comparison, the specific activity of the enzyme in the nervous or muscular tissues of other vertebrates is three or four orders of magnitude lower. The most detailed biochemical studies have, therefore, been done on the electric organ enzyme.

In the first section we will describe the structure of the molecular forms of acetylcholinesterase from electric organs and from higher vertebrae tissues. The structure and the regulation of the AChE activity and forms in many physiological and experimental conditions have been recently reviewed (Massoulie and Bon 1982). We will, therefore, briefly recall the major conclusions concerning the fine structure of this enzyme.

In the two following sections we will describe more extensively the location and the regulation of AChE in two systems: a murine neuroblastoma cell line and the rat skeletal muscle. We will focus our attention on the ectocellular location of the different molecular forms.

1 Molecular Forms of Acetylcholinesterase (see references in Massoulie and Bon 1982)

1.1 Electrophorus Acetylcholinesterase: Collagen-Tailed and Globular Forms

AChE may be solubilized from *Electrophorus* electric organs by homogenization in the presence of a high salt concentration. Such extracts contain three major enzyme forms

[1] Laboratoire de Neurobiologie, Ecole Normale Supérieure, 46, rue d'Ulm, 75005 Paris, France
[2] Biochimie Cellulaire, College de France, Paris Cedex 05, France
[3] Unite I.N.S.E.R.M.153.rue du Fer a Moulin, 75005 Paris, France

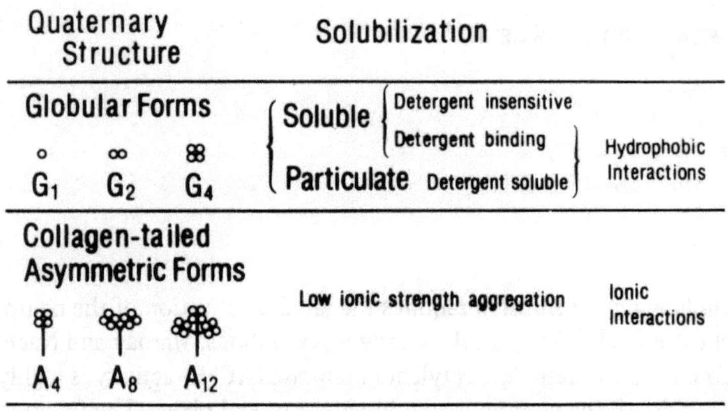

Fig. 1. The quaternary structure of the asymmetric (collagen-tailed) and globular forms of acetylcholinesterase. The proposed nomenclature distinguishes between the two types (*A* and *G*) and identifies each form by its number of catalytic subunits, hydrophobic interactions of the G forms, and ionic interactions of the A forms

as shown by sedimentation analysis in sucrose gradients. These forms are highly asymmetric molecules in which globular catalytic subunits are associated with a rod-shaped element or tail. They contain one, two, or three tetrameric groups of subunits corresponding to 4, 8, and 12 active sites, respectively. We call these forms A_{12}, A_8, and A_4 (Fig. 1). The mass of each catalytic subunit ist about 80,000.

The "tail" has a mass of approx. 100,000 daltons. It has a collagen-like structure as demonstrated by its sensitivity to collagenase and by the presence of hydroxyproline and hydroxylysine. The tail has a collagen-like triple helical structure. Each peptide of the three-stranded tail is covalently linked by disulfide bonds to a dimer which is itself associated through quaternary interactions with another disulfide-linked dimer.

Collagenase produces molecules that have lost most of the tail, so that the enzyme dissociates into individual tetramers. The collagen-tailed forms may also be dissociated by various proteases. These degradation procedures yield active tailless forms. Trypsin, for instance, converts the A_{12} form into A_8 and A_4 and releases tetramers. The tetramers represent particularly stable assembles of catalytic subunits. However, active dimers and monomers may be also obtained after proteolysis and disulfide bridge reduction.

Thus, the electric eel enzyme can exist as six distinct molecular forms, which are essentially equivalent in their catalytic activity.

We have proposed that the asymmetric collagen-tailed and the globular tailless molecules be named A and G forms, respectively, with a subscript indicating the number of catalytic subunits (see Fig. 1).

Acetylcholinesterase from *Electrophorus* electric organs reversibly aggregates in low salt conditions. The phenomenon is characteristic of the collagen-tailed forms and requires the integrity of the tail. After extensive purification the A forms no longer aggregate. Aggregation is, therefore, due to a distinct component, or aggregating agent. This agent is thermostable and is not a protein. It carries negative charges and we found that it could be replaced by various polyanions, such as chondroitin sulfate. This agent could correspond to glycosaminoglycans linked to protein cores in proteogylcan macromolecules.

In any case the collagen-tailed forms do not possess any obvious hydrophobic character. Thus, the A forms could be inserted in the tissues through ionic interactions. We discuss below the possible insertion of these forms in the basal lamina of the skeletal muscle. In the case of the electric organs they are probably located also in the same basal lamina extracellular structure which goes between the nerve endings and lines the postsynaptic membrane.

1.2 Torpedo Acetylcholinesterase

Torpedo electric organs contain not only collagen-tailed A forms, but also a large pool of acetylcholinesterase that interacts with nondenaturing detergents. The polymorphism and the location of the different molecular forms of *Torpedo* enzyme are described by C. Bon in this book.

1.3 Molecular Forms of Acetylcholinesterase in Other Vertebrates

Essentially all the acetylcholinesterase activity from nervous or muscular tissues of vertebrates can be recovered in a soluble extract after homogenization at high ionic strength (1 *M* NaCl) in the presence of a nonionic detergent.

The solubilized enzymes can be separated in a low salt aggregating fraction and a low salt soluble fraction. Each fraction contains distinct molecular forms.

The molecular parameters of the low salt soluble forms suggest that they represent monomers, dimers, and tetramers of catalytic subunits, which are generally in the range of 80,000 daltons. These forms occur both as soluble and hydrophobic molecules (Fig. 1). Schematically we can distinguish three classes of globular forms:

1. The forms which are soluble in absence of detergent and which do not interact with nondenaturing detergent, e.g., Triton X 100.
2. The forms which are soluble in absence of detergent, but which are modified by the Triton X 100.
3. The forms which are soluble only in the presence of detergent and aggregate in detergent-free buffers.

The low salt insoluble forms appear homologous to the *Electrophorus* collagen-tailed A_{12}, A_8, and A_4. This conclusion is based on several similarities. First they are asymmetric molecules and their hydrodynamic parameters are not affected by the presence of detergent. Second, these forms aggregate at low ionic strength and their aggregation is abolished by treatment with collagenase. Third, collagenase or trypsin degrade these forms in a sequential manner, producing G forms.

2 Polymorphism of Acetylcholinesterase in a Neuroblastoma Cell Line

Studies were carried out in a neuroblastoma X sympathetic ganglion cell hybrid cell line (T28). These cells produce mostly G_1 and G_4 forms, together with a very low level of G_2. They do not express the A forms.

2.1 Differential Induction of AChE Molecular Forms in T28 Cells

Cells grown in monolayer culture divide very intensively during the logarithmic phase, then at high cell density, the number of cells becomes constant (stationary phase). The beginning of the stationary phase coincides with an induction of total AChE activity, with a preferential accumulation of the G_4 form. After 2 days in the stationary phase there is a 6-fold increase in the total cellular AChE activity, reflecting 12-fold and 4-fold increases of the G_4 and G_1 forms, respectively (Table 1). It is worth noting that a similar induction of both the total activity and of the G_4 form occurs during the development of the rat brain in vivo. The activity released into the medium approximately equals the activity accumulated in the cells during the same period of time and presents a similar distribution of molecular forms.

2.2 Differential Subcellular Localization of AChE Forms in T28 Cells

We attempted to determine the ecto- or endocellular localization of the molecular forms which had been identified by sedimentation analysis. We have previously indicated that proteolysis or disulfide bridge reduction could induce the dissociation of the tetrameric form into dimers or monomers. Therefore, the relative proportion of the molecular forms obtained after extraction, even if it is highly reproducible might correspond to solubilization artifacts. In contrast, the physiological significance of ecto-/endocellular localization of the enzyme is obvious.

We used a quaternary phosphorylating inhibitor of AChE, phospholine, which is considered to be a nonpenetrating reagent because of its low lipid solubility. Treatment of intact cells with phospholine should, therefore, preferentially inhibit the AChE

Table 1. Total G_1 form, G_4 form, and ectoactivity of AChE in T28 cells treated or untreated with the nonpenetrating inhibitor phospholine in the logarithmic and stationary phases

Cell sample[a]	Total	AChE activity per dish (nmol/min at 37 °C)		
		G_4	G_1	Ectoactivity
Untreated logarithmic	18.3	4.2	14.1	4.0
Phospholine-treated + logarithmic	14.6	0.35	14.2	0.36
Untreated stationary	102.7	52.3	50.4	46.9
Phospholine-treated + stationary	47.5	6.5	41.0	4.4

[a] + in vivo at 10^{-6} *M* for 1.5 min

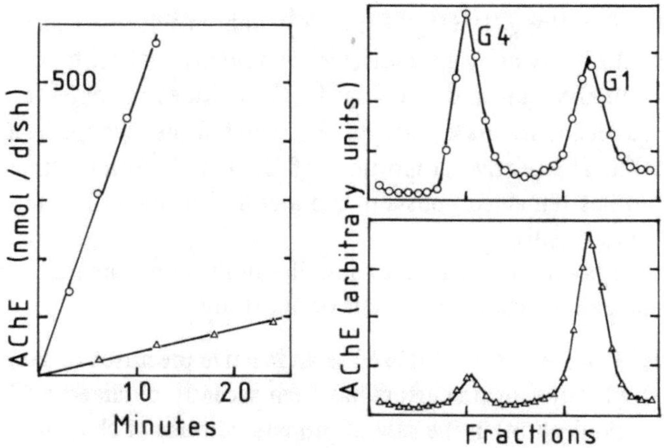

Fig. 2. Ectoactivity (*left panel:* intact cells) and sedimentation patterns (*right panel:* cell extracts) in stationary T28 cells, untreated (o) or treated (△) in vivo for 1.5 min with 10^{-6} *M* phospholine. The 5%–20% (wt/vol) sucrose gradients were in detergent (Triton X 100) containing buffer. Centrifugation was performed in a SW41 rotor at 36,000 rpm for 24 h at 4°

activity associated with the cell's surface. After 1.5 min in presence 10^{-6} *M* phospholine in the external medium the G_4 form was almost totally inhibited, while the G_1 component was essentially unaffected. We obtained similar results in the logarithmic and in the stationary phase (Table 1). Longer exposures resulted in substantial inhibition of the G_1 form, suggesting a time-dependent penetration of the inhibitor (Lazar and Vigny 1980).

The results indicated that in these cells G_1 is essentially intracellular, while the G_4 form is mostly exposed at the cell's outer surface. Thus, the intact cells should possess the ability to hydrolyze the acetylcholine present in the external medium. Acetylcholine or its analog, acetylthiocholine, does not penetrate the plasma membrane and remains confined outside the cell. Hydrolysis by intact cells is, therefore, due to the AChE activity located on the cell surface. We found that this ectoactivity roughly corresponds to the activity of the G_4 form. The inhibition of the G_4 form and of the ectoactivity in the presence of phospholine occur in a parallel manner (Table 1, Fig. 2).

2.3 Metabolism of the Molecular Forms of AChE in the T28 Hybrid Cell Line (Lazar et al. 1984)

We have used the method of heavy isotope labeling to study the metabolic turnover of AChE in the T28 cells in their differentiated state (stationary phase). This method was first introduced by Fambrough and collaborators in their studies of the acetylcholine receptor (Devreotes et al. 1977; Gardner and Fambrough 1979). This method is well adapted to the study of AChE which is a very minor component of total cellular proteins, but may be analyzed specifically with great sensitivity by its enzymatic activity.

The cells were transferred from the usual culture medium into a medium containing heavy isotope ([^{2}H], [^{13}C], [^{15}N]) amino acids. We achieved a satisfactory resolution of the heavy and light molecules by velocity sedimentation.

In the light and heavy media the cultures presented a constant and equal level of acetylcholinesterase activity, and secreted the enzyme at an identical rate for at least 50 h. The relative proportions of G_1 and G_4 forms estimated from the sedimentation profiles remained constant and identical in the cells after transfer into heavy experimental media.

It is not our purpose to describe all the experimental data. We would like only summarize the major conclusions of our study.

1. AChE is first synthesized as an inactive precursor of short half-life. Evidence for the existence of precursors has been recently obtained by R. Rotundo (personal communication) in the case of primary cultures of chick myotubes.
2. The monomeric G_1 form is renewed within a few hours, much faster than the tetrameric G_4 form.
3. Both G_1 and G_4 comprise several metabolic pools. For example, most of the synthesized enzyme (70%–80% according to our model) enters the fast G_1 pool and is rapidly degraded within the cells. Only a small pool of G_1 appears to be the precursor of the tetrameric form.

In conclusion, it is clear that in T28 cells the G_4 form is located on the surface of the cells with its catalytic sites directed towards the external medium. In contrast, the G_1 is mostly intracellular. Furthermore, the turnover rates of these forms are quite distinct and the metabolism of AChE in these cells appears very complex, thus, raising challenging problems.

3 Molecular Forms of Acetylcholinesterase in Rat Skeletal Muscle

3.1 Localization of A Forms in Rat Muscle

Neuromuscular end plates are characterized by a very high concentration of acetylcholinesterase. In adult rat muscles the A forms are exclusively localized at the end plate (Fig. 3) (Hall 1973; Vigny et al. 1976; Bon et al. 1978), but this is not the case in neonatal rats where they also exist in end plate-free segments (Sketelj and Brzin 1980; Koenig et al. 1980). Furthermore, in a recent paper, Fernandez and Seiter (1984) showed that during postnatal maturation of the muscle the A forms become concentrated and externalized at the end plates, but remain intracellular in noninnervated segments.

We have previously indicated that the A forms do not require detergent for quantitative solubilization, but are soluble only at high ionic strength. The aggregation of these forms at low ionic strength probably reflects the interaction of the tail with anionic components (Bon et al. 1978).

It has long been known that collagenase solubilizes AChE activity from the end plates of rat (Hall and Kelly 1971) and frog muscle (Betz and Sakmann 1973). This

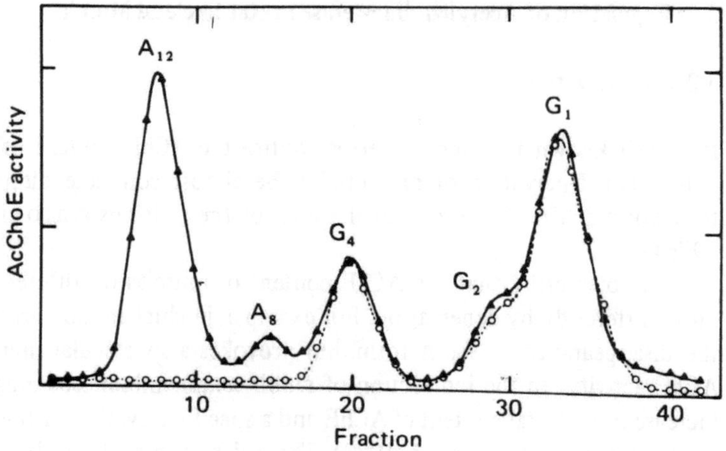

Fig. 3. AChE forms in neural and nerve-free regions of rat sternocleidomastoidian muscle. The 5%–20% (wt/vol) sucrose gradients were in 1 *M* NaCl/% Triton X 100 buffer. End plate-containing sections: o – o; end plate-free section: ▲ – ▲

observation suggested that the enzyme may be anchored within the extracellular basal lamina. A direct demonstration was given by McMahan and collaborators who observed the persistence of acetylcholinesterase activity in the end plate basal lamina of frog muscle after degeneration of both pre- and postsynaptic cellular elements (Marshall et al. 1977; McMahan et al. 1978). The basal lamina-linked activity probably corresponds to the asymmetric forms of acetylcholinesterase. The collagenous nature of the tail and their aggregation properties would be consistent with the solubilization of AChE activity by the collagenase treatment and with an attachment of these forms to acidic components of the basal lamina.

Recent data obtained by Inestrosa et al. in the case of myotubes of a mouse muscle cell line (C2) strengthens this hypothesis (Inestrosa et al. 1982). These cells express the A_{12} form of AChE. The active sites of the A_{12} form are accessible to the external medium and collagenase treatment removes specifically this form. Histochemical staining shows focal patches of AChE activity on the cell surface. They disappear after collagenase treatment, indicating that the patches consist of the A_{12} form of the enzyme.

We recently showed that in nerve-muscle cell cultures histochemically detectable patches of enzyme appear on the surface of the myotubes in the presence of the nerve. As in the C2 cell line these patches of enzyme are destroyed by collagenase treatment. Thus, in skeletal muscle and in muscle cells in culture the A forms are probably linked to the basal lamina. Because this enzyme is not directly anchored in the cellular membrane it probably cannot be regarded as an ectoenzyme, but rather as an immobilized, precisely located exoenzyme.

3.2 Regulation of Acetylcholinesterase in Rat Skeletal Muscle

3.2.1 Denervation

It is well known that motor nerves control the AChE content of the muscles (Guth 1968). On denervation of rat muscles the almost complete disappearance of the A forms occurs after 2 weeks, but the level of the G forms is also affected (Vigny et al. 1976).

It is noteworthy that the AChE content of muscles of different species is affected quite differently by denervation. For example, in chicken muscles denervation leads to the disappearance of the A form, but provokes a spectacular increase of the specific AChE activity. In the red section of rabbit semimembranosus muscle we observed an increase in the total content of AChE and a specific elevation of the relative proportion of the A forms (Bacou et al. 1982). These data demonstrate that the biosynthesis of the A forms in this case does not depend upon innervation.

After reinnervation of rat muscles the A forms reappear at the old end plates if the nerve has been implanted at these sites. If reinnervation occurs at an ectopic site the A forms appear at this level (Vigny et al. 1976). Moreover, in the ectopic reinnervation experiments Weinberg and Hall (1979) showed that the A forms are also restored at the original end plates. Therefore, the sites of neuromuscular contacts maintain the capacity of biosynthesis or accumulation of these forms. Specializations of the basal lamina are probably implicated in this process (for a review see Sanes 1983).

3.2.2 Expression of AChE Activity in Rat Myotubes

We previously indicated that in C2 myotubes the A forms are almost only distributed on the cell outer surface. In primary cultures of rat myotubes the A forms are distributed both inside and on the surface of the myotube (Brockman et al. 1982). In the presence of neurons synaptic contacts are made and patches or accumulations of AChE appear at these sites of contact. In our conditions, however, the total content of enzyme, the relative distribution of the different forms, and the cellular localization of each form are not modified in the nerve-muscle cultures as compared to myotubes cultured alone (unpublished result). The only effect of the nerve cell is to induce the appearance of patches which probably correspond to the A forms. Rat myotubes contract spontaneously. When action potentials and contractions are blocked by tetrodotoxin (TTX), a specific sodium channel blocker, the A forms disappear. This effect is completely reversible (Rieger et al. 1980). This result led us to analyze the effect of the antagonist drug, veratridine. Veratridine maintains the sodium channel in the open state and, therefore, depolarizes the cells, blocks the action potentials, and paralyzes the myotubes. This drug provoked a dramatic increase both of the total AChE activity and the relative proportion of the A forms. Depolarization induced by high K^+ concentration in the medium did not alter by itself the activity and the sedimentation pattern of AChE. Modifications of Ca^{++} permeability and/or the level of intracellular cGMP are probably implicated in the effect of veratridine (De la Porte et al., 1984). However, the mechanism of this effect will require further experimental analysis to be fully understood.

References

Bon S, Cartaud J, Massoulie J (1978) The dependence of acetylcholinesterase aggregation at low ionic strengh upon a polyanionic component. Eur J Biochem 85:1–14

Betz W, Sakmann B (1973) Effect of proteolytic enzymes on function and structure of frog neuromuscular junctions. J Physiol (Lond) 230:673–688

Brockman SK, Przybylski RJ, Younkin SG (1982) Cellular localization of the molecular forms of acetylcholinesterase in cultured embryonic rat myotubes. J Neurosci 2:1775–1785

Dale HH (1914) The action of certain esters of choline and their relation of muscarine. J Pharmacol Exp Ther 6:147–190

De la Porte S, Koenig J, Massoulie J, Vigny M (1984) Reversible action of veratridin on the acetylcholinesterase content in culture of rat muscle cells. Dev Biol 106:450–456

Fernandez HL, Seiter TC (1984) Subcellular distribution of acetylcholinesterase asymmetric forms during postnatal development of mammalian skeletal muscle. FEBS Lett 170:147–151

Guth L (1968) "Trophic" influences of nerve on muscle. Physiol Rev 48:655–687

Hall ZW (1973) Multiple forms of acetylcholinesterase and their distribution in endplate and non endplate regions of rat diaphragm muscle. J Neurol 4:343–361

Hall ZW, Kelly RB (1971) Enzymatic detachment of endplate acetylcholinesterase from muscle. Nature New Biol 232:262

Inestrosa NC, Silberstein L, Hall ZW (1982) Association of the synaptic form of acetylcholinesterase with extracellular matrix in cultured mouse muscle cells. Cell 29:71–79

Koenig J, Bournaud R, Rieger F (1980) Acetylcholinesterase and synaptic formation in striated muscle. In: Taxi J (ed) Ontogeneis and functional mechanism of peripheral synapses. Elsevier-North Holland, Amsterdam, pp 313–325

Lazar M, Vigny M (1980) Modulation of the distribution of acetylcholinesterase molecular forms in a murine neuroblastoma × sympathetic ganglion cell hybrid cell line. J Neurochem 35:1067–1079

Lazar M, Salmeron E, Vigny M, Massoulie J (1984) Heavy isotope labeling study of the metabolism of monomeric and tetrameric acetylcholinesterase in the murine neuronal-like T28 hybrid cell line. J Biol Chem 259:3703–3713

McMahan VJ, Sanes JR, Marshall L (1978) Cholinesterase is associated with the basal lamina at the neuromuscular junction. Nature (Lond) 271:172–174

Marnay A, Nachmansohn D (1937) Cholinesterase dans le muscle strie. C R Soc Biol 124:942–944

Marshal LM, Sanes JR, McMahan VJ (1977) Reinnervation of original synaptic sites on muscle fiber basement membrane after disruption of the muscle cells. Proc Natl Acad Sci USA 74:3073–3077

Massoulie J, Bon S (1982) Molecular forms of acetylcholinesterase. Annu Rev Neurosci 5:57–106

Rieger F, Koenig J, Vigny M (1980) Spontaneous contractile activity and the presence of the 16S form of acetylcholinesterase in rat muscle cells in culture. Reversible suppressive action of tetrodotoxin. Dev Biol 76:358–365

Sanes JR (1983) Roles of extracellular matrix in neural development. Annu Rev Physiol 45:581–600

Sketlej J, Brzin M (1980) 16S acetylcholinesterase in endplate free gerions of developing rat diaphragm. Neurochem Res 5:655–660

Vigny M, Koenig J, Rieger F (1976) The motor endplate specific form of acetylcholinesterase: Appearance during ambryogenesis and reinnervation of rat muscle. J Neurochem 27:1347–1353

Weinberg CG, Hall ZW (1979) Junctional form of acetylcholinesterase restored at nerve free endplate. Dev Biol 68:631–635

Acknowledgment. We are very grateful to Dr. J. Massoulie for his criticism of the manuscript.

An Ectocellular Form of Acetylcholinesterase on the Plasma Membrane of Nerve Terminals in *Torpedo* Electric Organ

C. Bon [1]

1 Introduction

In the present chapter, we will focus our attention on the presence of an amphiphilic, membrane-bound form of acetylcholinesterase which is localized at the external surface of nerve terminal plasma membranes in *Torpedo* electric organs. The experimental data have been described in detail (Li and Bon 1983) and similar results have been obtained by Morel and Dreyfus (1982). As described elsewhere in this volume by Vigny et al., acetylcholinesterase exists in multiple molecular forms. Therefore, we will first describe the polymorphism of acetylcholinesterase in *Torpedo* electric organs.

2 Polymorphism of Acetylcholinesterase in Torpedo Electric Organs

The various molecular forms of acetylcholinesterase may be classified as asymmetric and globular, depending upon the presence or absence of a collagen-like "tail" (review by Massoulié and Bon 1982). The globular forms may be operationally subdivided into amphiphilic forms that bind micelles of nondenaturing detergents, such as deoxycholate or Triton X 100, and nonamphiphilic forms that do not interact with such detergents. The nonamphiphilic forms are soluble and occur, for example, in the blood plasma and in the cerebrospinal fluid, whereas the amphiphilic forms may be either soluble or particulate (Massoulié 1984; Massoulié et al. 1984).

The main features of the model proposed for the quaternary structure of acetylcholinesterase forms from the electric organs of *Electrophorus* (Bon et al. 1976; Rosenberry and Richardson 1977; Anglister and Silman 1978) appear to be valid for the enzymes of other vertebrates, particularly mammals and birds (Bon et al. 1979; Allemand et al. 1981; see review by Massoulié and Bon 1982). This is also true for acetylcholinesterase from *Torpedo* electric organs, although biochemical studies have revealed that the asymmetric forms of *Torpedo* acetylcholinesterase are particularly complex, since in addition to the collagen-like tail and catalytic subunits, they contain noncatalytic globular subunits, which appear to randomly replace some of the tail-linked catalytic subunits (Lee and Taylor 1982; Lee et al. 1982).

[1] Unité des Venins, Unité associée Pasteur/I.N.S.E.R.M. 285, Institut Pasteur, 75015 Paris, France

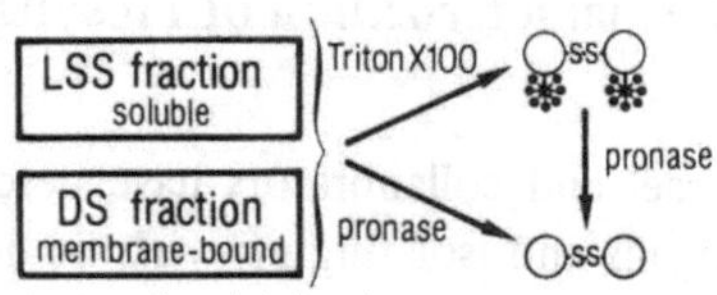

Fig. 1. Amphiphilic dimeric form of *Torpedo* acetylcholinesterase: effect of nondenaturing detergents and proteolytic cleavage. This figure has been modified from Massoulié and Bon (1982), assuming that each catalytic subunit binds one detergent micelle, as in the case of the membrane-bound dimeric form of acetylcholinesterase from human erythrocyte membranes (Rosenberry et al. 1984)

In contrast with *Electrophorus*, the electric organs of adult *Torpedo* contain a large proportion (about 60%) of globular, amphiphilic acetylcholinesterase (Bon and Massoulié 1980; Viratelle and Bernhard 1980). This enzyme corresponds to dimers (G_2) which exist as a soluble fraction (low salt soluble, LSS fraction) and a detergent-soluble (DS) fraction (Bon and Massoulié 1980). In the LSS fraction the dimers are noncovalently associated with noncatalytic proteins, from which they dissociate in the presence of Triton X 100, forming enzyme-detergent complexes. These complexes present the same hydrodynamic properties (sedimentation coefficient 6 S, Stokes radius 8 mm) as those obtained by detergent solubilization of the membrane-bound enzyme (DS fraction) (Fig.1).

The acetylcholinesterase catalytic subunits of the three enzyme fractions of *Torpedo* electric organs (asymmetric forms, LSS G_2 fraction, DS G_2 fraction) differ in their apparent polypeptide mass, as estimated by polyacrylamide gel electrophoresis in the presence of sodium dodecyl sulfate (SDS) (Witzemann and Boustead 1982; Massoulié et al. 1984). A chromatographic analysis of the peptides obtained after proteolytic cleavage showed a considerable similarity, but also some differences between the subunits of the asymmetric and LSS G_2 enzymes (Doctor et al. 1983). Monoclonal antibodies obtained by these authors against *Torpedo* acetylcholinesterase reacted, in general, equally well with all forms, but one clone was found to show a clear preference for the amphiphilic LSS enzyme, suggesting that it recognizes an antigenic site which is not present in the collagen-tailed molecules.

The amphiphilic G_2 acetylcholinesterase forms of both LSS and DS fractions may be converted into nonamphiphilic enzyme by proteolytic digestion with pronase or proteinase K (Massoulié and Bon 1982), as schematically illustrated in Fig. 1. The lytic enzyme sediments at 7.2 S, either in the presence or absence of Triton X 100, and the mass of the catalytic subunits appears only slightly inferior to that of the native enzymes. *Torpedo* amphiphilic acetylcholinesterase, thus, appears quite similar to the membrane-bound enzyme from human erythrocytes. This enzyme is also a dimer which may be solubilized in the form of a complex with Triton X 100 (Brodbeck and Ott 1984; Rosenberry et al. 1984) containing one detergent micelle bound to each subunit, and is converted by papain digestion into a soluble form, with cleavage of a 1,000 dalton C-terminal fragment (Rosenberry et al. 1984). However, Silman and his colleagues have recently found that both *Torpedo* and erythrocyte membrane-bound acetylcholinesterase dimers may be solubilized by phospholipase C, and suggested that these enzymes are anchored in the membranes not by a hydrophobic peptide, but rather by a covalently attached phospholipid (Futerman et al. 1984).

3 Characterization of Presynaptic Acetylcholinesterase

Israel and collaborators have described a method for fractionating *Torpedo* electric organs and isolating presynaptic nerve endings (Israel et al. 1976; Morel et al. 1977). In contrast with those obtained from mammalian central nervous system, these „synaptosomes" are detached from postsynaptic elements. Morel et al. (1977) have shown that the *Torpedo* synaptosomal fraction is, of course, enriched in acetylcholine and in choline acetyltransferase, but also contains a significant acetylcholinesterase activity. We have used a modification of this method to prepare a larger amount of presynaptic nerve terminals and we have studied their acetylcholinesterase content (Li and Bon 1983). Our preparation of cholinergic nerve terminals consisted of closed membranes, 1-5 μm in diameter, containing synaptic vesicles. It was able to synthesize acetylcholine and to release it when exposed to an extract of *Glycera convoluta* venom glands. This preparation contained a low concentration of acetylcholine receptors, indicating that it was not contaminated to a large degree by postsynaptic membranes. In contrast, the concentration of acetylcholine, related to protein content, was increased over ninefold compared to crude electric organ extracts; the specific activity of choline acetyltransferase was increased approx. sixfold and acetylcholinesterase specific activity by about the same factor.

We preincubated nerve terminals for 24 h in *Torpedo* physiological solution and in the same medium containing 0.3 *M* sucrose, so that their buoyant density was increased. We then equilibrated them in continuous sucrose gradients either from the top or from the bottom, and verified that we obtained essentially the same profiles in both cases. In these experiments, the distribution of acetylcholinesterase activity coincided with that of choline acetyltransferase activity, indicating that both enzymes were associated with the presynaptic structure.

4 Ectocellular Localization of Presynaptic Acetylcholinesterase

We compared the solubilization properties of acetylcholinesterase, choline acetyltransferase, and lactate dehydrogenase from the purified synaptosomal preparation. We used a three-step solubilization procedure, extracting the preparation first in a low salt, detergent-free buffer (low salt soluble, LSS fraction), then in a high salt, detergent-free buffer (high salt soluble, HSS fraction), and finally in the presence of 1% Triton X 100 (detergent-soluble, DS fraction). Table 1 shows that the major part of the cytoplasmic enzymes, choline acetyltransferase and lactate dehydrogenase, could be solubilized in the first step, indicating that the nerve terminals were efficiently lysed in our experimental conditions. Their solubilization was essentially completed in the second step, so that only residual amounts remained in the detergent-soluble fraction. In contrast, no acetylcholinesterase could be solubilized in the LSS or HSS media. Acetylcholinesterase was recovered quantitatively in the DS extraction medium, indicating that it is a membrane-bound enzyme. Therefore, our preparation does not contain LSS acetylcholinesterase which is probably solubilized during the first homogenization step.

Table 1. Solubilization of acetylcholinesterase from intact nerve endings by pronase treatment[a]

	AChE (mmol h^{-1})		LDH (μmol h^{-1})	
	Pellet	Supernatant	Pellet	Supernatant
Control	16.0	0.5	77	3
Pronase treated	0.3	17.0	67	6

[a] Nerve endings (1.6 mg of protein) were resuspended in 2 ml of *Torpedo* physiological solution supplemented with Tris–HCl, pH 8.0, at a final concentration of 50 μM. They were incubated for 1 h at 25 °C without (control) or with 10 μg per ml of pronase and centrifuged for 20 min at 17,000 g, at 4 °C

The absence of high-salt soluble enzyme is consistent with the absence of synaptic cleft components.

The detergent-solubilized acetylcholinesterase sedimented as a homogenous "6 S" peak in the presence of Triton X 100, and formed polydisperse aggregates in its absence. By polyacrylamide gel electrophoresis after specifically labeling the active site with the phosphorylating inhibitor [^{3}H]-diisopropylfluorophosphonate, we characterized this enzyme as a disulfide-linked, reducible dimer of 70,000 daltons catalytic subunits. It, therefore, corresponds to a G_2 amphiphilic form of acetylcholinesterase.

The purified preparation of nerve endings consists mainly of closed structures which retain their cytoplasmic components, particularly the synaptic vesicles. We showed in two different ways that presynaptic acetylcholinesterase is located on the external side of the plasma membrane. Firstly, we examined the hydrolysis of acetylthiocholine which does not penetrate the membranes, by intact and detergent-lysed synaptosomes. We found that Triton X 100 did not notably increase the acetylcholinesterase activity, suggesting that most of the active sites are accessible to the external medium. The

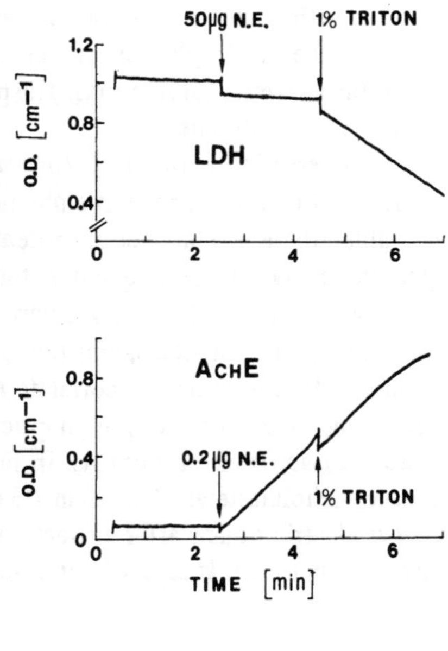

Fig. 2. External localization of acetylcholinesterase and internal localization of lactate dehydrogenase in *Torpedo* nerve endings. The external acetylcholinesterase activity of nerve endings (*N.E.*) was measured in *Torpedo* physiological solution supplemented with 50 m*M* Tris–CHl, pH 7.5, 0.1 m*M* DTNB and 0.7 m*M* acetylthiocholine, by recording the absorbance changes at 412 nm. The external lactate dehydrogenase activity of nerve endings was similarly determined in physiological solution supplemented with 0.25 m*M* $NADH_2$ and 1 m*M* sodium pyruvate, by following the absorbance changes at 340 nm. The total enzymatic activities were measured in the same experimental conditions, after addition of 1% Triton X 100. This figure is from Li and Bon (1983)

significance of this result was established by the fact that lactate dehydrogenase activity, assayed in the presence of $NADH_2$ and sodium pyruvate, was only revealed after addition of the detergent (Fig. 2). Secondly, we confirmed this result by showing that pronase was able to solubilize acetylcholinesterase quantitatively from intact synaptosomes, under conditions which did not lyse the membranes, since lactate dehydrogenase remained particulate (Table 1). The solubilized enzyme was detergent insensitive, sedimenting at 7.2 S either in the presence or absence of Triton X 100. This shows that the bulk of the protein is external to the plasma membrane, and may be detached by cleavage of a small peptide, which is perhaps covalently linked to a phospholipid (Futerman et al. 1984). These experiments prove that the amphiphilic G_2 form of acetylcholinesterase that is associated with nerve endings in *Torpedo* electric organs is an ectoenzyme.

5 Significance of Presynaptic, Ectocellular Acetylcholinesterase in Torpedo Electric Organs

We have clearly shown that in *Torpedo* electric organs, the plasma membrane of nerve terminals contains an externally accessible amphiphilic G_2 form of acetylcholinesterase. Similar results have been obtained by Morel and Dreyfus (1982).

The specific activity of this molecular form of the enzyme in the presynaptic synaptosomal preparation is increased ninefold when compared to the "6 S" acetylcholinesterase fraction (LSS and DS amphiphilic dimers), and about 12-fold when compared to the DS fraction alone, which represents approx. 45% of the total activity. The fact that 6 S acetylcholinesterase is more enriched than choline acetyltransferase over the corresponding pool of enzyme in the total electric organs, suggests, however, that this DS enzyme fraction must be largely, and perhaps entirely, presynaptic. Because we find only a very small proportion of soluble acetylcholinesterase in our preparation, it is likely that the LSS G_2 form is, on the contrary, contained principally in the cytoplasm of the electroplaques. If this is the case there is no metabolic relationship between the two enzyme fractions. The physiological significance of the intracellular LSS fraction is very obscure.

The innervation patterns of *Torpedo* and *Electrophorus* electric organs differ considerably; the postsynaptic membrane of *Torpedo* electroplaques is not electrically excitable; there are no action potentials, and the depolarization is entirely due to synaptic excitation. Consequently, the postsynaptic membrane is extensively covered by nerve terminals. In *Electrophorus* on the contrary, the postsynaptic membrane conducts action potentials, and the electroplaques are focally innervated by few nerve terminals. It is tempting to correlate this difference with the fact that a large proportion of membrane-bound, presynaptic acetylcholinesterase exists in *Torpedo*, but not in *Electrophorus* electric organs. In fact, a very small proportion of detergent-soluble G_2 acetylcholinesterase (less than 1% of the total activity) may be detected in *Electrophorus* electric organs after extraction of the bulk of the high salt soluble asymmetric forms (Bon 1982). It is, therefore, possible that an amphiphilic form of acetylcholinesterase is also present at the presynaptic level in *Electrophorus*, in a proportion consistent with the abundance of electromotor nerve terminals in the tissue.

The evolution of acetylcholinesterase forms has been studied in the electric organs of *Torpedo* embryos (Bon 1982; Witzemann and Boustead 1982) and correlated with the different stages of organogenesis (Fox and Richardson 1978, 1979). Two phases can be distinguished in the formation of the electric organs. The electrocytes first assume the morphology of normal spindle-shaped myotubes, forming vertically-oriented bundles (myogenic phase). They later become horizontally flattened into electroplaques, and are organized into vertical stacks or columns; the growing cones of the nerves then establish contacts with their ventral membrane and the electroplaques start to produce electric discharges (electrogenic phase). At birth, the electric organ is functional and morphologically similar to the adult. All forms of acetylcholinesterase, globular and asymmetric, already occur, although at a low level of total activity, at the early myogenic phase. The G_2 and A_{12} forms, which constitute the major components of acetylcholinesterase in the mature organ, accumulate rapidly, but not simultaneously. The accumulation of the A_{12} form occurs first, and is accompanied by a transient increase of A_8, perhaps representing a biosynthetic intermediate. This accumulation coincides with the formation of a basal lamina on the ventral side of the electrolytes, and precedes the formation of contacts with the nerve processes. The rapid accumulation of the 6 S G_2 form occurs later, and appears to coincide with the proliferation of nerve terminals, and with the establishment of morphologically recognizable synaptic contacts. Whereas the specific activity of the A_{12} form, per gram of tissue, reaches its adult value before birth, the specific activity of the G_2 form continues to increase after birth (Bon 1982; Witzeman and Boustead 1982).

In cholinergic synapses, acetylcholinesterase is usually considered to be associated with the postsynaptic plasma membrane, and is also known to exist in the basal lamina (McMahan et al. 1978), probably as collagen-tailed forms. In the case of rat skeletal muscles, acetylcholinesterase, particularly the A_{12} asymmetric form, is concentrated at the neuromuscular junctions, and decreases rapidly when the nerve is cut or paralyzed (Hall 1973; Vigny et al. 1976). In that case, most of the junctional enzyme appears to be provided by the muscle itself, since it returns to the normal level at denervated end plates if the muscle is reinnervated at an ectopic site (Weinberg and Hall 1979).

In the central nervous system, the presence of cholinesterase activity, which is easily detectable by cytochemical methods, has often been considered as a characteristic feature of cholinergic pathways (Silver 1974). This postulates a coincidence between cholinergic transmission and the presence of cholinesterase. In the case of *Torpedo* electric organ, we have presented evidence for the occurrence of cholinesterase activity in presynaptic plasma membranes of cholinergic nerve terminals. Such a situation may be the rule in cholinergic synapses of central nervous system. It is perhaps relevant that collagen-tailed acetylcholinesterase occurs only in trace amounts in the central nervous system of higher vertebrates, birds (Gomez-Barriocanal et al. 1981), and mammals (Grassi et al. 1982).

What is the physiological significance of ectocellular acetylcholinesterase at the presynaptic membrane? It is likely that because of the rapid diffusion of acetylcholine in the narrow synaptic cleft, this enzyme participates exactly like postsynaptic acetylcholinesterase in the hydrolysis of the neurotransmitter. It may be that at cholinergic synapses where asymmetric collagen-tailed forms are scarce or absent, transmission functions better if acetylcholinesterase is supplied at least in part by the nerve terminal, as a strategically located ectoenzyme.

6 Summary

In *Torpedo* electric organ about 40% of the acetylcholinesterase activity corresponds to collagen-tailed acetylcholinesterase forms which are thought to be anchored in the extracellular basal lamina through ionic interactions, and about 60% to hydrophobic dimers which associate with detergent micelles (Triton X-100). We have shown that a preparation of presynaptic nerve terminals, which does not contain postsynaptic elements, is considerably enriched in hydrophobic acetylcholinesterase. A comparison of the factor of enrichment of this form of acetylcholinesterase in such fractions with that of the specifically presynaptic enzyme, choline acetyltransferase, suggests that it is largely presynaptic. We have characterized the dimeric hydrophobic form of acetylcholinesterase as a membrane-bound, externally active enzyme in the nerve terminal preparation. The high level of this presynaptic ectoenzyme suggests that it plays an important role in cholinergic transmission in *Torpedo* electric organs. Presynaptic acetylcholinesterase may be physiologically active in other cholinergic synapses as well as suggested, for example, by the frequent abundance of acetylcholinesterase in cholinergic pathways in the central nervous system of vertebrates.

Acknowledgements. I gratefully acknowledge Drs. J. Massoulié and S. Bon for helpful discussions and criticism of the manuscript.

References

Allemand P, Bon S, Massoulié J, Vigny M (1981) The quaternary structure of chicken acetylcholinesterase and butyrylcholinesterase; effect of collagenase and trypsin. J Neurochem 38:860–867

Anglister L, Silman I (1978) Molecular structure of elongated forms of electric eel acetylcholinesterase. J Mol Biol 125:293–311

Bon S (1982) Molecular forms of acetylcholinesterase in developing *Torpedo* embryos. Neurochem Int 4:577–585

Bon S, Massoulié J (1980) Collagen-tailed and hydrophobic components in *Torpedo marmorata* electric organ. Proc Natl Acad Sci USA 77:4464–4468

Bon S, Huet M, Lemonnier M, Rieger F, Massoulié J (1976) Molecular forms of *Electrophorus* acetylcholinesterase; molecular weight and composition. Eur J Biochem 68:523–530

Bon S, Vigny M, Massoulié J (1979) Asymmetric and globular forms of acetylcholinesterase in mammals and birds. Proc Natl Acad Sci USA 76:2546–2550

Doctor BP, Camp S, Gentry MK, Taylor S, Taylor P (1983) Antigenic and structural differences in the catalytic subunits of the molecular forms of acetylcholinesterase. Proc Natl Acad Sci USA 80:5767–5771

Fox GQ, Richardson GP (1978) The developmental morphology of *Torpedo marmorata*: electric organ-myogenic phase. J Comp Neurol 179:677–697

Fox GQ, Richardson GP (1979) The developmental morphology of *Torpedo marmorata*: electric organ-electrogenic phase. J Comp Neurol 185:293–315

Futerman AH, Fiorini RM, Roth E, Michaelson DM, Low MG, Silman I (1984) Solubilization of membrane bound acetylcholinesterase by a phophatidylinositol-specific phospholipase C: enzymatic and physicochemical studies. In: Brzin M, Barnard EA, Sket D (eds) Cholinesterases. de Gruyter, Berlin, pp 99–113

Gomez-Barriocanal J, Barat A, Escudero E, Rodriguez-Borrajo C, Ramirez G (1981) Solubilization of collagen-tailed acetylcholinesterase from chick retina: effect of different extraction procedures. J Neurochem 37:1239–1249

Grassi J, Vigny M, Massoulié J (1982) Molecular forms of acetylcholinesterase in the bovine caudate nucleus and superior cervical ganglion: solubility properties and hydrophobic character. J Neurochem 38:457–469

Hall ZW (1973) Multiple forms of acetylcholinesterase and their distribution in endplate and non-endplate regions of rat diaphragm muscle. J Neurobiol 4:343–361

Israel M, Manaranche R, Mastour-Franchon P, Morel N (1976) Isolation of pure cholinergic nerve-endings from electric organ of *Torpedo marmorata*. Biochem J 160:113–115

Lee SL, Taylor P (1982) Structural characterization of the asymmetric (17 + 13) S forms of acetylcholinesterase from *Torpedo* component peptides obtained by selective proteolysis and disulfide bonds reduction. J Biol Chem 20:12292–12301

Lee SL, Heinemann S, Taylor P (1982) Structural characterization of the asymmetric (17 + 13) S forms of acetylcholinesterase from *Torpedo*; analysis of subunit composition. J Biol Chem 20: 12283–12291

Li ZY, Bon C (1983) Presence of a membrane-bound acetylcholinesterase form in a preparation of nerve-endings from *Torpedo marmorata* electric organ. J Neurochem 40:338–349

Massoulié J (1984) Polymorphism of cholinesterases. In: Serratrice et al. (eds) Neuromuscular diseases. Raven, New York, pp 425–429

Massoulié J, Bon S (1982) The molecular forms of cholinesterase and acetylcholinesterase in vertebrates. Annu Rev Neurosci 5:57–106

Massoulié J, Bon S, Lazar M, Grassi J, Marsh D, Meflah K, Toutant JP, Vallette F, Vigny M (1984) The polymorphism of cholinesterases: classification of molecular forms; interactions and solubilization characteristics; metabolic relationships and regulations. In: Brzin M, Bernard EA, Sket D (eds) Cholinesterases. de Gruyter, Berlin, pp 73–97

McMahan UJ, Sanes JR, Marshall LM (1978) Cholinesterase is associated with basal lamina at the neuromuscular junction. Nature (Lond) 271:172–174

Morel N, Dreyfus P (1981) Association of acetylcholinesterase with the external surface of the presynaptic plasma membrane in *Torpedo* electric organ. Neurochem Int 4:283–288

Morel N, Israel M, Manaranche R, Mastour-Frachon P (1977) Isolation of pure cholinergic nerve-endings from *Torpedo* electric organ. J Cell Biol 75:43–55

Rosenberry TL, Richardson JM (1977) Structure of 18 S and 13 S acetylcholinesterase. Identification of collagen-like subunits that are linked by disulfide bond to catalytic subunits. Biochemistry 16:3550–3558

Rosenberry TL, Scoggin DM, Dutta-Choudhury TA, Haas R (1984) Human erythrocyte acetylcholinesterase is an amphipathic form. In: Brzin M, Barnard EA, Sket D (eds) Cholinesterases. de Gruyter, Berlin, pp 155–172

Silver A (1974) The biology of cholinesterases. North-Holland, Amsterdam

Vigny M, Di Giamberardino L, Couraud JY, Rieger F, Koenig J (1976) Molecular forms of chicken acetylcholinesterase: effect of denervation. FEBS Lett 69:277–280

Viratelle OM, Bernhard SA (1980) Major component of acetylcholinesterases in *Torpedo* electroplax is not basal lamina associated. Biochemistry 19:4997–5007

Weinberg GG, Hall ZW (1979) Junctional forms of acetylcholinesterase restored at nerve-free endplate. Dev Biol 68:631–635

Witzemann V, Boustead C (1982) Change in acetylcholinesterase molecular forms during embryonic development of *Torpedo marmorata*. J Neurochem 39:747–755

Release of Acetylcholinesterase from Nigro-Striatal Neurons

SUSAN GREENFIELD[1]

1 Introduction

1.1 Acetylcholinesterase as a Secretory Neuroprotein

It is well established that acetylcholinesterase (AChE) is an ectoenzyme that hydrolyses acetylcholine (ACh) (see Bon, this volume). However AChE is polymorphic (see Vigny, this volume) and in both the periphery (Chubb and Smith 1975a) and in the brain (Chubb et al. 1976), some forms of the enzyme are soluble. Why should this be so? Since soluble AChE is intracellular (Somogyi et al. 1975), it could presumably have a function inside the cell, i.e. function as an endoenzyme. It would thus be playing a role other than its familiar one in cholinergic transmission (cf. Millar and Chubb 1984). Alternatively, soluble AChE could be released. Of the five soluble isoenzymes of AChE, one has been shown to be secreted from the adrenal gland in a K^+-evoked, Ca^{++}-dependent manner (Chubb and Smith 1975b). This same form of the enzyme is detectable in cerebrospinal fluid (CSF) and appears to be derived from the brain as a physiological process (Chubb et al. 1976). Hence, AChE not only exists as an ectoenzyme but perhaps also as an endoenzyme and secretory neuroprotein. Over the last few years we have been learning more about the release of AChE. Levels of AChE in CSF are markedly reduced following destruction of a particular region of the brain, the substantia nigra (Greenfield and Smith 1979). Hence, it is possible that AChE is normally released from nigral neurons. In this chapter I shall review the evidence that within the substantia nigra AChE is released specifically from the cells that project to the striatum (nigro-striatal neurons) and explore the possibility that the phenomenon is unrelated to cholinergic transmission.

1.2 Nigro-Striatal Neurons: Anatomy

Nigro-striatal neurons are best known as being the cells that degenerate in Parkinson's disease. Their somata lie in an area in the ventral mesencephalon, the substantia nigra, from which their axons project to the striatum. The striatum is classified into two spatially distinct but morphologically similar zones, caudate nucleus and putamen. The substantia nigra is also divided into two zones that are, in this case, morphologically distinguishable: 'pars compacta' i.e. the dorsal layer of the substantia nigra, con-

[1] University Laboratory of Physiology, Parks Road, Oxford OX1 3PT, UK

taining mainly the somata of nigro-striatal neurons; 'pars reticulata' i.e. the area ventral to the pars compacta, containing far more neuropil. The neuropil consists not only of afferent axons (including fibres from striatum, globus pallidus and raphe nuclei) but also of very long dendrites, up to 1,000 μ, extending down from nigro-striatal cell somata. For a more detailed description of the anatomical organization of the nigro-striatal pathway, see Dray (1979).

1.3 Nigro-Striatal Neurons: Transmitter Mechanisms

Nigro-striatal neurons contain the transmitter dopamine (DA), which is released in the familiar way from vesicles in the terminals in the striatum (Dray 1979). However DA is also present in the dendrites of these cells, within the substantia nigra (Bjorklund and Lindvall 1975). Dendritic DA is not stored in terminals, but in the smooth endoplasmic reticulum (SER) (McGeer et al. 1979; Wassef et al. 1981). Furthermore, the transmitter is released from dendrites in a fashion that is K^+-evoked, Ca^{++}-dependent but, curiously, resistant to blockade of Na^+ channels (Cheramy et al. 1981) and of impulse flow (Hefti and Lichtensteiger 1978). Dendritic release of DA thus differs from that seen at the axonal synapse, regarding both its presynaptic localization and its lack of relation to neuronal discharge. The functional significance of this phenomenon will be explored in Sect. 4, in the light of findings on the association of acetylcholinesterase with nigro-striatal neurons.

2 Nigro-Striatal Neurons and AChE

2.1 Cellular and Subcellular Localization

The fact that AChE is present in abundance in the substantia nigra, is puzzling. There are disproportionately low levels of its normal substrate acetylcholine (ACh) and the synthesizing enzyme choline acetyltransferase (ChAT) (Silver 1974; Kimura et al. 1981; Levey et al. 1983). Furthermore, no cholinergic input to the substantia nigra has been identified (Dray 1979; Butcher and Woolf 1982).

Within the substantia nigra, AChE is present in DA-containing nigro-striatal cells and also in some non-dopaminergic neurons in the pars reticulata (Butcher and Woolf 1982). The subcellular localization of AChE is not the same in these two populations of neurons, and is comparable to the subcellular localization of AChE in the striatum.

In non-dopaminergic neurons in the substantia nigra (which are also non-cholinergic, Levey et al. 1983) and non-cholinergic neurons in the striatum, AChE is present in the rough endoplasmic reticulum and nuclear envelope (Henderson 1981; Bolam et al. 1984). However in cholinergic neurons in the striatum and dopaminergic neurons in the substantia nigra, AChE is in addition localized in the Golgi apparatus (Henderson 1981; Bolam et al. 1984).

We know that AChE must pass through the Golgi region to acquire complex sugars prior to incorporation in the plasma membrane or release from the cell (Rotundo 1984). Hence in cells with a non-reactive Golgi apparatus, AChE may well play an intracellular role i.e. act as an 'endo-enzyme'. As we have seen, cells of this type, in both the striatum and substantia nigra, are not cholinergic. It seems likely therefore that in these neurons, AChE is playing a non-cholinergic, intracellular role. Precisely what this role might be is as yet unknown. Nevertheless it is noteworthy that in the retina, AChE has been shown to have an intracellular role in the processing of peptides (Millar and Chubb 1984).

In the cells in the striatum that are cholinergic, the presence of AChE in the Golgi region is not surprising, since the enzyme will be finally incorporated into the plasma membrane, in its familiar ectoenzyme position. But why should nigro-striatal neurons have an AChE-reactive Golgi apparatus? As mentioned in Sects. 1.2 and 2.1, these cells are not cholinergic, but dopaminergic, and do not receive a cholinergic input. In cultures of muscle two distinct pools of AChE exist: membrane-bound and secreted (Rotundo and Fambrough 1980). The released AChE is not derived from a sloughing off from the membrane and has a much higher turnover than the membrane-bound form (Rotundo and Fambrough 1980). In both the brain (Chubb et al. 1976) and in the adrenal gland (Chubb and Smith 1975b) released AChE carries more negative charge than the form functioning as an ectoenzyme.

Nigro-striatal neurons have much more RER than other nigral neurons, indicating a high synthesis rate for proteins such as AChE (Domesick et al. 1983). This observation, together with that of the strong AChE reactivity in the Golgi region of these cells and absence of cholinergic synapses, leads to the speculation that AChE might be released from nigro-striatal neurons.

2.2 Release of AChE

When the striatum and substantia nigra are each perfused locally in vivo, AChE is detectable in the perfusate i.e. it is released spontaneously from cells in both these structures (Greenfield et al. 1980). Release in both regions can be modified by depolarizing concentrations of KCl in the substantia nigra: in the substantia nigra there is a marked increase, accompanied by a decrease in release from the caudate nucleus (Greenfield et al. 1980). This finding implies that release of AChE is in some way related to acute events occurring in neuronal networks.

We have investigated the cellular origin of released AChE in both the striatum and substantia nigra, by destroying the nigro-striatal pathway with the neurotoxin, 6-hydroxydopamine (6-OHDA). In the striatum, lesions reduce the AChE released to a similar level to that seen following application of KCl, to the nigral somata (i.e. to about 30% control level) (Greenfield et al. 1983a). Hence stimulus-linked release of AChE in the striatum must occur almost exclusively from presynaptic terminals of dopaminergic nigro-striatal neurons.

In the substantia nigra, 6-OHDA lesions reduce spontaneous release of AChE by an amount corresponding to the DA itself depleted. In addition, potassium no longer evokes a release of the enzyme (Greenfield et al. 1983a). Hence it seems that AChE released locally from the substantia nigra, originates exclusively from dopaminergic nigro-striatal neurons, as suggested by ultrastructural studies (Sect. 2.1).

The terminals of nigro-striatal cell neurons are in the striatum, remote from the substantia nigra, which contain only dopaminergic somata and dendrites (Sects. 1.1

and 1.2). Since there is no evidence for axon collaterals, release of AChE must occur from nigro-striatal cell somata and/or dendrites. Dendritic release of AChE has already been studied in the guinea pig facial nucleus, where SER containing the enzyme was shown in apposition with the plasma membrane (Kreutzberg et al. 1975). In nigral cell dendrites also, AChE is present in the SER (Henderson and Greenfield 1984). Probably, therefore, AChE is released not only from the terminals, but also the dendrites, of nigro-striatal neurons.

3 Relation of AChE Release to Various Transmitter Systems

3.1 Cholinergic Transmission

It is conceivable that AChE released from terminals in the striatum will hydrolyze ACh. Although these terminals are dopaminergic, they possess pre-synaptic ACh receptors (Giorguieff et al. 1977). These receptors could be activated by ACh released from cholinergic interneurons, which are plentiful in the striatum (McGeer et al. 1971). Nevertheless we could question why, if this were the case, AChE would need to be released. Furthermore, striatal membrane-bound AChE is not distributed in parallel with releasable enzyme. There is far less (only 12%) striatal membrane-bound AChE located on pre-synaptic dopaminergic nerve terminals (Lehmann and Fibiger 1978).

In the substantia nigra, release of AChE is almost certainly not related to cholinergic transmission. Nigro-striatal cells are cholinoceptive i.e. excited by exogenous ACh (Walker et al. 1976). However, administration of cholinergic receptor blockers (Greenfield and Smith 1979) has no effect on release of AChE nor does stimulation with cholinomimetics enhance it (Cuello et al. 1981). In view of the overall low levels of ACh and ChAT in the substantia nigra, and the absence of cholinergic synapses, it would thus be hard to imagine how the release of AChE from the dendrites of nigro-striatal cells could be influenced by ACh.

3.2 Substance P Transmission

It has been shown in vivo that the secreted form of AChE can break down certain peptides, including substance P (Chubb et al. 1980). Perhaps then, AChE is released within the substantia nigra to perform an extracellular role in the degradation of this peptide. This idea is attractive as levels of substance P are high in the substantia nigra (Emson 1979) and there is a substance P input from the striatal cells (Dray 1979). However, application of exogenous AChE does not modify this excitation, although it can change signals mediated by other transmitters (Greenfield et al. 1981). Furthermore, exogenous AChE applied to the substantia nigra, has the *same* effect on motor behaviour as does substance P (Greenfield et al. 1984): if the former were breaking down the latter, this similarity would not hold. In addition, we have shown that an aminopeptidase electrophoretically distinct from AChE is in any case released from nigral neurons, and that the AChE itself released, cannot break down leucine-β-naphthylamide (Greenfield and Shaw 1982).

3.3 Dopamine Transmission

Release of AChE from nigro-striatal neurons appears not to interact with cholinergic or substance P transmitter systems. It does, however, seem closely related to nigral DA. As we have seen (Sect. 2.2) releasable AChE is present exclusively in dopaminergic nigral neurons. Furthermore, both AChE and DA are localized in the same type of subcellular compartment i.e. the SER of dendrites (Sects. 1.2, 2.1). The release of AChE can be evoked by the dopaminergic agonist, amphetamine (Greenfield and Shaw 1982) and the antagonist, haloperidol (Greenfield and Weston 1984). In addition release of AChE within the substantia nigra, resembles that of DA. Both compounds are released from dendrites in a K^+-evoked, Ca^{++}-dependent manner (Greenfield et al. 1983b).

However, perhaps the most intriguing similarity between dendritic release of DA and AChE is that neither phenomenon appears to be associated with the frequency of nigro-striatal cell firing (Table 1). This conclusion is drawn from studying the effects of transmitter agents on the firing rate of these cells and comparing it with the effects of the same substances on release. As can be seen from Table 1, no clear relationship emerges between firing rate and release of AChE or DA.

Tetrodotoxin (TTX), by blocking Na^+ channels, prevents neuronal discharge, as monitored by extracellular recording. However, when Llinas, Jahnsen and I recorded *intracellularly* from nigro-striatal neurons, we found conductances that were still operative in the presence of TTX (Llinas et al. 1984). These conductances were carried by Ca^{++} ions and occurred in the dendrites of the cells (Llinas et al. 1984). Since release of DA and AChE is unrelated to firing rate, is Ca^{++}-dependent and originates from dendrites, it is tempting to postulate that the Ca^{++} potentials we observed, underlie dendritic release of both the transmitter and protein.

Table 1. Effects of various substances on the firing rate of nigro-striatal cells and on release of AChE or DA in the substantia nigra

Substance	Firing rate[a]	Release DA[a]
5-HT	↓ (1)	↑ (5)
ACh	↑ (2)	↑ (5)
Substance P	↑ (2)	↓ (5)
GABA	↓ (3)	No effect (5)
TTX	↓ (4)	↑ (6)
		Release AChE
ACh	↑ (2)	No effect (8)
Amphetamine	↓ (7)	↑ (9)
Haloperidol	↑ (7)	↑ (10)

[a] *References:* (1) Dray et al. (1976); (2) Walker et al. (1976); (3) Scarnati and Pacitti (1982); (4) Llinas and Sugimori (1980); (5) Glowinski and Cheramy (1981); (6) Cheramy et al. (1981); (7) Groves et al. (1975); (8) Cuello et al. (1981); (9) Greenfield and Shaw (1982); (10) Greenfield and Weston (1984)

4 Function and Fate of Released AChE

If AChE is not released from nigro-striatal neurons to hydrolyze ACh, what functions could it have, if any? In an attempt to answer this question, we have infused exogenous AChE in the identified nigro-striatal cells and observed the effect on their discharge. AChE inhibits nigro-striatal cell firing rate: this effect is not via the hydrolysis of ACh, as butyrylcholinesterase is without effect (Greenfield et al. 1981). However whether AChE has a direct action on the membrane, or is acting as an enzyme towards an unknown substrate, has still to be decided.

DA agonists also inhibit nigro-striatal cell activity, probably via DA 'autoreceptors' located on the soma and dendrites of nigro-striatal cells (Groves et al. 1975). Thus, there is a further parallel between released AChE and DA in the substantia nigra, i.e. a similar effect on cell activity. Activation of DA autoreceptors results in membrane hyperpolarization (Grace and Bunney 1983). It is conceivable that by some means, AChE could also have this effect on the resting potential.

The inhibitory action of dendritically released AChE and DA would not be as fast or specific as that seen for a transmitter at the axonal synapse. There are no dendrodendritic synapses in the substantia nigra (Wassef et al. 1981). Thus both compounds would diffuse indiscriminately (Hefti and Lichtensteiger 1978) hence presumably affecting a large number of cells.

In addition, the lack of apposition of any specific 'post-synaptic' site would imply a far slower time course for dendritically released substances. DA, being a small molecule, would move faster than AChE, but probably be removed more rapidly from the extracellular space. On the other hand, AChE would diffuse more slowly but remain in its active form for a longer period. These features of dendritic release of AChE and DA, are summarized in Table 2. From them, we could perhaps formulate a possible function for dendritic release from nigro-striatal cells.

Given the slow time course and lack of specific target sites described above, it seems not unreasonable to postulate a 'modulatory' function for substances released from

Table 2. Features of dendritic release of DA and AChE in the substantia nigra

(1) Not stored in vesicles but in SER of dendrites
(2) Release not blocked by TTX or gamma-butyrolactone
(3) Release is Ca^{2+}-dependent
(4) Release is not related to firing frequency
(5) Paucity of dendro-dendritic synapses
(6) Approximate diffusion rates through extracellular space:
 a) DA (M.W.153.18), 17 min/1,000 μ
  4 min/ 500 μ
  10 s / 100 μ
 b) AChE 'one to two' mm/h (Kreutzberg and Kaija 1974)
(7) AChE and DA depress firing frequency
(8) Direct activation of DA receptors (apomorphine) hyperpolarizes cell membrane

dendrites. AChE and DA could well have the same function, in view of the close parallels in their storage and conditions governing their release. Since the half-life of AChE is probably much longer than that of DA, the advantage of a common function would be that AChE could amplify the DA 'signal' (Greenfield and Shaw 1982).

What would be the function of this 'modulatory' signal? As we have seen, the DA receptor agonist, apomorphine, hyperpolarizes nigro-striatal cells. We have shown that hyperpolarization may increase the sensitivity of the cell soma to signals received at distal dendrites (Llinas et al. 1984). Hence AChE, once released from a dendrite, could make both that and neighbouring neurons preferentially sensitive to signals from afferent inputs.

5 Summary

In certain parts of the brain AChE exists not as an ectoenzyme but as a secretory neuroprotein. AChE is released not only from the terminals but also the dendrites of nigro-striatal neurons. Dendritic release of AChE appears unrelated to cholinergic or substance P transmitter systems. Rather, there is a close association, within the substantia nigra, between DA and AChE, regarding both the origin and characteristics of their release from dendrites. Dendritic release of AChE and DA does not parallel nigro-striatal cell discharge, but is probably mediated by local Ca^{++} currents in the dendrites. Once released, both DA and AChE have an action on nigro-striatal cells that may involve changing the sensitivity of the soma to incoming signals. Hence DA and AChE may be acting as neuromodulators. Whether released AChE functions as an enzyme, or directly as a neuroprotein is, however, unknown.

References

Bjorklund A, Lindvall O (1975) Dopamine in dendrites of substantia nigra neurons: suggestions for a role in dendritic terminals. Brain Res 83:531–537

Bolam JP, Ingham C, Smith AD (1984) The section-golgi-impregnation procedure III. Combination of golgi-impregnation with enzyme histochemistry and electron microscopy to characterize acetylcholinesterase-containing neurons in the rat neostriatum. Neuroscience 12:687–709

Butcher LL, Woolf NJ (1982) Monoaminergic-cholinergic relationships in the chemical communication matrix of the substantia nigra and neostriatum. Brain Res Bull 9:485–492

Cheramy A, Leviel V, Glowinski J (1981) Dendritic release of dopamine in the substantia nigra. Nature (Lond) 289:537–542

Chubb IW, Smith AD (1975a) Isoenzymes of soluble and membrane bound acetylcholinesterase in bovine splanchnic nerve and adrenal medulla. Proc R Soc Lond B Biol Sci 191:245–261

Chubb IW, Smith AD (1975b) Release of acetylcholinesterase into the perfusate from the ox adrenal gland. Proc R Soc Lond B Biol Sci 191:263–269

Chubb IW, Goodman S, Smith AD (1976) Is acetylcholinesterase secreted from central neurons into the cerebrospinal fluid? Neuroscience 1:57–62

Chubb IW, Hodgson AJ, White GH (1980) Acetylcholinesterase hydrolyses substance. P. Neuroscience 5:2065–2072

Cuello AC, Romero E, Smith AD (1981) In vitro release of acetylcholinesterase from the rat substantia nigra. J Physiol (Lond) 312:14P

Domesick VB, Stinus L, Paskevich PA (1983) The cytology of dopaminergic and nondopaminergic neurons in the substantia nigra and ventral tegmental area of the rat: a light- and electron-microscope study. Neuroscience 8:743–765

Dray A (1979) The striatum and the substantia nigra: commentary on their relationships. Neuroscience 4:1407–1439

Dray A, Gonye TJ, Oakley NR, Tanner T (1976) Evidence for the existence of a raphe projection to the substantia nigra in rats. Brain Res 113:45–57

Emson PC (1979) Peptides as neurotransmitter candidates in the mammalian CNS. Prog Neurobiol 13:61–116

Fonnum F, Grofova I, Rinvik E, Storm-Mathisen J, Walberg F (1974) Origin and distribution of glutamate decarboxylase in substantia nigra of the cat. Brain Res 71:77–92

Giorguieff MF, Le Floc'h ML, Glowinski J, Besson MJ (1977) Involvement of cholinergic presynaptic receptors of nicotinic and muscarinic types in the control of the spontaneous release of dopamine from striatal dopamine terminals in the rat. J Pharmacol Exp Ther 200:535–544

Glowinski J, Cheramy A (1981) Dendritic release of dopamine; its role in the substantia nigra. In: Stjarne L, Hedquist P, Lagercrantz H, Wennmalm A (eds) Chemical neurotransmission: 75 years. Academic Press, New York, pp 285–299

Grace AA, Bunney BS (1983) Effects of apomorphine on nigral dopamine neurons recorded intracellularly. Prog Neuro-Psychopharmacol Biol Psychol (Suppl) 185

Greenfield SA, Smith AD (1979) The influence of electrical stimulation of certain brain regions on the concentration of acetylcholinesterase in rabbit cerebrospinal fluid. Brain Res 177:445–459

Greenfield SA, Shaw SG (1982) Release of acetylcholinesterase and aminopeptidase in vivo following infusion of amphetamine into the substantia nigra. Neuroscience 7:2883–2893

Greenfield SA, Weston J (1984) Acetylcholinesterase release from rat nigro-striatal cells following infusion of haloperidol into the substantia nigra. J Physiol (Lond) 351–23P

Greenfield SA, Cheramy A, Leviel V, Glowinski J (1980) In vivo release of acetylcholinesterase in the cat substantiae nigrae and caudate nuclei. Nature (Lond) 284:355–357

Greenfield SA, Stein JF, Hodgson AJ, Chubb IW (1981) Depression of nigral pars compacta cell discharge by exogenous acetylcholinesterase. Neuroscience 6:2287–2295

Greenfield SA, Grunewald RA, Foley P, Shaw SG (1983a) Origin of various enzymes released from the substantia nigra and caudate nucleus: effects of 6-hydroxydopamine lesions of the nigro-striatal pathway. J Comp Neurol 214:87–92

Greenfield SA, Cheramy A, Glowinski J (1983b) Evoked release of proteins from central neurons in vivo. J Neurochem 40:1048–1057

Greenfield SA, Chubb IW, Grunewald RA, Henderson Z, May J, Portnoy S, Weston J, Wright MC (1984) A non-cholinergic function for acetylcholinesterase in the substantia nigra: behavioural evidence. Exp Brain Res 54:513–520

Groves PM, Wilson CJ, Young SJ, Rebec GV (1975) Self-inhibition by dopaminergic neurons. Science (Wash DC) 190:522–529

Hefti F, Lichtensteiger W (1978) Subcellular distribution of dopamine in substantia nigra of the rat brain: effects of gamma-butyrolactone and destruction of noradrenergic afferents suggest formation of particles from dendrites. J Neurochem 30:1217–1230

Henderson Z (1981) Ultrastructure and acetylcholinesterase content of neurons forming connections between the striatum and substantia nigra of rat. J Comp Neurol 197:185–196

Henderson Z, Greenfield SA (1984) Ultrastructural localization of acetylcholinesterase in the substantia nigra: a comparison between rat and guinea pig. J Comp Neurol 230:278–286

Kimura H, McGeer PL, Peng JH, McGeer EG (1981) The central cholinergic system studied by choline acetyltransferase immunohistochemistry in the cat. J Comp Neurol 200:157–201

Kreutzberg GW, Kaija H (1974) Exogenous acetylcholinesterase as tracer for extracellular pathways in the brain. Histochimie 42:233–237

Kreutzberg GW, Toth L, Kaija H (1975) Acetylcholinesterase as a marker for dendritic transport and dendritic secretion. Adv Neurol 12:269–281

Lehman J, Fibiger HC (1978) Acetylcholinesterase in substantia nigra and caudate-putamen of the rat: properties and localisation in dopaminergic neurons. J Neurochem 30:615–624

Levey AI, Wainer BH, Mufson EJ, Mesulam M-M (1983) Co-localization of acetylcholinesterase and choline acetyltransferase in the rat cerebrum. Neuroscience 9:9–22

Llinas R, Sugimori M (1980) Electrophysiological properties of in vitro Purkinje cell somata in mammalian cerebellar slices. J Physiol (Lond) 305:171–195

Llinas R, Greenfield SA, Jahnsen H (1984) Electrophysiology of pars compacta cells in the in vitro substantia nigra – a possible mechanism for dendritic release. Brain Res 294:127–132

McGeer PL, McGeer EG, Fibiger HL, Wickson V (1971) Neostriatal choline acetylase and cholinesterase following selective brain lesions. Brain Res 35:308–314

McGeer L, Del Fiaco M, Cuello AC (1979) The smooth endoplasmic reticulum as a possible storage site for dendritic dopamine in substantia nigra neurons. Experientia (Basel) 35:101–103

Millar TJ, Chubb IW (1984) Sections of chick retinae treated with acetylcholinesterase show enhanced enkephalin and substance P immunoreactivity. Neuroscience 12:441–452

Rotundo RL (1984) Synthesis, assembly and processing of AChE in tissue cultured muscle. In: Brzin M, Kiauta T, Barnard EA (eds) Cholinesterases; fundamental and applied aspects. de Gruyter, Berlin, in press

Rotundo RL, Fambrough DM (1980) Synthesis, transport and fate of acetylcholinesterase in cultured chick embryo muscle cells. Cell 22:583–594

Scarnati E, Pacitti C (1982) Neuronal responses to iontophoretically applied dopamine, glutamate and GABA of identified dopaminergic cells in the rat substantia nigra after kainic acid-induced destruction of the striatum. Exp Brain Res 46:377–342

Silver A (1974) The biology of the cholinesterases. Elsevier, Amsterdam

Somogyi P, Chubb IW, Smith AD (1975) A possible structural basis for extracellular release of acetylcholinesterase. Proc R Soc Lond B Biol Sci 191:271–283

Walker RJ, Kemp JA, Yajima H, Kitagawa K, Woodruff GN (1976) The action of substance P on mesencephalic reticular and substantia nigra neurons of the rat. Experientia (Basel) 32:214–215

Wassef M, Berod A, Sotelo C 91981) Dopaminergic dendrites in pars reticulata of the rat substantia nigra and their striatal input. Combined immunocytochemical localization of tyrosine-hydroxylase and anterograde degeneration. Neuroscience 6:2125–2139

Ectopeptidases

Endopeptidase-24.11: An Ectoenzyme Capable of Hydrolysing Regulatory Peptides at the Surface of Many Different Cell Types

A. J. Kenny [1]

1 Introduction: Renal Brush Border Peptidases

The brush border membrane which lines the renal proximal convoluted tubule is richly endowed with hydrolases capable of hydrolysing glycosidic, ester and peptide bonds. The peptidases of this membrane constitute a well-studied group of enzymes and several of them are known to be among the major proteins of the membrane (Table 1). They form a functional group that, from their various modes of attack, can be assumed to be capable of hydrolysing most peptides to their constituent amino acids. Only one (in

Table 1. Pig kidney brush border peptidases

Enzyme	Subunit M_r ($\times 10^{-3}$)	Proportion of membrane protein (%)	Active site	Specificity[a]	Specific inhibitors
Endopeptidase-24.11 (E.C. 3.4.24.11)	90	4	Zn^{2+}	–o–o↓●–o– (Hydrophobic)	Phosphoramidon Thiorphan
Aminopeptidase N	160	8	Zn^{2+}	●↓o–o–o– (Many)	Amastatin
Aminopeptidase A	170	1	Ca^{2+}	●↓o–o–o– (Glu/Asp)	
Aminopeptidase P	?	?	?	o↓●–o–o– (Pro)	
Proline carboxypeptidase	?	?	Me^{2+}	–o–o–●↓o (Pro)	
Dipeptidyl peptidase IV	130	4	Serine	o–●↓o–o– (Pro)	Dip–F
Peptidyl dipeptidase	180	?	Zn^{2+}	–o–o↓●–● (non-specific)	Captopril
γ-Glutamyl transferase	60,25	?	?	●↓o–o–o– (γ-Glu)	

[a] Residue shown ● indicated in parentheses

[1] MRC Membrane Peptidase Research Group, Department of Biochemistry, University of Leeds, Leeds LS2 9JT, UK

most species) acts as an endopeptidase capable of hydrolysing internal bonds of peptides, the rest are exopeptidases attacking from either the N- or C-terminus of the peptide. The endopeptidase (E.C. 3.4.24.11) is uniquely placed to initiate the attack on a peptide, thus generating smaller fragments accessible for hydrolysis by the exopeptidases. It is this endopeptidase that forms the subject of this chapter.

2 Endopeptidase-24.11: Historical Background

About 15 years ago, we realised that rat kidney contained a particulate enzyme that hydrolysed [^{125}I]-insulin B chain at neutral pH (Wong-Leung and Kenny 1968). Further study showed that this activity was associated with the brush border membranes (George and Kenny 1973) and we then embarked upon its purification from rabbit kidney (Kerr and Kenny 1974a,b). The enzyme was difficult to release from the membrane, requiring a rather severe treatment with toluene and trypsin before it could be obtained in a hydrophilic form. When purified, this hydrophilic protein was shown to be a monomer of M_r 93,000, containing one Zn atom and about 15% carbohydrate. Its specificity, determined with insulin B chain and other small peptides, was similar to that of the chelator-sensitive microbial endopeptidases typified by thermolysin. Both types of enzyme hydrolyse peptide bonds involving the amino group of hydrophobic residues, provided they are not C-terminal, nor adjacent to the N-terminus. The kidney enzyme was thus the first example of this type of enzyme to be isolated from a vertebrate source. The similarity in its active site was further confirmed by the finding that phosphoramidon, a natural product from *Streptomyces tanashiensis* (Suda et al. 1973) that strongly inhibited thermolysin also inhibited the kidney enzyme (Kenny 1977). The Enzyme Commission classification places it in the same group as the microbial enzymes, listing it as E.C. 3.4.24.11, kidney brush border neutral proteinase, a name that seemed appropriate at the time, but became unsatisfactory as it became clear that the enzyme was neither limited to the kidney, nor to brush border membranes, but had a wider distribution on the surfaces of many cell types (see Sect. 6). We have recently proposed an unambiguous and non-restrictive name: endopeptidase-24.11 (Matsas et al. 1983).

3 Endopeptidase-24.11 is an Ectoenzyme

When visualised in negatively-stained electron micrographs, the surfaces of kidney microvilli are seen to be coated with particles (Fig. 1). These are not organised in an ordered monolayer, but appear as a confused array comprising more than one layer. The effect of papain treatment of the membranes is to convert the pattern into an ordered monolayer. Essentially, papain removes those particles attached to the membrane by long stalks, leaving only the short-stalked proteins in situ, as shown in the morphometric analysis (Fig. 2) (Kenny et al. 1983). Among the papain-releasable peptidases are aminopeptidases A and N, dipeptidyl peptidase IV and gamma-glutamyl transferase (Kenny

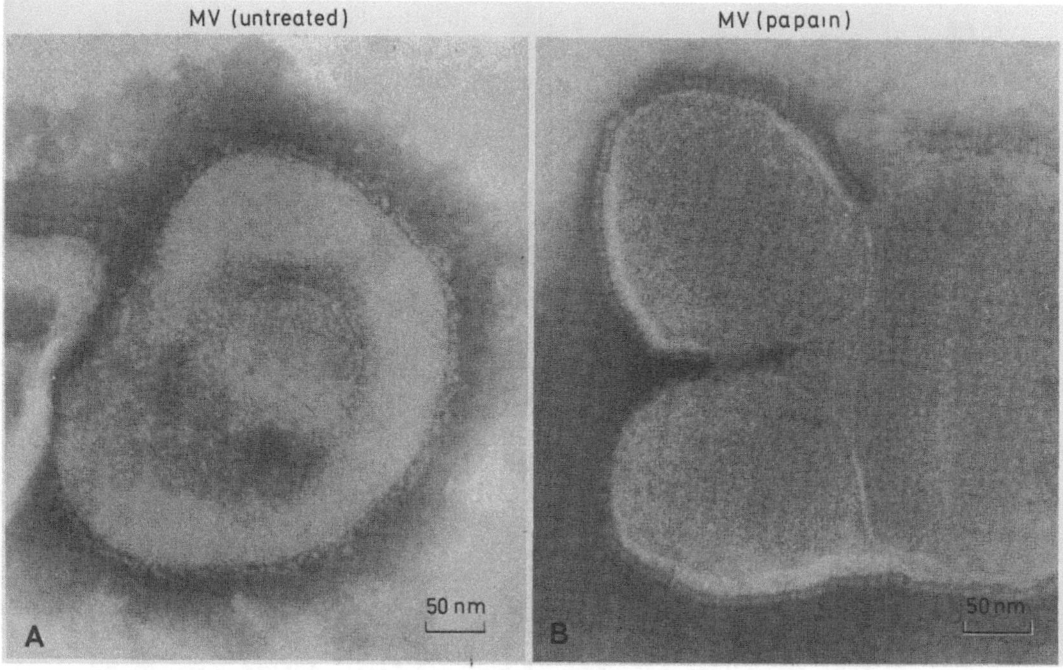

Fig. 1A,B. Negatively-stained electron micrographs of microvilli. **A** Untreated; **B** papain treated. (Kenny et al. 1983)

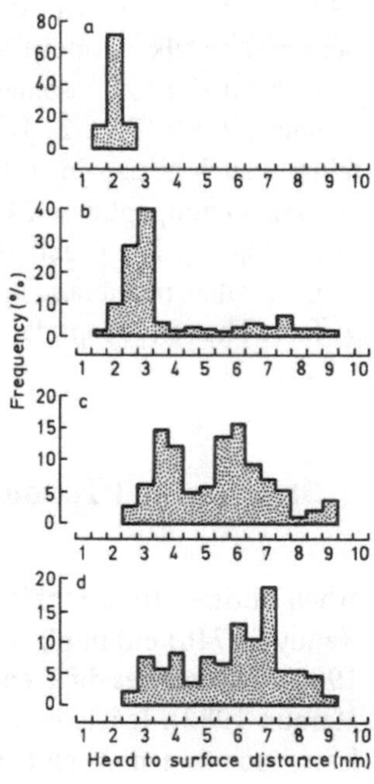

Fig. 2A–D. Morphometric analysis of the length of the junctional (stalk) peptide. **A** Endopeptidase-24.11 reconstituted in liposomes; **B** microvilli, papain treated; **C** microvillar proteins, reconstituted in liposomes, **D** microvilli, untreated (Kenny et al. 1983)

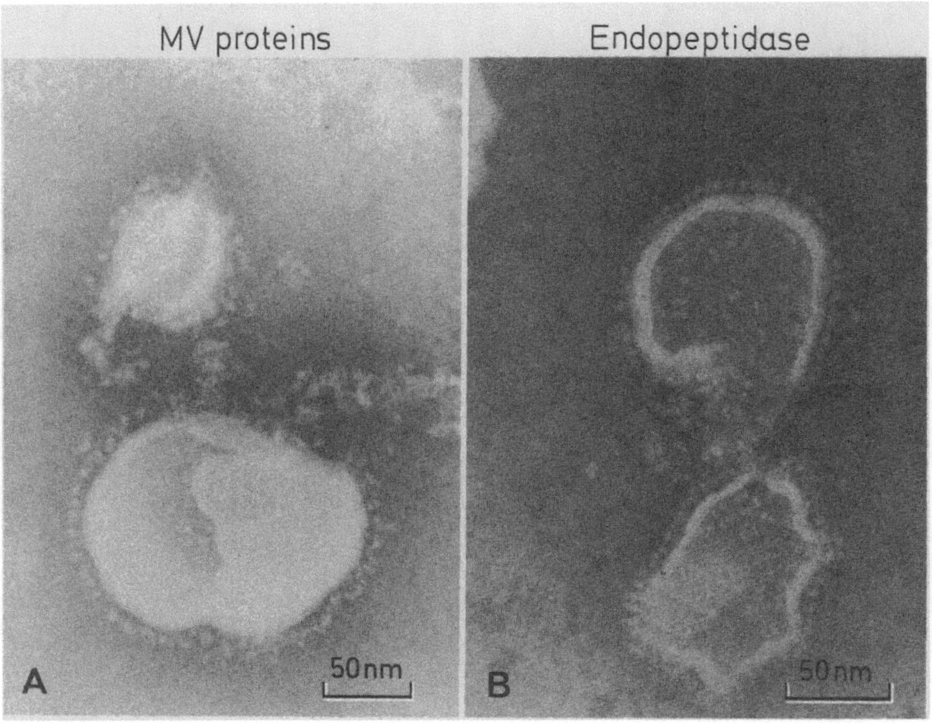

Fig. 3A,B. Negatively-stained electron micrographs of liposomes. **A** Reconstituted with microvillar proteins; **B** reconstituted with endopeptidase-24.11 (Kenny et al. 1983)

1977), and it is known that when aminopeptidase N is reconstituted into liposomes, it possesses a stalk of about 5 nm (Hussain et al. 1981). When endopeptidase-24.11 was reconstituted into liposomes, it formed an ordered array of particles with a stalk length of about 2 nm (Figs. 2, 3), which is too small a gap to allow access of even small proteinases such as papain. All the peptidases that have been reconstituted in liposomes, including endopeptidase-24.11, do so in a consistently right-side-out orientation, none exhibiting any latent activity when detergent is present. The same conclusion applies to microvillar membrane vesicles: all the peptidase activity is accessible at the external surface. Thus endopeptidase-24.11 is demonstrably an ectoenzyme.

4 Structure and Topology of Kidney Endopeptidase-24.11

When purified from rabbit kidney, the enzyme is a monomer of M_r 93,000 (Kerr and Kenny 1974b) but in pig it exists as a dimer of subunit M_r 89,000 (Fulcher and Kenny 1983). This species difference is common to other brush border peptidases (Kenny and Maroux 1982). It is a homodimeric protein, unlike certain other brush border hydrolases, such as gamma-glutamyl transferase. Both subunits are believed to be anchored

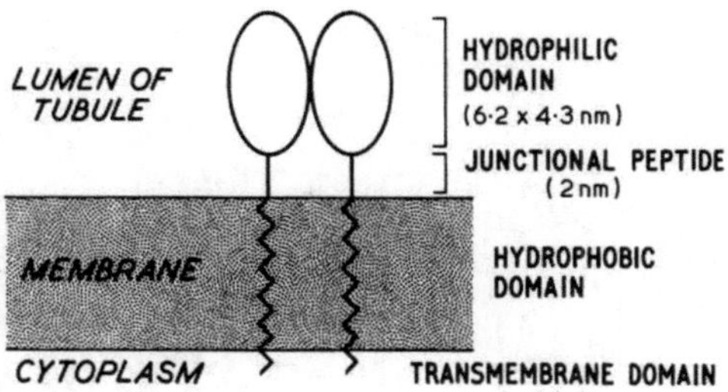

Fig. 4. Model showing the structure and topology of endopeptidase-24.11

to the membrane. In the case of other hydrolases that are readily released from the membrane by proteinase treatment, it is possible to assess (by difference in subunit M_r) the size of the peptide bearing the hydrophobic anchor (see Kenny and Maroux 1982), but endopeptidase-24.11 is not so easily assessed in this respect and we can only infer that it is a relatively small portion of the molecule. This conclusion is based on the M_r of the purified rabbit enzyme, which was released by toluene-trypsin treatment and was nevertheless comparable in M_r to the detergent-released subunits of the pig enzyme. Furthermore, the reconstitution experiments (Kenny et al. 1983) show by electron microscopy that the bulk of the protein lies external to the membrane surface.

Asymmetric labelling experiments using a photo-activated compound 'DIABS' (3,5-di[^{125}I]iodo-4-azidobenzenesulphonate) have established that the endopeptidase is a transmembrane protein (Booth and Kenny 1980). But it follows from the comments above that very little of the polypeptide chain can be exposed at the cytoplasmic surface. The location of the peptide chain that includes the cytoplasmic domain, the hydrophobic domain and the stalk (or junctional) peptide, remains uncertain in the case of endopeptidase-24.11. Where the N-terminal sequence of other brush border enzymes has been determined it appears that this peptide is included in the N-terminal sequence (for review, see Kenny and Maroux 1982; Semenza et al. 1983; Matsuda et al. 1983). Preliminary work on the N-terminal sequence of endopeptidase-24.11 has also implied that a hydrophobic sequence is located near the N-terminus (I.S. Fulcher and A.J. Kenny, unpublished work). It has been suggested that this topological feature might arise from the retention rather than cleavage of the signal peptide (Kenny and Booth 1978) but this hypothesis awaits unequivocal experimental proof by direct comparison of the N-terminal sequence of the in vitro translated product with that of the mature enzyme. Our present knowledge of the structure and topology of the endopeptidase is summarised in the model (Fig. 4).

5 Biosynthesis of Endopeptidase-24.11

The biosynthesis of the endopeptidase has recently been investigated by pulse-chase labelling experiments using slices of piglet kidney cortex (Stewart and Kenny 1984a,b). Immediately after a 15 min pulse of [^{35}S]methionine, a precursor form was detected in a Mg^{2+}-precipitated subcellular fraction that contained intracellular membranes

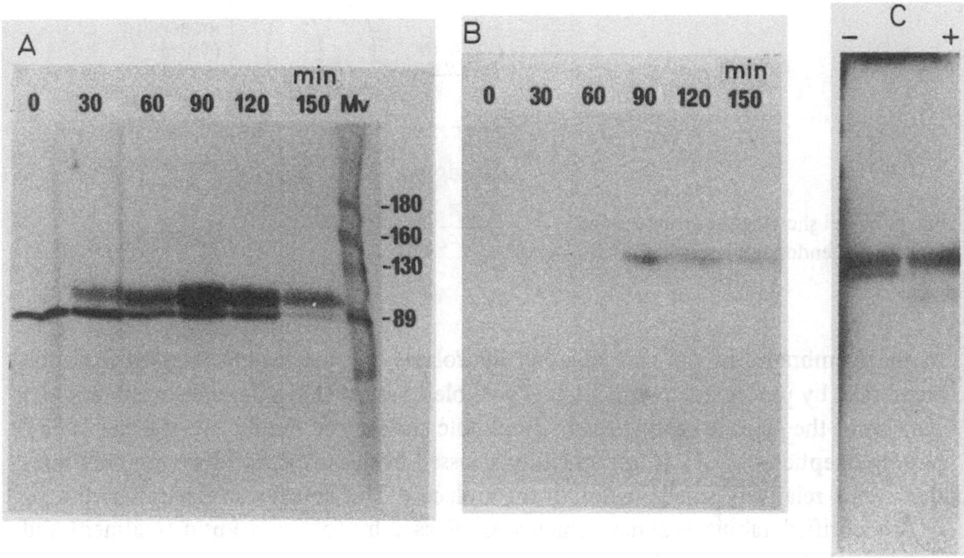

Fig. 5A–C. Biosynthesis of endopeptidase-24.11 in pig kidney slices. Slices were pulsed with [^{35}S]-methionine and chased for the periods indicated. Membranes were treated with detergent and immunoprecipitated before analysis by SDS-PAGE and fluorography. **A** Mg^{2+} pellet containing intracellular membranes; **B** microvillar fraction; **C** total homogenate showing effect of endoglycosidase H on the faster migrating glycoprotein. (Data from Stewart and Kenny 1984a)

(Fig. 5A). Thirty minutes later, a second form about 4 kD larger was detected in the same fraction. While the first precursor was sensitive to endoglycosidase-H, the larger glycoprotein was resistant (Fig. 5C) indicating the conversion of the high mannose- to the complex glycosidated-form. This mature form of the protein reached a maximum after 90 min of chase and then diminished. The microvillar fraction revealed only the mature form and this was not detected until 90 min of the chase had elapsed – representing the time required for intracellular transport of the newly synthesised glycoprotein to the apical membrane (Fig. 5B). No precursor forms were detected in the cytosol: biosynthesis and post-translational processing were restricted to membrane compartments of the cell. Monensin, an ionophore that causes distension of Golgi elements, held up the processing at the high-mannose stage and prevented onward movement to the microvillar membrane. It is clear, therefore, that the earliest precursor to be recognised corresponds to the monomer and that unlike the heterodimeric enzymes of the brush border, e.g. gamma-glutamyl transferase (Nash and Tate 1982, 1984) that are synthesised as single chain precursors, these subunits of endopeptidase-24.11 are not generated by a proteolytic step. In this respect it is typical of other homodimeric enzymes (Danielsen et al. 1984; Stewart and Kenny 1984a,b).

6 Distribution of Endopeptidase-24.11

6.1 Endopeptidase-24.11 Activity in Tissues

The enzyme was first recognised in kidney brush border membranes (George and Kenny 1973) and only later identified in intestinal microvilli (Danielsen et al. 1980), from which source it was later purified and found to be differently glycosylated from the kidney enzyme (Fulcher et al. 1983). The search for the enzyme in other cell types was first attempted by exploiting the inhibitor, phosphoramidon, which, so far, appears to be specific for this particular enzyme in mammals. Phosphoramidon-sensitive activity hydrolysing [^{125}I]insulin B chain was observed in several other tissues including spleen, aorta, myocardium, and lung (Kenny and Fulcher 1983). At about the same time we realised that another substance, thiorphan, that had been synthesised as a specific inhibitor of the activity in brain referred to as 'enkephalinase', was indistinguishable from phosphoramidon in its ability to inhibit purified kidney endopeptidase-24.11. This observation and the ability of phosphoramidon to inhibit the hydrolysis of [Leu]-enkephalin at the Gly^3–Phe^4 bond by brain synaptic membranes (Fulcher et al. 1982) suggested that endopeptidase-24.11 was also present in pig brain. The later development of this research is described in Sect. 7.

6.2 Immunoradiometric Assay (IRMA) of Endopeptidase-24.11 in Pig Tissues

A monoclonal antibody to the endopeptidase has been produced (Gee et al. 1983). It has been useful in immuno-affinity purification of the enzyme as well as in immunofluorescent microscopy to show the localisation of the antibody to the brush borders of kidney and intestine. More recently, we have set up an IRMA, which appears to have certain advantages over enzyme assay when applied to crude tissue homogenates (Gee et al. 1985; Matsas et al. 1985). The enzymic assay using [^{125}I]insulin B chains is amplified by other endogenous peptidases and may be inhibited by endogenous peptides that compete with the labelled substrate. Moreover, we cannot yet exclude the possibility that unrelated endopeptidases may be sensitive to phosphoramidon. On the other hand, the monoclonal antibody employed in the IRMA, GK7C2, seems to be specific for endopeptidase-24.11, from experiments in which it has been used in Western blotting and as a means of purifying the enzyme from various tissues. Generally the two assays have correlated extremely well when membrane pellets were compared, but with some tissues the correlation was weaker when unfractionated homogenates were used. The results of the IRMA survey of tissues and organs, other than brain and spinal cord, are shown in Fig. 5. No tissue exceeded kidney in specific activity; in the gastrointestinal tract, the enzyme was mainly located in the jejunum and the stomach was devoid of activity. However, lymph nodes proved to be very rich in the endopeptidase. The values in Fig. 6 are from analyses of new-born piglet nodes and generally higher values have been obtained from adult pigs. Preliminary studies (M.A. Bowes and A.J. Kenny, unpublished work) have suggested that reticular cells, especially in the T-cell areas, are rich in the enzyme. Significant activities were demonstrated in various glandular tissues. Among isolated cell preparations, only chondrocytes were rich in the endopeptidase and the amount diminished during culture.

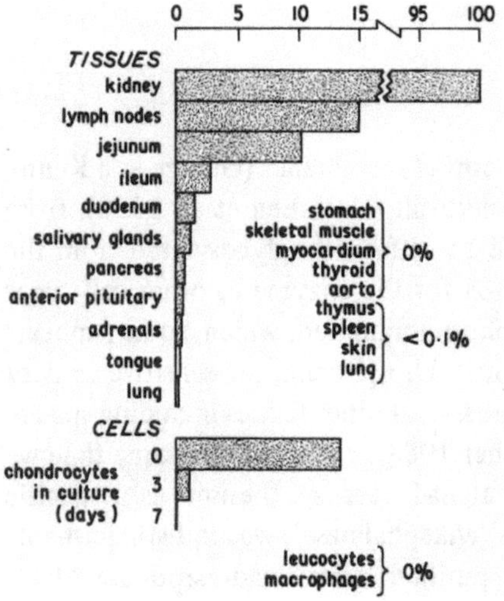

Fig. 6. Distribution of endopeptidase in pig tissues. An IRMA was used to determine the endopeptidase content in unfractionated homogenates of pig tissues. The results are specific activities (U mg^{-1} of protein relative to kidney homogenate) (Gee et al. 1985)

7 Identity of Endopeptidase-24.11 with So-Called Enkephalinase in Brain

The enzymes/enkephalin concerned in the biological inactivation of the enkephalin have received much attention, not least because compounds capable of specifically inhibiting their hydrolysis might lead to the development of a new class of analgesics. Enkephalins (Tyr–Gly–Gly–Phe–Leu and Tyr–Gly–Gly–Phe–Met) are readily hydrolysed by aminopeptidases at the Tyr-Gly bond. However, greater interest has been focused on the nature of the peptidase attacking the Gly^3–Phe^4 bond, an activity misleadingly called 'enkephalinase' (Swerts et al. 1979; Schwartz et al. 1983), implying an exclusive role for this enzyme in the CNS. Various strands of experimental evidence have led to the unequivocal conclusion that this activity is attributable to endopeptidase-24.11. The initial finding that thiorphan and phosphoramidon were identical in their action on kidney endopeptidase-24.11 has been mentioned above (Sect. 6.1). Our early experiments revealed only partial inhibition when synaptic membranes were incubated with [Leu]enkephalin. This was partly because the endopeptidase is not uniformly distributed in all regions of the brain. The IRMA values for synaptic membrane preparations from regions of pig brain are shown in Fig. 7. In the brain, the striatum contained the highest specific activity and the spinal cord was an even richer source. When striatal synaptic membrane preparations were tested with [D–Ala^2,Leu^5]enkephalin (an analogue that resists aminopeptidase attack), phosphoramidon fully inhibited hydrolysis with an IC_{50} value identical to that observed with the purified kidney enzyme (Matsas et al. 1983). Substance P (Arg–Pro–Lys–Pro–Gln–Gln–Phe–Phe–Gly–Leu–Met–NH_2) was attacked at the same three bonds (Gln–Phe, Phe–Phe, and Gly–Leu) by both synaptic membranes and purified enzyme and in the case of the membranes the hydrolysis was inhibited by a polyclonal antiserum raised to the kidney enzyme (Matsas et al. 1983). This work established that the enzyme in pig caudate synaptic membrane preparations was identical in bond specificity, inhibitor sensitivity and was immunologi-

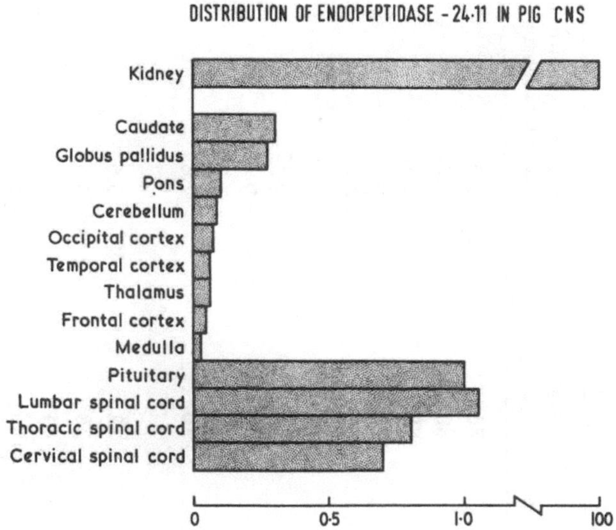

Fig. 7. Distribution of endopeptidase-24.11 in pig central nervous system. See legend to Fig. 2 (Matsas et al. 1985)

cally identical to the pig kidney enzyme. By using the monoclonal antibody as an immuno-affinity absorbent column, it was possible to purify to homogeneity endopeptidase-24.11. It had a subunit M_r of 87,000, slightly less than the renal form and contained less carbohydrate. It had the same specific activity and showed immunological identity with purified preparations from kidney and intestine (Relton et al. 1983).

8 The Hydrolysis of Neuropeptides by Endopeptidase-24.11

Once the complete identity of the brain and kidney enzymes had been established, it became logical to use the latter, which can be obtained in milligram quantities by immuno-affinity purification, to define the kinetic repertoire of endopeptidase-24.11. The primary specificity is readily defined as the cleavage of bonds involving the amino groups of hydrophobic amino acid residues. But peptidase action is often markedly influenced by residues in positions somewhat removed from the scissile bond. The sequence of the peptide substrate around the scissile bond (P_1-P_1') is conventionally described as

$$P_5-P_4-P_3-P_2-P_1-P_1'-P_2'-P_3'-P_4'-P_5' .$$

For this enzyme, the first requirement is that P_1' should be hydrophobic and not C-terminal or adjacent to the N-terminus. But it is now clear that residues in other positions may have major influences on both K_m and k_{cat} values. Some results (from Matsas et al. 1984a,b) are given in Table 2. Among the enkephalins, certain effects may be discerned. Amidation of the C-terminal (P_2') carboxyl group impairs the efficiency of hydrolysis, as noted by Malfroy and Schwartz (1982), but extension by the addition of Arg^6 or by a dipeptide Arg^6-Phe^7 either did not impair or actually improved hydrolysis. It was also apparent that methionine, unlike other hydrophobic residues, did not permit hydrolysis at the adjacent bond. The tachykinins, however,

Table 2. Hydrolysis of neuropeptides by endopeptidase-24.11[a]

Enkephalins and analogues	Bonds hydrolysed	K_m (m*M*)	k_{cat}/K_m (relative to [Leu[5]]enkephalin)
[Leu[5]]enkephalin	Y G G↓F L	86	1.00
[Leu[5]]enkephalinamide	Y G G↓F LNH_2	680	0.04
[D–Ala[2],Leu[5]]enkephalin	Y dA G↓F L	88	1.46
[D–Ala[2],Leu[5]]enkephalinamide	Y dA G↓F LNH_2	773	0.05
[Leu[5]]enkephalin–Arg[6]	Y G G↓F↓L R	111	0.95
Dynorphin (1–9)	Y G G↓F⇣L R R↓I R	119	0.56
[Met[5]]enkephalin	Y G G↓F M	62	0.95
[D–Ala[2],Met[5]]enkephalinamide	Y dA G↓F MNH_2	432	0.32
[Met[5]]enkephalin–Arg[6]	Y G G↓F M R	39	1.22
[Met[5]]enkephalin–Arg[6] Phe[7]	Y G G↓F M R F	50	2.67
Tachykinins and analogues			
Substance P	R P K P Q Q↓F↓F G↓L MNH_2	32	3.62
Substance P (deamidated)	R P K P Q Q↓F↓F G↓L M	38	4.62
[NorLeu[11]]substance P	R P K P Q Q↓F↓F G↓L $NorLNH_2$	84	1.93
Physalaemin	E A D P N K↓F↓Y G↓L MNH_2	106	1.59
DiMeC7	E Q F (MeF (MeG)↓L MNH_2	2,567	0.10
Others			
Bradykinin	R P P G⇣F S P↓F R	92	1.57
Cholecystokinin (CCK-8) (sulphated)	D Y M G⇣W M D↓FNH_2	296	0.68

[a] Solid and dotted arrows represent sites of cleavage under initial rate conditions and after extensive substrate hydrolysis, respectively. The one letter code for amino acids is used and NH_2 at the C-terminus indicates an amidated residue. The abbreviations dA and NorL refer to D-alanine and norleucine, respectively. DiMeC7 is an analogue of substance P (5–11), methylated at peptide bonds 7–8 and 8–9. (Data from Matsas et al. 1984a, b)

show that an amidated C-terminal methionine in the P_2' position was by no means incompatible with efficient hydrolysis, indeed substance P had the lowest K_m of any of this group of peptides. Cholecystokinin (CCK-8) is of interest in showing the ability of the enzyme to release a C-terminal Phe–NH_2 residue efficiently, suggesting that phenylalanine may be the preferred residue in the P_1' position. Orlowski and his co-workers have explored a series of synthetic peptide substrates and their results show major influences of residues in P_2, P_3, and P_4 positions on K_m and k_{cat} (Orlowski and Wilk 1981; Almenoff and Orlowski 1983). It is apparent that endopeptidase-24.11 has an extended active site, a property that seems to justify the much larger subunit size of this molecule compared with the microbial enzymes of the same E.C. class. Thermolysin has distinctly different specificity requirements (Pozsgay et al. 1985). However, it must be emphasised that it is not yet possible to formulate hard and fast rules that would enable a sound prediction of the efficiency by which any given peptide might be hydrolysed by this enzyme.

9 Role of Endopeptidase-24.11 in Metabolising Peptides at Cell Surfaces

9.1 General Considerations

The physiological role of this enzyme is likely to be determined by five main considerations. First, and most importantly, there is the precise cellular location in the tissues and organs in which it has been found. In the kidney, for example, its actions, in company with the other microvillar hydrolases, can only be effected on the constituents of the luminal fluid. The second consideration concerns the identity and the concentrations of peptide substrates present in the microenvironment at that particular cell surface. In some instances these may be predictable or even experimentally determinable, but generally these will be uncertain parameters. Thirdly, the presence of other peptidases at the same location will be important. For example, the enkephalins are very susceptible to aminopeptidase attack, and if this activity predominates in some locations, the action of the endopeptidase may be of minor significance. Fourthly, the individual susceptibility of the relevant peptides may be crucial. Here it is possible to draw valid conclusions if the kinetic parameters are known. It is apparent from Table 2 that there is a large variation in K_m and k_{cat} values for the group of peptides studied. Another peptide, not listed in Table 2, luliberin (LHRH), is an extremely poor substrate giving a k_{cat}/K_m value two orders of magnitude less than substance P. The extended active site of the endopeptidase therefore generates some degree of peptide specificity, even though its primary specificity, for hydrophobic residues in the P_2' position, might suggest an indiscriminate action. Finally, we must consider the possibility of endogenous inhibitors. Many intracellular proteinases appear to be regulated by the association of a specific peptide or protein inhibitor. There may be similar inhibitors at the cell surface, but so far no experiments have suggested that this phenomenon occurs. Abnormally high recoveries of the enzyme that might occur if an inhibitor was displaced have never, at least in this laboratory, been observed. Moreover, the final specific activities of many batches of purified enzyme from various tissue sources have all been of similar values, which again argues against the existence of an endogenous inhibitor. So, in summary, the main considerations concern the identity of both the peptidases and the peptides in the immediate vicinity at the external surface of the membrane.

9.2 Kidney

The battery of peptidases on the luminal surface of the proximal tubule is listed in Table 1. Peptides present in the glomerular filtrate will encounter these peptidases and are likely to be wholly degraded before entering the distal tubule and loop of Henle. While we may admire the efficiency of the process, it is difficult to guess why it should be mandatory that all biological activity of such peptides should be extinguished in the proximal tubule. The salvaging of the very small quantities of amino acids, which would be efficiently reabsorbed by the same membrane, seems an inadequate explanation. It is more likely that the raison d'être lies in the possibility that one or more of the filtered peptides might cause undesirable, nonphysiological effects if presented at the luminal surface of cells in the distal nephron or indeed in the remainder of the urinary tract. But this suggestion is no more than speculation: we do not know which peptides might be harmful, nor what the nature of the harm might be.

9.3 Intestine

The concentration of the enzyme in the jejunum is logical if it, together with the other microvillar peptidases, plays a role in the final stages of hydrolysis of peptides generated by the action of gastric and pancreatic proteinases. In this respect, the endopeptidase has a role in digestion and nutrition rather than in regulation of the action of functional peptides.

9.4 Central Nervous System

Although the amount of endopeptidase-24.11 in even the richest regions of the brain is extremely small compared with kidney (Fig. 7), we can argue a convincing case for its role in terminating the signal arising from the release of some neuropeptides. While the enzyme is enriched in preparations of synaptic membranes, we still await unequivocal evidence that it is a component of the synapse rather than some other membrane that co-purifies with synaptosomes. One report has identified this enzyme on membranes of glial cells (Lentzen and Palenker 1983). But this observation does not exclude a neuronal localisation as well, nor if it should be exclusively on glial cells would it eliminate it from neuropeptide metabolism, since these may diffuse from their point of release towards an adjacent glial membrane.

9.5 Other Tissues and Organs

The demonstration of high concentrations of endopeptidase-24.11 in lymph nodes and isolated chondrocytes is novel and at present provokes more questions than answers. The cells in lymph nodes that express the enzyme have yet to be defined, as are the possible peptide substrates. Chondrocytes not only lay down the matrix of cartilage but, if appropriately stimulated, can resorb it. One of the agents that can stimulate is a product of the synovium, a 21 kD protein, which has been also isolated from monocyte culture medium and named catabolin, but may also be homologous to interleukin-1 (Saklatvala et al. 1983a,b). It remains to be seen whether catabolin is a susceptible substrate. In lymph nodes there are many cell types which exchange information by a variety of humoral factors, many of which are known to be peptides. Among these are lymphokines including interleukins and interferons. Others include chemotactic factors of undefined structure. However, a synthetic chemotactic peptide (f–Met–Leu–Phe) is a substrate for a phosphoramidon-sensitive peptidase (Yuli and Snyderman 1984). The complex interrelationships of the neuro-endocrinological and immune systems have been recently reviewed by Blalock (1984) and it is clear that there is no lack of candidates for the lymph node endopeptidase to hydrolyse.

10 Summary

Endopeptidase-24.11 is a widely distributed ectoenzyme which is capable of hydrolysing a variety of peptides at bonds adjacent to hydrophobic residues. In the pig kidney, it is a dimeric glycoprotein (subunit M_r 89,000) attached to the microvillar membrane by 2 nm stalks. Its biosynthesis, distribution and functional roles are discussed.

References

Almenoff J, Orlowski M (1983) Membrane-bound kidney neutral metalloendopeptidase: interaction with synthetic substrates, natural peptides and inhibitors. Biochemistry 22:590–599

Blalock JE (1984) The immune system as a sensory organ. J Immunol 132:1067–1070

Booth AG, Kenny AJ (1980) Proteins of the kidney microvillar membrane. Asymmetric labelling of the membrane by lactoperoxidase-catalysed radioiodination and by photolysis of 3,5-Di-[^{125}I]iodo-4-azidobenzenesulphonate. Biochem J 187:31–44

Danielsen EM, Vyas JP, Kenny AJ (1980) A neutral endopeptidase in the microvillar membrane of pig intestine. Partial purification and properties. Biochem J 191:645–648

Danielsen EM, Cowell GM, Norén O, Sjostrom H (1984) Biosynthesis of microvillar proteins. Biochem J 221:1–14

Fulcher IS, Kenny AJ (1983) Proteins of the kidney microvillar membrane. The amphipathic forms of endopeptidase purified from pig kidneys. Biochem J 211:743–753

Fulcher IS, Matsas R, Turner AJ, Kenny AJ (1982) Kidney neutral endopeptidase and the hydrolysis of enkephalin by synaptic membranes show similar sensitivity to inhibitors. Biochem J 203:519–522

Fulcher IS, Chaplin MF, Kenny AJ (1983) Endopeptidase-24.11 purified from pig intestine is differently glycosylated. Biochem J 215:317–323

Gee NS, Matsas R, Kenny AJ (1983) A monoclonal antibody to kidney endopeptidase-24.11. Its application in immunoadsorbent purification of the enzyme and immunofluorescent microscopy of kidney and intestine. Biochem J 214:377–386

Gee NS, Bowes MA, Buck P, Kenny AJ (1985) An immunoradiometric assay for endopeptidase-24.11 shows it to be a widely distributed enzyme in pig tissues. Biochem J 228:119–126

George SG, Kenny AJ (1973) Studies on the enzymology of purified preparations of brush border from rabbit kidney. Biochem J 134:43–57

Hussain MM, Tranum-Jensen J, Norén O, Sjostrom H, Christiansen K (1981) Reconstitution of purified amphiphilic pig intestinal microvillus aminopeptidase. Biochem J 199:179–186

Kenny AJ (1977) Proteinases associated with cell membranes. In: Barrett AJ (ed) Proteinases in mammalian cells and tissues. Elsevier/North-Holland Biomedical, Amsterdam, pp 393–444

Kenny AJ, Booth AG (1978) Microvilli: their ultrastructure, enzymology and molecular organization. Essays Biochem 14:1–44

Kenny AJ, Fulcher IS (1983) Microvillar endopeptidase, an enzyme with special topological features and a wide distribution. Ciba Found Symp 95:12–33

Kenny AJ, Maroux S (1982) Topology of microvillar membrane hydrolases of kidney and intestine. Physiol Rev 62:91–128

Kenny AJ, Fulcher IS, McGill KA, Kershaw D (1983) Proteins of the kidney microvillar membrane. Reconstitution of endopeptidase in liposomes shows that it is a short-stalked protein. Biochem J 211:755–762

Kerr MA, Kenny AJ (1974a) The purification and specificity of a neutral endopeptidase from rabbit kidney brush border. Biochem J 137:477–488

Kerr MA, Kenny AJ (1974b) The molecular weight and properties of a neutral metallo-endopeptidase from rabbit kidney brush border. Biochem J 137:489–495

Lentzen H, Palenker J (1983) Localization of the thiorphan-sensitive endopeptidase, termed enkephalinase A, on glial cells. FEBS Lett 153:93–97

Malfroy B, Schwartz J-C (1982) Purification and substrate specificity of rat kidney 'enkephalinase'. Life Sci 31:1745–1748

Matsas R, Fulcher IS, Kenny AJ, Turner AJ (1983) Substance P and [Leu]enkephalin are hydrolyzed by an enzyme in pig caudate synaptic membranes that is identical with the endopeptidase of kidney microvilli. Proc Natl Acad Sci USA 80:3111–3115

Matsas R, Kenny AJ, Turner AJ (1984a) The metabolism of neuropeptides. The hydrolysis of peptides, including enkephalins, tachykinins and their analogues, by endopeptidase-24.11. Biochem J 223:433–440

Matsas R, Turner AJ, Kenny AJ (1984b) Endopeptidase-24.11 and aminopeptidase activity in brain synaptic membranes are jointly responsible for the hydrolysis of cholecystokinin octapeptide (CCK-8). FEBS Lett 175:124–128

Matsas R, Rattray M, Kenny AJ, Turner AJ (1985) The metabolism of neuropeptides. Endopeptidase-24.11 in human synaptic membrane preparations hydrolysis Substance P. Biochem J 228:487–492

Matsuda Y, Tsuji A, Katunuma N (1983) Studies on the structure of gamma-glutamyl transpeptidase III. Evidence that the amino terminus of the heavy subunit is the membrane binding segment. J Biochem (Tokyo) 93:1427–1433

Nash B, Tate SS (1982) Biosynthesis of rat renal gamma-glutamyl transpeptidase. Evidence for a common precursor of the two subunits. J Biol Chem 257:585–588

Nash B, Tate SS (1984) In vitro translation and processing of rat kidney gamma-glutamyl transpeptidase. J Biol Chem 259:678–685

Orlowski M, Wilk S (1981) Purification and specificity of a membrane-bound metalloendopeptidase from bovine pituitaries. Biochemistry 20:4942–4950

Pozsgay M, Michaud C, Orlowski M (1985) The active site of endopeptidase-24.11: substrate and inhibitor studies. Biochem Soc Trans 13:44–47

Relton JM, Gee NS, Matsas R, Turner AJ, Kenny AJ (1983) Purification of endopeptidase-24.11 ('enkephalinase') from pig brain by immunoadsorbent chromatography. Biochem J 215:317–323

Saklatvala J, Sarsfield SJ, Pilsworth LMC (1983a) Characterization of proteins from human synovium and mononuclear leucocytes that induce reabsorption of cartilage proteoglycan in vitro. Biochem J 209:337–344

Saklatvala J, Curry VA, Sarsfield SJ (1983b) Purification to homogeneity of pig leucocyte catabolin, a protein that causes cartilage reabsorption in vitro. Biochem J 215:385–392

Schwartz J-C, Malfroy B, d la Baume S (1983) Biological inactivation of enkephalin and the role of enkephalin-dipeptidyl-carboxypeptidase ('enkephalinase') as neuropeptidase. Life Sci 29: 1715–1740

Semenza G, Brunner J, Wacker H (1983) Biosynthesis and assembly of the largest and major intrinsic polypeptide of the small intestinal brush borders. Ciba Found Symp 95:92–112

Swerts JP, Perdrisot R, Patey G, De la Baume S, Schwartz J-C (1979) Enkephalinase is distinct from brain angiotensin-converting enzyme. Eur J Pharmacol 57:279–281

Stewart JR, Kenny AJ (1984a) Proteins of the kidney microvillar membrane. Biosynthesis of endopeptidase-24.11, dipeptidylpeptidase IV and aminopeptidases N and A in pig kidney slices. Biochem J 224:549–558

Stewart JR, Kenny AJ (1984b) Proteins of the kidney microvillar membrane. Effects of monensin, vinblastine, swainsonine and glucosamine on the processing and assembly of endopeptidase-24.11 and dipeptidyl peptidase IV in pig kidney slices. Biochem J 224:559–568

Suda H, Aoyagi T, Tacheuchi T, Umezawa H (1973) A thermolysin inhibitor produced by actinomycetes: phosphoramidon. J Antibiot (Tokyo) 26:621–623

Wong-Leung YL, Kenny AJ (1968) Some properties of a microsomal peotidase in rat kidney. Biochem J 110:5P

Yuli I, Snyderman R (1984) Extensive hydrolysis of f–Met–Leu–[^{3}H]Phe by a receptor-associated metalloprotease on human neutrophils, a potential mechanism for perception of a temporal chemotactic signal. Fed Proc 43:1418

Ectopeptidases Responsible for the Inactivation of Enkephalins

J.-C. Schwartz, C. Gros, B. Giros, C. Llorens, B. Malfroy, C. Rose, K. Zuzel, H. Pollard, and I. Pachot [1]

1 Introduction

Our current views about the enzymes responsible for the inactivation of "classical" neurotransmitters, like the monoamines, can be summarized as follows. In some cases an ectoenzyme, like acetylcholinesterase, efficiently cleaves the neurotransmitter shortly after its release because of its high catalytic rate and adequate localization on membranes of, or in the vicinity of, cholinergic neurons. In other cases (the monoamine methyltransferases, for instance) the responsible enzymes are intracellularly localized in a variety of cells (including the neurons of origin) into which the released neurotransmitter is taken up by more or less selective transport systems.

It is unlikely that the second scheme can apply to neuropeptides for which rapid uptake systems have so far not been established. On the other hand, there is already strong evidence in the case of the best studied neuropeptides, i.e., the enkephalins, that diffusion alone cannot account for their inactivation and that ectoenzymes play a crucial role.

The enkephalins can be cleaved by a variety of peptidases, but it is increasingly clear that two enzyme activities are mainly if not solely involved in their physiological inactivation (for reviews see Schwartz et al. 1981, 1982, 1983; Hui and Lajtha 1983; Schwartz 1983). The first, responsible for the cleavage of the Gly^3–Phe^4 amide bond of the opioid pentapeptides, is generally termed "enkephalinase" (Malfroy et al. 1978). It is primarily associated with synaptic membranes (De la Baume et al. 1980) and is potently and selectively inhibited by Thiorphan (Roques et al. 1980). Its inhibition allows an increased recovery of enkephalins released from striatal slices and, in vivo, a diminished responsiveness to various nociceptive stimuli (Patey et al. 1981; De la Baume et al. 1983; Chaillet et al. 1983a). In addition, an aminopeptidase activity sensitive to bestatin, an inhibitor of microbial origin (Umezawa et al. 1976) seems to be physiologically involved in the hydrolysis of endogenous enkephalins as shown by the strong potentiation of the effects of Thiorphan as regards both protection of the peptide and antinociceptive effect (De la Baume et al. 1983; Chaillet et al. 1983a).

Are these enzymes true "ectopeptidases" in the CNS? Although direct morphological studies are just starting there is strong indirect evidence that both enkephalinase and the aminopeptidase are located within plasma membranes in a position allowing them

[1] Unité 109 de Neurobiologie, Centre Paul Broca de l'INSERM, 2 ter rue d'Alésia, 75014 Paris, France

to catalyze the transformation of substrates present in the brain extracellular fluid. This is mainly indicated by experiments with brain slices: on this preparation which maintains to a large extent its cellular integrity, the two enzyme activities were shown to be responsible for the rapid hydrolysis of peptides in the extracellular fluid, i.e., either added into the bathing medium or released by depolarization (Patey et al. 1981; De la Baume et al. 1983). We report here recent data concerning the identity and properties of these two enzyme activities which strengthen the view that they indeed play a "neuropeptidase" role.

2 Enkephalinase: Substrates and Inhibitors

The first studies of enkephalinase specificity were conducted on crude cerebral membranes, using [^{3}H]enkephalins as substrates in the presence of inhibitors of interfering peptidases, like ACE or the aminopeptidases (Malfroy et al. 1978, 1979; Llorens et al. 1980). Although these studies allowed the establishment of the main structural features of peptides required for interaction with the enzyme in an inhibitory fashion (like the necessity of an aromatic or hydrophobic residue in P′1), they did not allow the definition of the true substrate specificity. The purification of the enzyme nearly to homogeneity showed (Almenoff et al. 1981; Fulcher et al. 1982; Malfroy and Schwartz 1982) that it was identical or closely similar to a so-called "neutral proteinase from rabbit kidney brush border" (E.C. 3.4.24.11) whose presence in brain had not been previously shown and whose function had remained unknown. The substrate specificity of the latter was mainly studied on a few natural peptides and it was characterized as a "thermolysin-like" enzyme because like this bacterial enzyme, (1) it is a Zn-containing peptidase, (2) it is strongly inhibited by phosphoramidon, and (3) it hydrolyses the amide bond comprising the amino group of an aromatic amino acid (Kenny 1977). In contrast, our previous studies with cerebral enkephalinase had underlined its similarities with another Zn-containing peptidase, angiotensin-converting enzyme (ACE) considered as a dipeptidylcarboxypeptidase and, therefore, classified (E.C. 3.4.15.1) as an exopeptidase. Indeed like enkephalinase, ACE is able to cleave, the enkephalins at the level of the Gly^3–Phe^4 amide bond, although with a lower affinity (Erdös et al. 1978). The rate of (Met^5)enkephalin hydrolysis by semipurified ACE is strongly decreased (by more than 30-fold) by amidification of the terminal carboxylate (Malfroy and Schwartz 1985; Schwartz et al. 1984) and, similarly this modification clearly reduces the specificity constant for enkephalinase (Malfroy and Schwartz 1982; see also Table 1). Hence, although ACE displays some endopeptidase activity (Inokuchi and Nagamatsu 1981; Bünning et al. 1983) it acts considerably more efficiently as a dipeptidylcarboxypeptidase and the same seems true for enkephalinase, even if the difference is less marked.

This latter view is further illustrated when the rate of hydrolysis of various opioid peptides is compared: extension of the C-terminal end of enkephalins progressively leads to a less effective recognition by enkephalinase so that the hydrolysis of β-endorphin (31 amino acids) or of dynorphin 1–9 becomes negligible as compared to that of the parent pentapeptides (Table 1). This strongly suggests that enkephalinase does not

Table 1. Kinetic parameters for the hydrolysis of various opioid peptides by enkephalinase[a]

	(1) Km (or Ki) (μM)	(2) Kcat (min)	Specificity constant [ratio (1)/(2)]
(Met^5)Enkephalin	1.3	329	253
(D–Ala^2, Met^5)Enkephalin	2.0	257	125
(D–Ala^2, Met^5–NH_2)Enkephalin	56	671	12
(Met^5, Arg^6, Phe^7)Enkephalin	20	362	187
β-Endorphin		< 20	
(Leu^5)Enkephalin	12	897	75
Dynorphin 1–6	38	332	9
Dynorphin 1–9	95	< 20	< 0.2

[a] The hydrolysis of the various peptides at the level of their Gly^3–Phe^4 bond was quantified by the appearance of the tripeptide Tyr–Gly–Gly (HPLC, UV absorbance) using enkephalinase highly purified from rat kidney (Malfroy and Schwartz 1982)

play any major role in the inactivation of opioid peptides of the β-endorphin and dynorphin families. In support of this view the analgesic effect of β-endorphin in mice is not significantly modified by co-administration of Thiorphan (J. Costentin, personal communication).

Regarding enkephalinase inhibitors, a series of transition state analogs have been developed, all based on an experimental model of the enzyme active site (Llorens et al. 1980; Schwartz et al. 1981; Malfroy and Schwartz 1982). The various compounds mainly differ in the nature of the Zn-chelating group which may be a thiol as in Thiorphan (Roques et al. 1980), a phosphoramidate as in phosphoramidon, a natural inhibitor (Kenny 1977; Malfroy and Schwartz 1982) or phosphoryl-Leu-Phe (Altstein et al. 1982), an hydroxamate (Hudgin et al. 1981) or a carboxyl as in the series of N-(l-carboxy, 3-phenylpropyl) Phe–X dipeptide derivatives (Mumford et al. 1982). The affinity of most compounds lies in the low nanomolar range and they constitute useful tools to evaluate the functional role of the enzyme.

3 Production of Monoclonal Antibodies Against and Localization of Enkephalinase in Ependymal Cells, and in Soluble Form in Cerebrospinal Fluid

Initial neurochemical studies in this laboratory on the cerebral enzyme indicated (1) its marked regional variations, (2) its association with subcellular fractions containing synaptic membranes, and (3) its decrease following administration of neurotoxins (Schwartz et al. 1981). All these data are consistent with the localization of at least a fraction of the cerebral activity within neuronal membranes. However, for a more precise localization, we have recently developed monoclonal antibodies (Mab) against the renal enzyme purified nearly to homogeneity (Malfroy and Schwartz 1982, 1984).

BALB/c mice received 30 μg antigen in 0.2 ml of complete Freund's adjuvant three times at 3 week intervals. Fusion of SP_2/oAg myeloma cells with BALB/c spleen cells was performed using PEG (Köhler and Milstein 1975). Hybrids, clones, and ascitic fluids were tested for the presence of antibodies using an ELISA test against the immunogen. Finally, confirmation of the presence of antibodies was checked by enkephalinase activity measurements. Although the Mab did not, at any dilution, modify the enzyme activity, indicating that they did not recognize the active site, the activity could be immunoprecipitated, using a double antibody system; immunoprecipitation occurred with the purified enzyme and, using a [^{125}I] labeled sheep anti-mouse γ-globulin antibody, recognition of the enzyme in crude membranes from rat kidney could be established. With the cerebral membranes the recognition by the Mab was less apparent, due to a relatively high background, a question under current investigation. Initial immunohistochemical studies using the PAP method with the Mab have already allowed us (1) to confirm the localization of enkephalinase in very high density in the brush border of rat *kidney tubuli* and in the *intestinal mucosa*, (2) to establish that the relatively high enkephalinase activity found in rat lung (Llorens and Schwartz 1981) corresponds to its high localization in *bronchial epithelium*, and (3) to evidence the presence of the enzyme in *ependymal cells* all along the cerebral ventricles, in *tanycytes* and in *choroid plexuses* (Fig. 1). From these initial data, it is interesting to note similarities and differences as compared with the related peptidase, angiotensin-converting enzyme (E.C. 3.4.15.1) which is also concentrated in the kidney and lungs (apparently in endothelial cells) and, finally, in choroid plexuses. The high enkephalinase level in choroid plexus was confirmed by a biochemical enzyme assay which indicated that the activity represented about fourfold that found in striatum. Current studies are aimed at the immunohistochemical localization of enkephalinase within the brain parenchyma.

Enkephalinase has been so far known as a strictly membrane-bound peptidase requiring relatively potent detergent treatments for solubilization. However, using a sensitive two-step fluorometric assay (Mumford et al. 1980; Spillantini et al., submitted), we were recently able to detect a soluble enkephalinase activity in several human body fluids, like the blood plasma, amniotic fluid, and CSF. In the latter the activity was, indeed, extremely low, but could still be characterized with regard to the apparent kinetic constants of substrates, like the enkephalins, or inhibitors like Thiorphan or phosphoramidon. These values did not differ significantly from those found with the highly purified or membrane-bound enzymes, indicating their close similarity from a catalytic point of view. There are, indeed, other examples of mainly membrane-bound enzymes also detectable in a soluble form as in the case, for instance, of acetylcholinesterase (Chubb et al. 1976). It is likely that the presence of soluble enkephalinase in CSF reflects its high localization in cells producing this fluid and lining the ventricular spaces in which it circulates. Current studies are aimed at determining whether its level in CSF constitutes a marker for pathological changes.

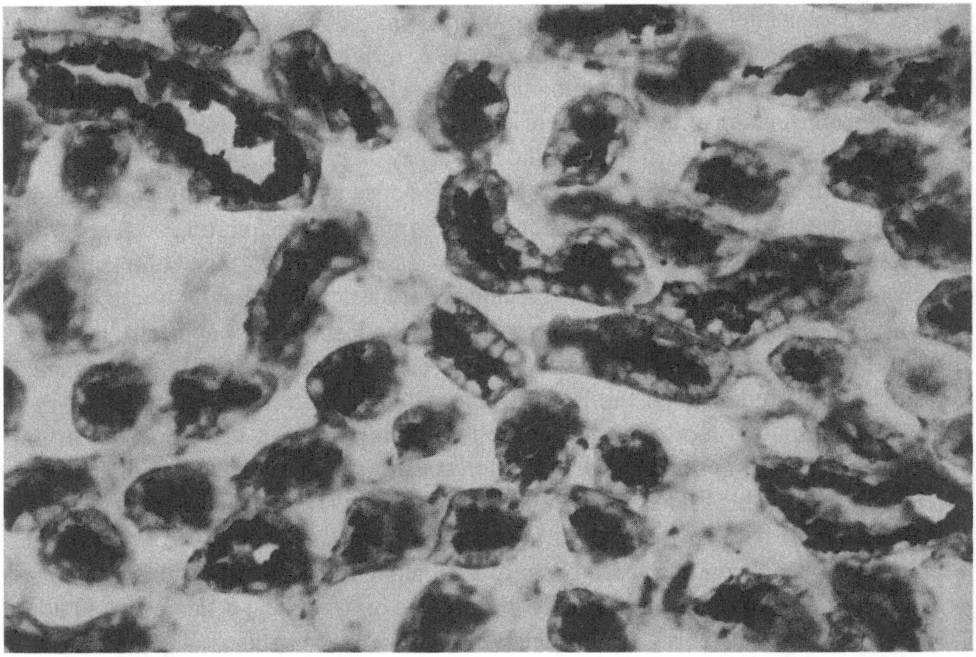

A

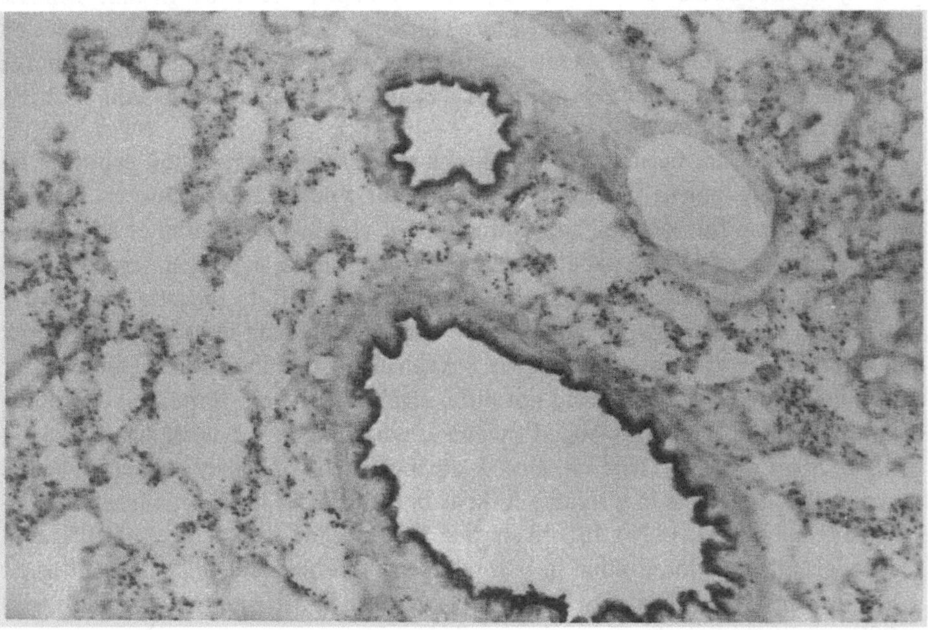

B

Fig. 1A–C. Immunohistochemical localization of enkephalinase with monoclonal antibodies using the PAP method in rat tissues. **A** Kidney cortex; **B** lung; **C** third cerebral ventricle

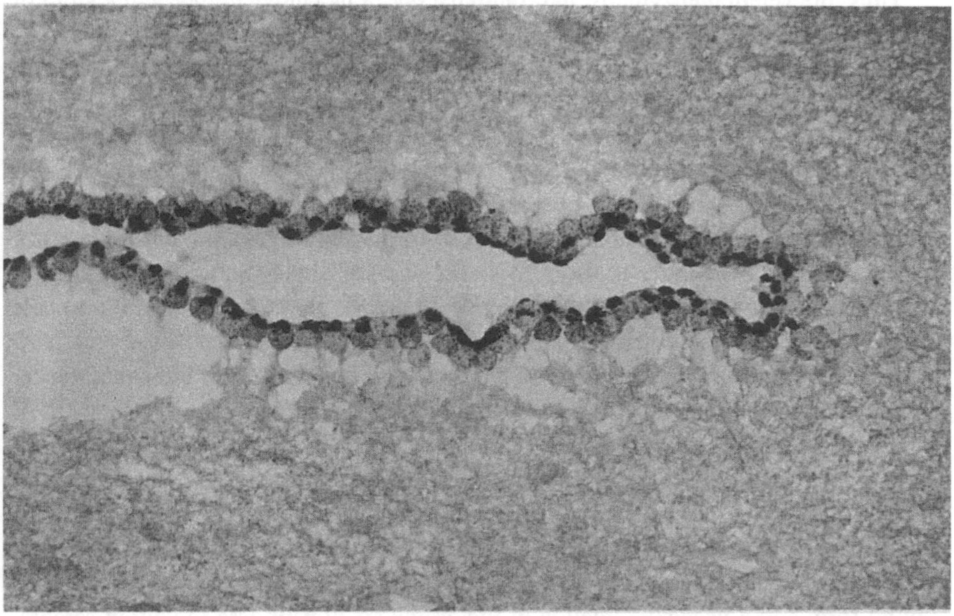

C

4 Isolation and Identification of Membrane-Bound Enkephalin-Hydrolyzing Aminopeptidases

The recovery of endogenous enkephalins released from brain slices can be improved when bestatin, an aminopeptidase inhibitor (Umezawa et al. 1976) is present in the medium (De la Baume et al. 1983), whereas puromycin, another aminopeptidase inhibitor, has no such effect (Patey et al. 1981). This may suggest that several enkephalin-hydrolyzing aminopeptidases with different sensitivities to inhibitors may be present in cerebral membranes, an hypothesis consistent with a preliminary study indicating that rat cerebral membranes contain two unidentified aminopeptidases for which puromycin displays a 1,000-fold difference in IC_{50} value (Hersh 1981). Recently an enkephalin-hydrolyzing aminopeptidase was purified almost to homogeneity from rat cerebral membranes (Hui et al. 1983), but its sensitivity to puromycin was not assessed and its relationships with the two above mentioned peptidases has not been clearly established.

Starting also from rat cerebral membranes, we have recently purified the solubilized aminopeptidase activity using a series of chromatographic steps (Gros et al. 1985). In a parallel manner the aminopeptidase M activity was purified from rat kidney membranes and conventional antibodies raised in rabbits; these were apparently directed against the active site as shown by the strong inhibition of enzyme activity they elicited. A series of data suggest that the "puromycin-insensitive" aminopeptidase from rat cerebral membranes corresponds to aminopeptidase M (E.C. 3.4.11.2) an enzyme bound to the membrane of kidney microvilli.

Thus, the cerebral enzyme is, like the renal one, a metallo-peptidase and a series of substrates, including the enkephalins and various naphthylamides, display similar affinities towards the renal and cerebral enzymes. In addition the immunotitration curves of the aminopeptidase activity from the two preparations are superimposable. Finally, the Ki values of inhibitors are the same on both enzymes, i.e., 100 μM for puromycin and 4 μM for bestatin (Gros et al. 1985).

These Ki values should be compared with those for the puromycin-sensitive aminopeptidase from cerebral membranes, i.e., 1 μM for puromycin and 0.5 μM for bestatin which shows that the latter is also a "bestatin-sensitive" aminopeptidase, even slightly more sensitive to this inhibitor than aminopeptidase M itself. This observation raises the possibility that the much better protection of endogenous enkephalins by bestatin than by puromycin could be due to the physiological involvement of either aminopeptidase M selectively or both peptidases from cerebral membranes. The development of inhibitors selectively directed against aminopeptidase M might help to select between these two hypotheses. In any case these data demonstrate the crucial role of aminopeptidase M in the inactivation of endogenous enkephalins, a rather remarkable result since this enzyme, the presence of which is in brain demonstrated for the first time, does not represent more than 10% of the overall aminopeptidase activity in membrane-bound form and a negligible fraction of the overall aminopeptidase activity in cerebral homogenates.

5 Neurochemical and Pharmacological Effects of Inhibitors of Enkephalin Metabolism

The hydrolysis of exogenous enkephalins in vitro by striatal slices (De la Baume et al. 1983; Hui et al. 1983) as well as in vivo after intraventricular administration (Chaillet et al. 1983a) is largely prevent by a combination of Thiorphan and bestatin. Even more interestingly, each of the two compounds partially protects endogenous enkephalins released from depolarised brain slices from degradation which, in their absence, occurs to the extent of more than 80% (Patey et al. 1981; De la Baume et al. 1983). These are the main neurochemical data which led to the proposition that two peptidase activities were equally important in the metabolism of endogenous enkephalins (Schwartz 1983).

In view of a large number of studies indicating that enkephalinergic neurons may impinge upon and modify the activity of monoaminergic neurons in CNS, it was of interest to establish whether the administration of inhibitors of enkephalin degradation may result in a modified turnover of catecholamines and serotonin (5-HT) in brain (Llorens and Schwartz 1984).

In the case of noradrenaline from mouse cerebral cortex, Thiorphan, bestatin, or the combination of the two inhibitors induced a significant decrease in the rate of α-methylparatyrosine (α-MPT)-induced decline (Table 2). The latter effect was completely prevented by naloxone, indicating that it presumably resulted from the stimulation of opiate receptors by endogenous opioid peptides. Synthetic opiates or opioid

Table 2. Effects of Thiorphan and bestatin on the α-methylparatyrosine decline in noradrenaline levels in mouse cerebral cortex

	Cortical noradrenaline		
	Levels[a] (ng g^{-1})		α-MPT-induced change (%)
Controls	246 ± 7	(26)	
α-MPT (200 mg kg^{-1}, i.p.)			
+ Solvent (i.c.v.)	153 ± 6	(33)	–38%
+ Thiorphan (50 μg, i.c.v.)	190 ± 7*	(8)	–23%
+ Bestatin (50 μg, i.c.v.)	197 ± 14**	(8)	–20%
+ Thiorphan (25 μg, i.c.v.) + Bestatin (50 μg, i.c.v.)	196 ± 9**	(19)	–20%
+ Solvent (i.c.v.) + Naloxone (10 mg kg^{-1}, i.p.)	138 ± 15[N.S.]	(15)	–44%
+ Thiorphan (25 μg) + Bestatin (50 μg) + Naloxone (10 mg kg^{-1})	133 ± 9[N.S.+]	(9)	–46%

[a] N.S. nonsignificant; * $P < 0.05$, ** $P < 0.001$ as compared to solvent. $P^{+} < 0.001$ as compared to Thiorphan + bestatin. All injections were performed 70 min prior to sacrifice. Means ± SEM of number of experiments indicated in parentheses

peptides are known to inhibit the firing of noradrenergic neurons (Korf et al. 1974) and the release of noradrenaline (Montel et al. 1974) presumably through opiate receptors borne by noradrenergic neurons (Llorens et al. 1978). In the same animals, we found that the α-MPT-induced decline of striatal dopamine was increased significantly (and in a naloxone-reversible manner) by the combination of Thiorphan and bestatin (Llorens and Schwartz 1984). Similarly, enkephalinase inhibitors increase striatal DOPAC levels in rat striatum (Algeri et al. 1981; Wood 1982). Finally, the two peptidase inhibitors increase 5-HT turnover in mouse cerebral cortex as judged from changes in the levels of 5-HIAA and 5-HT, again in a naloxone-reversible manner.

The various neurochemical effects of the two peptidase inhibitors, presumably mediated by opiate receptors, are likely to underlie the various behavioral responses already observed upon administration of these agents. Among the latter an antinociceptive activity has been described using a variety of tests (Roques et al. 1980; Schwartz et al. 1982; Chipkin et al. 1982; Yaksh and Harty 1982; Kayser and Guilbaud 1983; De la Baume et al. 1983; Chaillet et al. 1983a). Moreover, a Thiorphan analog was recently found to display significant analgesic activity in a double-blind test performed in human headache sufferers (Floras et al. 1983).

These two peptidase inhibitors also reproduce other behavioral responses known to occur following administration of exogenous opioids, such as in catatonic immobility (Chaillet et al. 1983b) or "behavioral despair" (Ben Natan et al. 1984).

6 Evaluation of the Role of Enkephalinase in the Inactivation of Cholecystokinin

As a rule peptidases display a limited substrate specificity generally determined by a few amino acid residues in the vicinity of the susceptible amide bond. Enkephalinase obeys this rule, being able to hydrolyze, in addition to enkephalins, a variety of neuropeptides containing aromatic or hydrophobic amino acid residues, such as oxytocin, bradykinin, neurotensin (Almenoff et al. 1981), or substance P (Matsas et al. 1983). Recently, Deschodt-Lanckman and Strosberg (1983) described the hydrolysis of cholecystokinin-8 (CCK-8) by a partially purified preparation from cerebral membranes resembling enkephalinase.

Using the purified enzyme (Malfroy and Schwartz 1982, 1984) a progressive disappearance of CCK-8 accompanied by a complex pattern of hydrolysis products can be demonstrated (Zuzel et al. 1985; Vargas et al. 1985; Schwartz et al. 1984). They indicate that the hydrolysis of CCK-8 mainly occurs at the two levels indicated below:

$$\text{Asp–Tyr}(SO_3H)\text{–Met–Gly}\overset{\downarrow}{-}\text{Trp–Met–Asp}\overset{\downarrow}{-}\text{Phe–}NH_2 \,.$$

However, since the appearance of Phe–NH_2 entirely accounts for the disappearance of intact CCK-8, this seems the primary cleavage by enkephalinase. The kinetic parameters of Phe–NH_2 formation are the following: K_m value of 60 μM and V_{max} of 6 μmol min^{-1} mg^{-1} to be compared to respective values of 1 μM and 3 μmol min^{-1} mg^{-1} for (Met^5)enkephalin which, therefore, displays a 25-fold higher "specificity constant" than CCK-8.

Table 3. Effect of Thiorphan (0.1 μM), an enkephalinase inhibitor, on the recovery of endogenous cholecystokinin-8 immunoreactivity (CCK-IR) released by depolarization of slices of rat cerebral cortex[a]

Condition	CCK-IR levels (fmol mg^{-1} protein)	
	Slices	Medium
2 mM K^+ (20)	2,930 ± 110	12 ± 3
50 mM K^+ (20)	2,370 ± 90**	101 ± 11**
Difference	560 ± 140	89 ± 11[bb] (15.9%)
2 mM K^+ + Thiorphan (15)	2,940 ± 160[N.S.]	17 ± 2[N.S.]
50 mM K^+ + Thiorphan (15)	2,320 ± 170*[N.S.]	117 ± 14**[N.S.]
Difference	620 ± 230[N.S.]	100 ± 14[N.S.b] (16.1%)

[a] After 5 min preincubation in the presence or absence of Thiorphan, slices were exposed to 50 mM K^+ for 5 min then separated from the media. Values in parentheses (left) represent number of incubations. Percent values in parentheses (right) represent the ratios of difference in medium to difference in slices (i.e., recovered/released). * $p < 0.05$, ** $p < 0.001$ when compared with 2 mM K^+ , bb: $p < 0.005$; b: $p < 0.05$ when compared to amount lost from slices. N.S. Not significantly different to corresponding value without Thiorphan

Furthermore, the possible role of enkephalinase in the inactivation of *endogenous* CCK-8 was assessed using a model which had allowed us previously to evaluate its role in the inactivation of endogenous enkephalins (Patey et al. 1981; De la Baume et al. 1983). Depolarized slices from rat cerebral cortex (or striatum) release a certain amount of CCK-8 of which less than 20% is recovered in the incubation medium in intact (immunoreactive) form, indicating that an extensive degradation takes place in the extracellular space of the tissue preparation before the neuropeptide reaches the medium in which it is diluted. However, Thiorphan, whose presence ensures, in the case of enkephalins, a marked increase (+100%) in recovery, does not significantly protect endogenous CCK-8 from hydrolysis (Table 3). This may suggest that although CCK-8 clearly is a substrate of enkephalinase, *endogenous* CCK-8 may not be inactivated mainly by this enzyme, due to either inappropriate localization or an insufficient hydrolysis rate compared with other (still largely unidentified) peptidases.

7 Conclusions

The role of enkephalinase (E.C. 3.4.24.11) in the inactivation of endogenous enkephalins proposed when the enzyme was demonstrated in cerebral membranes (Malfroy et al. 1978) has been largely confirmed. When inhibited together with a membrane-bound aminopeptidase sensitive to bestatin, i.e., aminopeptidase M (E.C. 3.4.11.2), a variety of neurochemical, electrophysiological, and behavioral responses mediated by opiate receptors occur. This leaves little doubt about the mode of termination of enkephalinergic signals, i.e., hydrolytic cleavage by two enzyme activities and not simple diffusion. Indeed, both enzymes are able to cleave in vitro a large variety of peptides besides the enkephalins, but their participation in the physiological inactivation of other neuropeptides remains to be established as illustrated in the case of cholecystokinin.

It is interesting to note that both enkephalinase and aminopeptidase M are typical peptidases from the brush borders of the intestinal mucosa and the renal proximal tubules, leaving little doubt about their ectoenzymatic nature. We had already postulated several years ago (Malfroy et al. 1978) that ectoenzymes were probably responsible for the "turning off" of neuropeptides and recent developments at least in the case of enkephalins seem to have largely substantiated this view.

8 Summary

A major role of ectopeptidases might be the inactivation of endogenous neuropeptides as illustrated in the case of enkephalins. Two enzyme activities, from cerebral membranes recently characterized in this laboratory, i.e., enkephalinase (E.C. 3.4.24.11; also designated neutral metalloendopeptidase, neutral proteinase from rabbit brush border membranes) and aminopeptidase M (E.C. 3.4.11.2; also designated microsomal aminopeptidase) seem to be involved in the inactivation of the endogenous opioid peptides. This is mainly shown by the effects of the two inhibitors Thiorphan and bestatin which protect the endogenous enkephalins from degradation, increase their cerebral level and induce a series of neurochemical and behavioral responses similar to those of opiates and reversed by opiate receptor antagonists.

Like most other peptidases, these two enzyme activities display limited substrate specificity, being able to hydrolyze a variety of neuropeptides. However, this does not automatically mean that they are physiologically involved in their inactivation as illustrated in the case of cholecystokinin, a substrate for enkephalinase. Factors like the specificty constants and the localization of the peptidase might be important to consider. Initial results on the immunohistochemical localization of enkephalinase with monoclonal antibodies are reported.

References

Algeri S, Altstein M, De Simone GM, Guardabasso V (1981) In vivo potentiation of (D-Ala^2)met-enkephalinamide central effects after administration of an enkephalinase inhibitor. Eur J Pharmacol 74:261–262

Almenoff J, Wilk S, Orlowski M (1981) Membrane bound pituitary metalloendopeptidase: apparent identity to enkephalinase. Biochem Biophys Res Commun 102:206–214

Altstein M, Blumberg S, Vogel Z (1982) Phosphoryl-Leu-Phe: a potent inhibitor of the degradation of enkephalins by enkephalinase. Eur J Pharmacol 76:299–300

Ben Natan L, Chaillet P, Lecomte JM, Marcais-Collado H, Huchida G, Costentin J (1984) Involvement of endogenous enkephalins in the mouse "behavioral despair" test. Eur J Pharmacol 97:301

Bünning P, Holmquist B, Riordan JF (1983) Substrate specificity and kinetic characteristics of angiotensin converting enzyme. Biochemistry 22:103–116

Chaillet P, Marcais-Collado H, Costentin J, Yi CC, De la Baume S, Schwartz JC (1983a) Inhibition of enkephalin metabolism by, and antinociceptive activity of, bestatin, an aminopeptidase inhibitor. Eur J Pharmacol 86:329–336

Chaillet P, Marcais-Collado H, Costentin J (1983b) Catatonic or hypotonic immobility induced in mice by intracerebroventricular injections of mu and kappa opioid receptor agonists as well as enkephalines or inhibitors of their degradation. Life Sci 33:2105–2111

Chipkin RE, Iorio LC, Barnett A, Berger J, Billard W (1982) In: Costa E, Trabucchi M (eds) Regulatory peptides: from molecular biology to function. Raven, New York, pp 235–242

Chubb IN, Goodman S, Schmith AD (1976) Is acetylcholinesterase secreted from central neurons into the cerebrospinal fluid. Neuroscience 1:57–62

De la Baume S, Patey G, Schwartz JC (1980) Subcellular distribution of enkephalin dipeptidylcarboxypeptidase ("enkephalinase") in rat brain. Neuroscience 6:315–321

De la Baume S, Yi CC, Schwartz JC, Chaillet P, Marcais-Collado H, Costentin J (1983) Participation of both "enkephalinase" and aminopeptidase activities in the metabolism of endogenous enkephalins. Neuroscience 8:143–151

Deschodt-Lanckman M, Strosberg D (1983) In vitro degradation of the C-terminal octapeptide of cholecystokinin by enkephalinase A. FEBS Lett 152:109–113

Erdös EG, Johnson AR, Boyden NT (1978) Hydrolysis of enkephalin by cultured human endothelial cells and by purified peptidyl dipeptidase. Biochem Pharmacol 27:843–848

Floras P, Bidabé AM, Caillé JM, Simonnet G, Lecomte JM, Sabathie M (1983) Double-blind study of effects of enkephalinase inhibitor on adverse reactions to myelography. Am J Neuroradiol 4:653–655

Fulcher IS, Matsas R, Turner AJ, Kenny AJ (1982) Kidney neutral endopeptidase and the hydrolysis of enkephalins by synaptic membrane show similar sensitivity to inhibitors. Biochem J 203:519–522

Gros C, Giros B, Schwartz JC (1985) Indentification of aminopeptidase M as an enkephalin inactivating enzyme in rat cerebral membranes. Biochemistry 24:2179–2185

Hersh LB (1981) Solubilization and characterization of two rat brain membrane-bound aminopeptidases active on Met-enkephalin. Biochemistry 20:2345–2350

Hudgin RL, Charleson SE, Zimmerman M, Mumford R, Wood PL (1981) Enkephalinase: selective peptide inhibitors. Life Sci 29:2593–2601

Hui KS, Lajtha A (1983) Neuropeptidases. In: Lajtha E (ed) Handbook of neurochemistry, vol 4. Plenum, New York, pp 1–19

Hui KS, Wang YJ, Lajtha A (1983) Purification and characterization of an enkephalin aminopeptidase from rat brain membranes. Biochemistry 22:1062–1067

Inokuchi JI, Nagamatsu A (1981) Tripeptidyl carboxypeptidase activity of kininase II (angiotensin converting enyzme). Biochem Biophys Res Commun 662:300–307

Kayser V, Guilbaud G (1983) The analgesic effects of morphine but not those of the enkephalinase inhibitor, Thiorphan, are enhanced in arthritic rats. Brain Res 267:131–138

Kenny AJ (1977) Proteinases associated with cell membranes. In: Barrett AJ (ed) Proteinases in mammalian cells and tissues. North-Holland, Amsterdam, pp 393–444

Köhler G, Milstein C (1975) Continuous cultures of fused cells secreting antibody of predefined specificity. Nature (Lond) 256:495–497

Korf J, Bunney BS, Aghajanian GK (1974) Noradrenergic neurons: morphine inhibition of spontaneous activity. Eur J Pharmacol 251:65–169

Llorens C, Schwartz JC (1981) Enkephalinase activity in rat peripheral organs. Eur J Pharmacol 69:113–116

Llorens C, Schwartz JC (1984) Changes in turnover of cerebral monoamines following inhibition of enkephalin metabolism by Thiorphan and bestatin. Eur J Pharmacol 104:369–374

Llorens G, Martres MP, Baudry M, Schwartz JC (1978) Hypersensitivity to noradrenaline in cortex after chronic morphine: relevance to tolerance and dependance. Nature (Lond) 274:603–605

Llorens C, Gacel G, Swerts JP, Perdrisot R, Fournie-Zaluski MC, Schwartz JC, Roques BP (1980) Rational design of enkephalinase inhibitors: substrate specificity of enkephalinase studied from inhibitory potency of various dipeptides. Biochem Biophys Res Commun 961:710–1717

Malfroy B, Schwartz JC 91982) Properties of "enkephalinase" from rat kidney: comparison of dipeptidyl-carboxypeptidase and endopeptidase activity. Biochem Biophys Res Commun 106: 276–285

Malfroy B, Schwartz JC (1984) "Enkephalinase" from rat kidney: purification, characterisation and study of substrate specificity. J Biol Chem 259:14365–14370

Malfroy B, Schwartz JC (1985) Comparison of dipeptidylcarboxypeptidase and endopeptidase activities in the three enkephalin-hydrolysing metallopeptidases: "angiotensin-convertiop enzyme", thermolysin and "enkephalinase". Biochem Biophys Res Commun 130:372–378

Malfroy B, Swerts JP, Guyon A, Roques BP, Schwartz JC (1978) High affinity enkephalin-degrading peptidase in brain is increased after morphine. Nature (Lond) 276:523–526

Malfroy B, Swerts JP, Llorens C, Schwartz JC (1979) Regional distribution of a high-affinity enkephalin-degrading peptidase ("enkephalinase") and effects of lesions suggest localization in the vicinity of opiate receptors in brain. Neurosci Lett 11:329–334

Matsas R, Fulcher IS, Kenny AJ, Turner AJ (1983) Substance P and (Leu)-enkephalin are hydrolysed by an enzyme in pig caudate synaptic membranes that is identical with the endopeptidase of kidney microvilli. Proc Natl Acad Sci USA 80:3111–3115

Montel H, Starke K, Weber F (1974) Influence of morphine and naloxone on the release of noradrenaline from brain cortex slices. Naunyn-Schmiedebergs Arch Pharmakol 283:357–359

Mumford RA, Strauss AW, Powers JC, Pierzchala PA, Nishino N, Zimmerman M (1980) A zinc metallo-endopeptidase associated with dog pancreatic membranes. J Biol Chem 255:2227–2230

Mumford RA, Zimmerman M, Broeke JT, Taub D, Joshua H, Rothrock SW, Hirshfield JM, Springer JP, Patchett AA (1982) Inhibition of porcine kidney "enkephalinase" by substituted-D-carboxymethyl dipeptides. Biochem Biophys Res Commun 109:1303–1309

Patey G, De la Baume S, Schwartz JC, Gros C, Roques BP, Fournie-Zaluski MC, Soroca-Lucas E (1981) Selective protection of methionine enkephalin released from brain slices by enkephalinase inhibition. Science (Wash DC) 212:1153–1155

Roques BP, Fournie-Zaluski MC, Soroca E, Lecomte JM, Malfroy B, Warens C, Schwartz JC (1980) The enkephalinase inhibitor Thiorphan shows antinociceptive activity in mice. Nature (Lond) 288:286–288

Schwartz JC (1983) Metabolism of enkephalins and the concept of neuropeptidase. Trends Neurosci 6:15–18

Schwartz JC, Malfroy B, De la Baume S (1981) Biological inactivation of enkephalins and the role of enkephalin-dipeptidyl-carboxypeptidase ("enkephalinase") as neuropeptidase. Life Sci 29: 1715–1740

Schwartz JC, De la Baume S, Chaillet P, Marcais-Collado H, Costentin J (1982) Enkephalin metabolism in brain and its inhibition. Prog Neurol Psychiatry 6:665–671

Schwartz JC, De la Baume S, Chaillet P, Marcais-Collado H, Costentin J (1983) Peptidases involved in the inactivation of exogenous and endogenous enkephalins. J Neurol Trans 18:235–243

Schwartz JC, Gros C, Giros B, Llorens C, Malfroy B, Rose C, Zuzel K (1984) Metabolic inactivation of enkephalins. In: Vizi ES, Magyar K (eds) Regulation of transmitter function: basic and clinical aspects. Akadémai, Budapest, pp 217–229

Umezawa H, Aoyagi T, Suda H, Hamada M, Takeuchi T (1976) Bestatin an inhibitor of aminopeptidase B, produced by actinomycetes. J Antibiot (Tokyo) 29:97–99

Vargas F, Frerot O, Dam Trung Tuong M, Zuzel K, Rose C, Schwartz JC (1984) Sulfation and sefulfation of cerebral cholecystokinin. In: Vanderhaeghen JJ, Crawley JN (eds) Neuronal cholecystokinin. Annals of the New York academy of Sciences Vol 448, New York, pp 110–120

Wood PL (1982) Phasic enkephalinergic modulation of nigrostriatal dopamine metabolism: potentiation with enkephalinase inhibitors. Eur J Pharmacol 82:119–120

Yaksh TL, Harty GM (1982) Effects of thiorphan on the antinociceptive actions of intrathecal ($Dala^2$, Met^5)enkephalin. Eur J Pharmacol 79:293–300

Zhang AZ, Yang HYT, Costa E (1982) Nociception, enkephalin content and dipeptidyl carboxypeptidase activity in brain of mice treated with exopeptidase inhibitors. Neuropharmacology 21: 625–630

Zuzel K, Rose C, Schwartz JC (1985) Assessment of the role of "enkephalinase" in cholecystokinin inactivation. Neuroscience 15:149–158

Cell Surface-Bound Leucine Aminopeptidase: Target of the Immunomodulator Bestatin

W. E. G. Müller [1], D. K. Schuster [1], G. Leyhausen [1], C. Sobel [1], and H. Umezawa [2]

1 Introduction

The study of low molecular weight enzyme inhibitors of microbial origin was initiated by Umezawa in 1965 (see Umezawa 1972). Since the discovery of an inhibitor of tyrosine hydroxylase, nearly 50 inhibitors of various enzymes have been found by him; their structures were elucidated and most of the compounds were chemically synthesized (Umezawa 1982). Among them one inhibitor of both aminopeptidase B and the ectoenzyme, leucine aminopeptidase was found in 1976 and was termed bestatin (Fig. 1), [(2S,3R)-3-amino-2-hydroxy 4-phenyl-butanoyl]-(S)-leucine (Umezawa et al. 1976).

2 Subcellular Aspects

Umezawa et al. (Ishizuka et al. 1981) studied the inhibitory activities of bestatin against purified aminopeptidase and found that the compound is a potent competitive inhibitor of leucine aminopeptidase(s) (Leu–AP) and aminopeptidase B (AP–B). He already established that at least one of these enzymes has the characteristics of ectoenzymes. In order to obtain further evidence as to which of these aminopeptidases are present on the cell surface, both biochemical and immunological studies were performed.

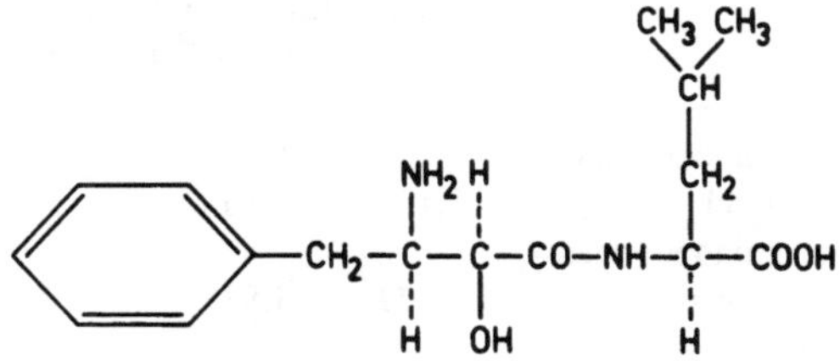

Fig. 1. Chemical structure of bestatin (2S,3R)-3-Amino-2-hydroxy-4-phenylbutanoyl-L-leucine (Bestatin)

[1] Institut für Physiologische Chemie, Universität, Abteilung für „Angewandte Molekularbiologie", Duesbergweg, 6500 Mainz, FRG

[2] Institute of Microbial Chemistry, 3-14-23 Kamiosaki, Shinagawa-ku, Tokyo 141, Japan

2.1 Biochemical Identification of Cell Surface Associated Leucine Aminopeptidase

The identification of the target enzyme for bestatin on the cell surface was approached by comparison with the purified enzymes cytosolic leucine aminopeptidase (Leu–APc; E.C. 3.4.11.1; swine kidney, Boehringer) and microsomal leucine aminopeptidase [Leu–APm; E.C. 3.4.11.2; swine kidney, Boehringer (Leyhausen et al. 1983a)]. The cell surface associated aminopeptidase (AP) was solubilized from cell membranes obtained from L5178y mouse lymphoma cells and subsequently partially enriched (Table 1). The two Leu–APs were found to be competitively inhibited by bestatin if Leu–NA was used as substrate; the stronger effect has been determined for Leu–APc ($K_i:K_m = 6.3 \times 10^{-6}$) compared to Leu–APm ($K_i:K_m = 4.2 \times 10^{-2}$) (Table 2). The

Table 1. Enrichment of cell membrane associated leucine aminopeptidase from L5178y cells[a]

Step	Total protein (mg)	Total enzyme activity (μmol min^{-1})	Specific activity (μmol mg^{-1} min^{-1})	Yield (%)	Purification -fold
Membrane fraction	57	1.85	0.0324	100	1
Nonidet treatment	2.8	1.31	0.47	71	14.5
Sephadex G-100	0.64	1.09	1.70	59	52.5

[a] Cell membranes were isolated from 7×10^{10} cells according to the Tris method, described by Warren et al. (1967). Subsequently the membranes were treated (3 h; 2 °C) with a Nonidet-Tris system (25 m*M* Tris-HCl, 1 *M* NaCl, 1% Nonidet NP-40; pH 7.3) and the particle free supernatant was obtained by centrifugation (100,000 × *g*; 60 min; 2 °C). Then the Nonidet fraction was passed through a Sephadex G-100 column. The leucine aminopeptidase activity appeared between Ve/Vo values (Determann 1969) of 1.0–1.22. The fractions were pooled and assayed for enzyme activity using Leu–NA as substrate (Leyhausen et al. 1983a)

Table 2. Kinetic properties of isolated leucine aminopeptidases and cell surface aminopeptidase(s)[a]

Enzyme preparation	K_m (μM)	V_{max} (μmol mg^{-1} min^{-1} or pmol 10^{-5} cells min^{-1}	$\frac{V_{max}}{K_m}$	K_i bestatin (μM)	$\frac{K_i}{K_m}$ ratio
Leucine aminopeptidase (cytosolic)	1,420	4.26	0.003	0.009	6.3×10^{-6}
Leucine aminopeptidase (microsomal)	73	5.84	0.08	3.03	4.2×10^{-2}
Aminopeptidase (cell surface bound)	63	330.2	5.24	13.6	2.2×10^{-1}
Aminopeptidase (cell membrane bound)					
a) particle associated	58	0.0324	0.0006	4.1	7.1×10^{-2}
b) solubilized	69	1.70	0.02	2.3	3.3×10^{-3}

[a] The procedures for the evaluation of the kinetic properties have been given earlier (Leyhausen et al. 1983a). The solubilized cell membrane associated aminopeptidase fraction was obtained as described in Table 1. For the determination of the aminopeptidase activity(ies) on the surface of intact cells, 6×10^5 L5178y cells were added to the 1 ml assays

cell surface associated AP activity was also inhibited by bestatin. Using Leu–NA as substrate, the inhibition is of the competitive type with a potency comparable to that determined for isolated Leu–APm (K_i:K_m = 7.1×10^{-2}). After fractionation and partial purification the cell membrane bound AP retained its enzymatic activity and its sensitivity to bestatin action. The catalytic activity (V_{max}:K_m ratio) of the solubilized cell membrane associated AP is almost identical with the one determined for Leu–APm and is distinct from the catalytic activity of Leu–APc.

A direct identification of the cell surface associated AP as Leu–APm was achieved firstly by in situ detection of the solubilized enzyme in polyacrylamide gels (Fig. 2) and by immunological techniques (see Sect. 2.2). It was found that the activity released from cell membrane preparations comigrated with the isolated Leu–APm, while its mobility in the gel differed from that characteristic for Leu–APc. Furthermore, this activity was proven to be inhibited by bestatin (Leyhausen et al. 1983a).

The determination of the AP activity detectable on the surface of a series of mammalian cells revealed the lowest activity on normal lymphocytes obtained from mouse spleen (Table 3). Higher activities have been measured on the surface of two tumor cell lines (L5178y and P815); impressively high are the activities on macrophages and

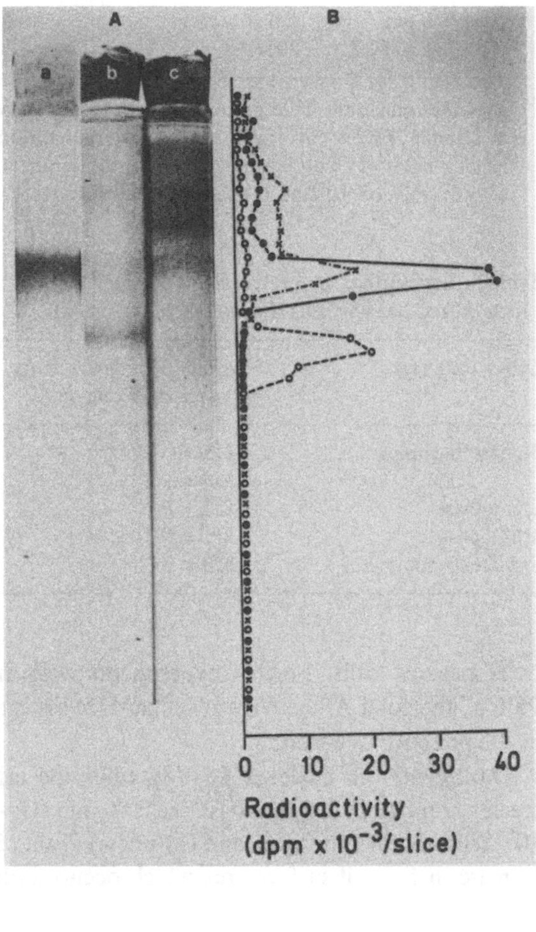

Fig. 2A,B. Identification of cell membrane-associated aminopeptidase by gel electrophoresis. **A** Separation of purified Leu–APm (10 µg gel^{-1}) (*a*), of purified Leu–APc (40 µg gel^{-1}) (*b*), and of solubilized membrane fraction [supernatant obtained after Triton extraction (Leyhausen et al. 1983a)] (*c*). The gels were incubated with 300 *µM* Leu–NA to trace the enzyme activity. **B** In situ binding of [^{3}H]-bestatin to Leu–APm (•–•), to Leu–APc (o– –o), and membrane-associated AP (x –.–x). Enzyme preparations, at the same concentrations as used under **A**, were separated by polyacrylamide gel electrophoresis and subsequently incubated with [^{3}H]-bestatin as described in detail (Leyhausen et al. 1983b)

Table 3. Activity of cell surface bound aminopeptidase and binding capacity for [^{3}H]-bestatin of different mouse cells (Müller et al. 1982; Leyhausen et al. 1983a)

Cells	Aminopeptidase activity (pmol 10^{-5} cells min^{-1}		Binding of [^{3}H]-bestatin (molecules/cell)
	Minus bestatin	32 μM Bestatin	
L5178y cells	115.1 ± 9.9	93.8 ± 5.4	134 ± 9 × 10^4
P815 cells	113.2 ± 8.1	88.2 ± 4.7	187 ± 11 × 10^4
T-lymphocytes (spleen)	33.8 ± 2.3	20.8 ± 2.6	254 ± 17 × 10^4
B-lymphocytes (spleen)	49.9 ± 2.8	30.9 ± 2.2	92 ± 6 × 10^4
Macrophages	791.3 ± 23.4	205.7 ± 8.7	1,459 ± 95 × 10^4
Bone marrow	330.4 ± 13.8	240.9 ± 9.3	Not determined

Table 4. Aminopeptidase activity in macrophages, as determined with Leu–NA as substrate under assay conditions described earlier (Leyhausen et al. 1983a)[a]

Compound added	Aminopeptidase activity (pmol 10^{-5} cells min^{-1}		
	Cell surface associated	Homogenate	Homogenate sonicated
Bestatin; 0 μM	791.3 ± 23.1	829.4 ± 23.9	824.6 ± 24.2
32 μM	205.7 ± 8.4	219.5 ± 7.4	228.3 ± 6.7

[a] The determinations were performed with (*a*) intact cells, (*b*) homogenates (cells homogenized in 20 m*M* HEPES buffer pH 7.2 and 100 m*M* NaCl using a Dounce potter), and (*c*) sonicated homogenates (15 s at setting 2 of a Branson sonifier). The activities are referred to 10^5 cells or cell equivalence. 10^6 macrophages were used for each set of experiments

Table 5. Cell surface associated aminopeptidase activity and bestatin binding during cell cycle of synchronized L5178y cells (Leyhausen et al. 1983a)

Cell cycle phase	Aminopeptidase activity (pmol 10^{-5} cells min^{-1}	[^{3}H]-Bestatin binding (molecules/cell)
S-phase; beginning	115 ± 6	172 ± 10 × 10^4
mid-	153 ± 8	328 ± 18 × 10^4
S/G_2-phase	151 ± 7	484 ± 26 × 10^4
M/G_1-phase	111 ± 5	299 ± 19 × 10^4
S-phase, beginning	129 ± 7	183 ± 12 × 10^4

bone marrow cells. Further experiments with macrophages revealed (Table 4) that 95% of the total AP activity is cell surface associated; 75% of this activity is inhibited in the presence of bestatin.

During the cell cycle of L5178y cells, the cell surface associated AP activity increases from 111–115 pmol/10^5 cells × min (G_1- and onset of S-phase) to 150 pmol/10^5 cells × min (during S- and G_2-phase) (Table 5). This change is not only due to an increase in the cell surface area which occurs during the S- and G_2-phase, but must be attributed to a higher enzyme density on the surfaces of S- and G_2-cells as well (Leyhausen et al. 1983a).

2.2 Immunochemical Identification of Cell Surface Associated Leucine Aminopeptidase

To support the biochemical data, we performed immunochemical studies to identify the cell surface associated AP (Leyhausen et al. 1983b). Antibodies were prepared against purified Leu–APm (swine kidney, Boehringer) and AP–B [E.C. 3.4.11.6; purified from mouse liver by the method of Hopsu et al. (1966)]. They were determined to function monospecifically with respect to the enzyme spectrum Leu–APm, Leu--APc, and AP–B. Using permeabilized L5178y cells fixed with ethanol, it was established that both Leu–APm and AP–B are distributed intracellularly. However, in nontreated cells, fluorescence could be developed with anti-Leu–APm only; anti-AP–B failed to label the cells (Leyhausen et al. 1983b). Moreover, it was demonstrated that fluorescence was restricted to the cell surface after incubation with anti-APm (Fig. 3A).

The cell surface associated AP was further characterized by immunodiffusion experiments using rabbit anti-Leu–APm. They revealed (Fig. 3B) that the antiserum reacted both with the purified kidney Leu–APm and with the cell surface associated AP with

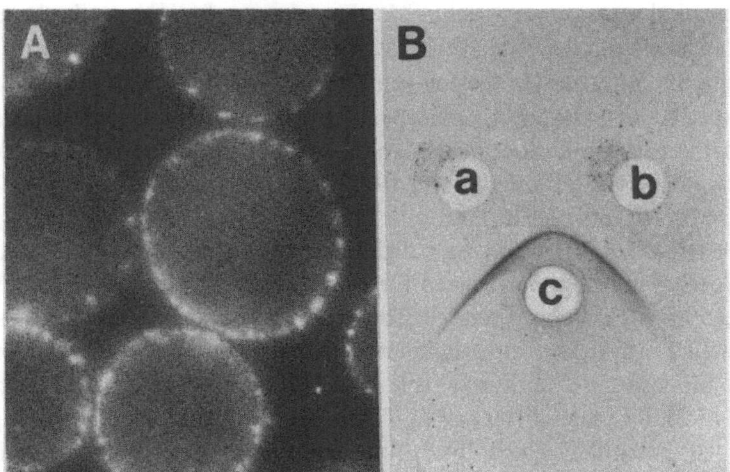

Fig. 3A,B. Identification of the cell surface-associated AP as Leu–APm. **A** Nonfixed L1578y cells were incubated with antibodies against Leu–APm (anti-Leu-APm; from rabbit) and after washing in PBS, treated with fluorescein-conjugated goat and anti-rabbit IgG (Leyhausen et al. 1983b). *Bar* 10 μm. **B** Immunodiffusion experiments with anti-Leu–APm. The center well (*c*) contained anti-Leu–APm in a dilution of 1:128. The outer wells contained purified swine kidney Leu–APm (*a*) or cell membrane-associated AP, solubilized and enriched from L5178y cells as summarized in Table 1 *(b)*

formation of a strong precipitin line in a reaction of identity. This means that the antiserum cannot distinguish the two antigens from each other. Anti-AP–B was determined not to react with the membrane derived preparation (data not shown).

2.3 Binding Studies with [^{3}H]-Bestatin

In a further series of studies, we prepared [^{3}H]-bestatin to demonstrate the binding of this compound to the surface of L5178y cells (Müller et al. 1979). Scatchard plots indicated that only one form of "receptor" exists; the K_a was determined to be $0.8 \times 10^5\ M^{-1}$. The rate of binding was temperature dependent with an optimum at 37 °C and it, therefore, appears to be dependent on cell metabolism. At a higher temperature (45 °C) bestatin which was bound at "physiological" temperature to the cells and was released into the medium without having been degraded. This result together with earlier inhibition studies (Umezawa 1981; Ishizuka et al. 1981; Aoyagi et al. 1977) strongly indicates that bestatin is not sequestered inside the cell. Based on the finding that L-leucine-β-naphthylamide inhibited the binding of bestatin to the cell membrane we propose that the bestatin receptor is identical with the Leu–APm.

As shown above (Sect. 2.1) the cell surface associated AP was found to comigrate in polyacrylamide gels with Leu–APm, but not with Leu–APc. This finding already implied that the solubilized membrane associated AP is the target enzyme for bestatin. To support this assumption in a direct way, in situ binding studies in polyacrylamide gels with [^{3}H]-bestatin were performed. After incubation of the gels (after the chromatographical separation procedure) in the presence of [^{3}H]-bestatin the radioactivity peak coincided with the band of enzyme activity (Fig. 2B). On the basis of these biochemical data we could answer the following two questions (Müller et al. 1982); is the binding capacity for bestatin (1) cell type dependent and (2) cell cycle dependent? The results of this study indicated that B-lymphocytes (from mouse spleen) have only relatively few binding sites for bestatin (92×10^4/cell), while L5178y cells (T-lymphoma; 134×10^4/cell), P815 mast cells (187×10^4/cell), and T-lymphocytes (from mouse spleen; 254×10^4/cell) have a slightly higher binding capacity for bestatin. The highest capacity has been determined for macrophages (from nonirritated CBA/J mice) with $1{,}459 \times 10^4$ binding sites/cell. The values for the number of binding sites per cell type parallel with those for the aminopeptidase activity, were determined on the cell surface of the respective cell type (Table 3): L5178y cells (115.1 pmol/10^5 cells × min) P815 (113.2), B-Lymphocytes (49.9), T-lymphocytes (33.8), and macrophages (791.3).

Of potentially chemotherapeutic importance are the studies which revealed that both the number of bestatin binding sites and the activity of the cell surface bound Leu–AP activity changes during the cell cycle (Müller et al. 1982; Leyhausen et al. 1983a). The experiments were performed with synchronized L5178y cells (Table 5). The lowest levels of bestatin binding are observed during the G_1-phase (1.4×10^6 molecules/cell) and early S-phase ($1.3-2.0 \times 10^6$), while the highest level was determined during the transition from S- to G_2-phase (4.8×10^6). Again parallel with this change, a fluctuation of Leu–AP activity occurs. The lowest activity was determined during G_1-phase (111–115 pmol/10^5 cells × min); significantly higher activities were measured at mid-S-phase (153 pmol) and at S/G_2-phase (151 pmol).

The conclusion that one of the cell surface associated APs is identical with Leu–APm (E.C. 3.4.11.2) can be drawn from both biochemical and immunochemical data summarized above. Moreover, the in vitro (intact cell studies) and in situ (polyacrylamide gel experiments) binding studies strongly suggest that the cell surface associated Leu–APm is the target enzyme (receptor) for bestatin. This deduction is additionally supported by a stoichiometric comparison revealing that the values for the bestatin binding sites and for the Leu–APm molecules (on the surface) per cell are very similar (e.g., for macrophages around 5×10^6) (Leyhausen et al. 1983b).

3 Cellular Aspects

Based on the finding that bestatin augmented establishment of delayed-type hypersensitivity against red blood cells (Aoyagi et al. 1977), we studied the activation of DNA metabolism by this compound in vivo (Müller et al. 1979; Müller 1981). Choosing CBA/J mice, we showed that bestatin caused both an increase of [^{3}H]-thymidine incorporation rate and a stimulation of the DNA-synthesizing systems (DNA polymerase α and terminal deoxynucleotidyl transferase) only in lymphoid tissue, in particular in the T-cells from spleen and thymus, as well as in lymphocytes from bone marrow.

The same effect of bestatin on lymphocytes was observed also in in vitro studies (Ishizuka et al. 1981; Müller 1981). The experiments revealed that the ectoenzyme inhibitor caused an increase of [^{3}H]-thymidine incorporation into lymphocytes in vitro only under such conditions at which macrophages are present in the cultures.

4 Summary

Based on the reported in vivo and in vitro results, the following series of events which might be triggered by the immunomodifier bestatin can be formulated (Fig. 4). The AP inhibitor bestatin has been shown to bind to the surface of mammalian cells, and especially to those of macrophages. The receptor for this microbial product was identified as the ectoenzyme Leu–APm. In response to the competitive inhibition of this enzyme, the uptake rate of Urd increases (Koren et al. 1981), resulting in an activation of macrophages. An induction and maintaining of naive T-cell proliferation to natural killer cells occurs (Aoike et al. 1981) very likely through an increased production of interleukin 1 and 2 (Ishizuka and Umezawa 1983). Moreover, bestatin causes an induction of DNA polymerase α and TdT in lymphocyte-precursor cells both in vivo (bone marrow) and in vitro (Molt-4 cells), suggesting that this microbial product positively influences the growth capacity of these primordial cells of the immune system.

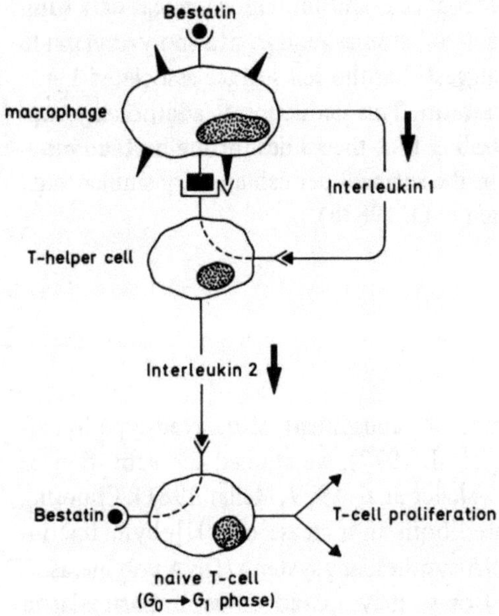

Fig. 4. Proposed mode of action of bestatin on macrophage-lymphocyte system in vivo

Acknowledgment. This work was supported by a grant from the Deutsche Krebshilfe e.V. (W.E.G. M.).

References

Aoike A, Tanaka Y, Hosokawa T, Yamaguchi N, Kawai K (1981) Effect of bestatin on natural killer activity. In: Umezawa H (ed) Small molecular immunodifiers of microbial origin. Pergamon, Oxford, pp 101–108

Aoyagi T, Ishizuka M, Takeuchi T, Umezawa H (1977) Enzyme inhibitors in relation to cancer therapy. Jpn J Antibiot 33 S:S121–S132

Determann H (1969) Gel chromatography. Springer, Berlin Heidelberg New York

Hopsu VK, Mäkinen KK, Glenner GG (1966) Purification of a mammalian peptidase selective for N-terminal arginine and lysine residues: aminopeptidase B. Arch Biochem Biophys 114:557–566

Ishizuka M, Umezawa H (1983) Effect of bestatin on experimental animal tumors. 13th Int Congr Chemother Wien 1983. Abstract book SY 67, part 204, pp 204/3–204/8

Ishizuka M, Aoyagi T, Takeuchi T, Umezawa H (1981) Activity of bestatin: enhancement of immune responses and antitumor effect. In: Umezawa H (ed) Small molecular immunomodifiers of microbial origin. Pergamon, Oxford, pp 17–38

Koren R, Bercovitz H, Umezawa H, Müller WEG (1981) The role of bestatin, an inhibitor of cell surface proteases, in the interaction of serum with untransformed cells in culture. J Antibiot (Tokyo) 34:90–94

Leyhausen G, Schuster DK, Vaith P, Zahn RK, Umezawa H, Falke D, Müller WEG (1983a) Identification and properties of the cell membrane bound leucine aminopeptidase interacting with the potential immunostimulant and chemotherapeutic agent bestatin. Biochem Pharmacol 32: 1051–1057

Leyhausen G, Gramzow M, Steffen R, Zahn RK, Umezawa H, Müller WEG (1983b) Immunochemical identification of the cell surface bound leucine aminopeptidase, the target enzyme for the immunostimulant bestatin. J Antibiot (Tokyo) 36:728–734

Müller WEG (1981) Effect of bestatin on DNA and RNA metabolism in T-cells. In: Umezawa H (ed) Small molecular immunomodifiers of microbial origin. Pergamon, Oxford, pp 39–58

Müller WEG, Zahn RK, Arendes J, Musch N, Umezawa H (1979) Activation of DNA metabolism in T-cells by bestatin. Biochem Pharmacol 28:3131–3137

Müller WEG, Schuster DK, Zahn RK, Maidhof A, Leyhausen G, Falke D, Koren R, Umezawa H (1982) Properties and specificity of binding sites for the immunomodulator bestatin on the surface of mammalian cells. Int J Immunopharmacol 4:393–400

Umezawa H (1972) Enyzme inhibitors of microbial origin. University of Toyko Press, Tokyo, pp 1–114

Umezawa H (ed) (1981) Small molecular immunomodifiers of microbial origin. Pergamon, Oxford, pp 1–254

Umezawa H (1982) Low molecular-weight enzyme inhibitors of microbial origin. Annu Rev Microbiol 36:75–99

Umezawa H, Aoyagi T, Suda H, Hamada M, Takeuchi T (1976) Bestatin, an inhibitor of aminopeptidase B, produced by actinomycetes. J Antibiot (Tokyo) 29:97–99

Warren L, Glick MC, Nass NK (1967) Methods of isolation of the surface membrane. J Cell Physiol 68:269–288

The Intramolecular Turnover Heterogeneity, a New Property of Plasma Membrane Glycoproteins. Dipeptidylaminopeptidase IV as a Model

W. KREISEL [1], R. BÜCHSEL [1], W. REUTTER [2], and W. GEROK [1]

1 Introduction

The glyoproteins of the hepatocyte plasma membrane are involved in a variety of functions, such as enzymatic activity, antigenicity, receptor and transport function, and cellular communication (Hughes 1976). Their actual concentration in the membrane is regulated by de novo synthesis, insertion into and removal from the membrane, including an eventual recycling, and degradation (for reviews on membrane glycoprotein turnover see Doyle et al. 1982; Hubbard 1978; Blobel 1980; Struck and Lennarz 1980). The different steps of this turnover are schematically depicted in Fig. 1 (taken from Kreisel et al. 1983). Most membrane glycoproteins are synthesized on membrane-bound polysomes and inserted into the endoplasmatic reticulum membrane. Via dolichol phosphate a preliminary carbohydrate chain is added, which is further processed. The glycoproteins are finally glycosylated in the Golgi complex. The membrane vesicles are transported and fuse with the plasma membrane (pathway 1). The glycoproteins are removed from the membrane mainly by internalization of membrane units. These vesicles may contain a random distribution of all membrane glycoproteins (pathway 2, 3, 4); some special molecules, namely receptors, are concentrated in coated pits and

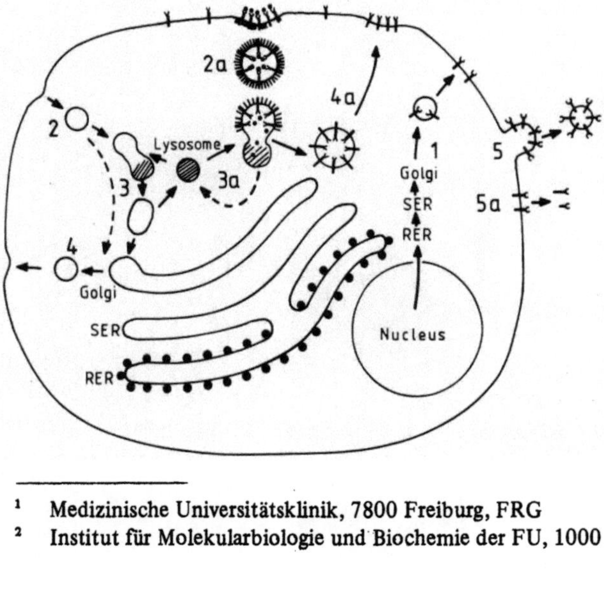

Fig. 1. Pathways of plasma membrane glycoprotein turnover (for details see text). *SER*, smooth endoplasmatic reticulum; *RER*, rough endoplasmatic reticulum

[1] Medizinische Universitätsklinik, 7800 Freiburg, FRG
[2] Institut für Molekularbiologie und Biochemie der FU, 1000 Berlin, FRG

vesicles (pathway 2a, 3a, 4a). After fusion with lysosomes the glycoproteins can be degraded. For a variety of membrane glycoproteins, a recycling back to the plasma membrane has been proven. Examples for this process are the receptors for low density lipoprotein, insulin, lysosomal enzymes, EGF, desialylated serum glycoproteins, and 5′-nucleotidase (for reviews see Steinman et al. 1983; Ashwell and Harford 1982; Dunn and Hubbard 1984). Additionally some glycoproteins can be released from the membrane by a shedding (pathway 5, 5a) (Doljanski 1982).

The glycoproteins of liver and hepatoma plasma membrane are not turned over in bulk, but exhibit a marked intermolecular heterogeneity with respect to their individual turnover rates and half-lives, i.e., different glycoproteins of the plasma membrane turn over at individual rates (Doyle et al. 1982; Vischer and Reutter 1980; Tauber and Reutter 1978; Büchsel et al. 1980; Tanabe et al. 1979; Bock et al. 1970; Elovson 1980; Kreisel et al. 1980). We report on a further aspect of turnover kinetics of plasma membrane glycoproteins, the intramolecular turnover heterogeneity. This term describes the observation that some glycoproteins are not turned over as a whole, but there exists an intramolecular turnover of terminal carbohydrates: terminal carbohydrates turn over several times during the life span of these membrane glycoproteins of the liver. This effect was first demonstrated for dipeptidylaminopeptidase IV isolated from rat liver plasma membranes (Kreisel et al. 1980).

Dipeptidylaminopeptidase IV (DPP IV, EC 3.4.14.5) is a dimeric enzyme consisting of two identical subunits. It splits off dipeptides from the N-terminus of polypeptides with specificity for proline as the penultimate amino acid. This enzyme has been found in a variety of organs: liver, kidney, intestinal microvilli, T-helper lymphocytes, pancreas, brain, fibroblasts, endothelia of blood vessels, and serum. Its biological function remains to be established. Probably it plays a role in the regulation of the regional blood flow and of the blood clotting. Another possibility could be an involvement in dipeptide transport. Until now not all criteria for its being an ectoenzyme have been applied to this enzyme. One hint in this direction is the fact that the membrane-bound enzyme acts on the external substrate, fibrinogen (Heymann and Mentlein 1984) and the neurotransmitter and gut peptide, substance P (Conlon and Sheehan 1983).

As outlined below, the intramolecular turnover of terminal carbohydrates seems to be a general phenomenon of plasma membrane glycoproteins. Preliminary unpublished data suggest that this effect can also be shown for the ectoenzyme 5′-nucleotidase. Since nearly all ectoenzymes are glycosylated plasma membrane proteins, we feel it is justified to add a chapter to this book on ectoenzymes, which deals with a new aspect of plasma membrane glycoprotein turnover.

2 Turnover of Carbohydrate and Polypeptide Moieties of DPP IV

Using a four-steep procedure, we isolated a glycoprotein from rat liver plasma membrane. It proved to be pure on SDS gels and on isoelectric focusing gels. The relative molecular weight is 110,000, the isoelectric point about 4.5. It contains fucose, sialic acid, galactose, mannose, and N-acetylglucosamine. The molar ratio of the carbohydrates and the lectin-binding capacity of this glycoprotein suggest a biantennary complex type

Table 1. Glycoprotein employed for turnover studies of carbohydrate and protein moiety

Isolation	Characterization
1. Extraction of plasma membranes with Triton X-100	M_r: 110,000
	Isoelectric point: 4.5
2. Affinity chromatography on Concanavalin A	Carbohydrate chain(s): fucose, sialic acid, galactose, mannose, N-acetylglucosamine
3. Affinity chromatography on wheat germ lectin	probably complex type structure, partly core-fucosylated
4. Preparative SDS-gel electrophoresis	
Identification	
Monomer of dipeptidylaminopeptidase IV (DPP IV, E.C. 3.4.14.5)	

structure of the carbohydrate chain(s) which is partly core-fucosylated (Kreisel et al. 1980). By immunoprecipitation and electrophoretic methods, we identified this glycoprotein as dipeptidylaminopeptidase IV (Kreisel et al. 1982; Table 1).

In pulse-chase experiments we measured the half-lives of the respective components of DPP IV (Table 2):

1. The terminal carbohydrates: the fucose half-life is 12 h, galactose 20 h, and sialic acid 30 h.
2. The inner core carbohydrates: the half-life of mannose amounts to 56 h, of N-acetylglucosamine 42 h.
3. The polypeptide chain: after labeling with ^{35}S-methionine we obtained 70 h and after ^{14}C-guanidolabeled arginine, 78 h.

From these results one must conclude that terminal carbohydrates turn over several times during the life cycle of DPP IV. An intramolecular turnover of terminal carbohydrates takes place (Kreisel et al. 1980). In contrast the inner core carbohydrates behave similarly to the polypeptide chain (Volk et al. 1983).

This intramolecular heterogeneity of turnover is not restricted to DPP IV. Tauber et al. (1983) and Hanski et al. (1984) found similar results for additional liver plasma

Table 2. Half-lives of the components of dipeptidylaminopeptidase IV isolated from rat liver plasma membrane

Component		Half-live (h)
Terminal carbohydrates	Fucose	12
	Sialic acid	30
	Galactose	20
Inner core carbohydrates	Mannose	56
	N-Acetylglucosamine	42
Polypeptide chain	Methionine	70
	Arginine	78

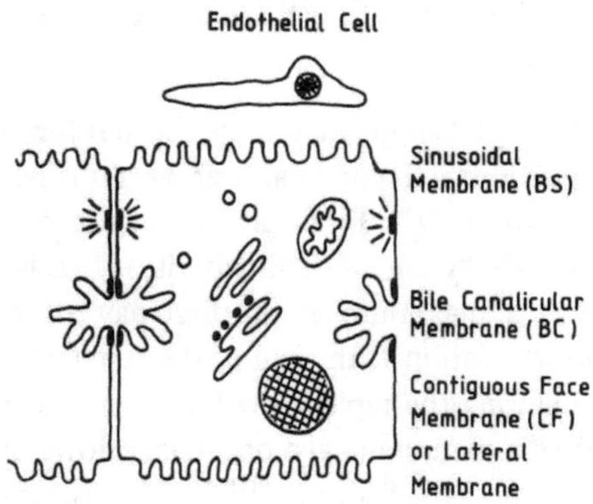

Fig. 2. The three domains of the hepatocyte plasma membrane

membrane glycoproteins, whose functions have not yet been identified. Further confirmation of these observations came from Baumann et al., who analyzed a hepatoma cell line (1983). Thus, intramolecular turnover heterogeneity might be a general phenomen of liver and hepatoma plasma membrane glycoproteins.

3 Turnover of DPP IV in Different Hepatocyte Plasma Membrane Domains

The results mentioned above have been obtained from the total liver plasma membrane. However, the hepatocyte plasma membrane is not a uniform organelle, but consists of three functionally and morphologically different domains: (1) the membrane lining the blood sinus; (2) the lateral or contiguous face membrane; and (3) the bile canalicular membrane (Evans 1980; Wisher and Evans 1975) (Fig. 2). We tried to answer the question if the intramolecular turnover heterogeneity concerns all three domains, or if it is restricted to one or two of them. DPP IV occurs in all three domains. The enrichment factor in comparison to the homogenate is 15–20. The yield of DPP IV is lowest in the blood sinus membrane. In Table 3 the ratios of fucose and methonine half-lives are given for the three plasma membrane domains. The lower this ratio, the faster the intramolecular fucose turnover and vice versa. The ratio is low in sinusoidal and bile canalicular membranes. This demonstrates a more rapid intramolecular fucose turnover than in the lateral membrane which possesses a higher ratio of the two half-lives (Büchsel et al., in preparation).

Table 3. Ratios of half-lives of fucose and methionine of dipeptidylaminopeptidase IV in the three rat hepatocyte plasma membrane domains

Sinusoidal membrane	< 0.28
Bile canalicular	0.20
Lateral membrane	0.47

4 Turnover of DPP IV in Regenerating Liver

After partial hepatectomy the residual liver weight rapidly increases. At about the eigthth day the original liver weight is reached again. During this condition the specific activity of DPP IV in plasma membranes falls to about 80% of the original level at the second day and then slowly rises. This is accompanied by an increase in serum DPP IV with a maximum at the first day after partial hepatectomy followed by a gradual normalization at the end of the regeneration phase.

During the rapid growth in liver regeneration the degradation of proteins, including plasma membrane glycoproteins, slows down (Tauber and Reutter 1978; Scornik and Botbol 1976; Scornik 1984). The determination of a true half-life of an isolated protein is difficult to perform in this non-steady-state system as the increase in the amount of total protein essentially leads to an underestimation of half-lives as derived from the loss of protein-bound radioactivity. The ratio of two half-lives of a glycoprotein obtained after labeling with two different precursors is a reliable parameter to compare the intramolecular turnover of terminal carbohydrates in normal and regenerating liver. We calculated an approximate half-life of the protein component of DPP IV in regenerating liver from the apparent methionine half-life, assuming that the increase in DPP IV is parallel to the increase in total liver protein. Thus, we obtained a value of more than 110 h. This means that the degradation rate of the polypeptide chain of DPP IV is reduced during liver regeneration.

As indicated in Table 4, the discrepancy of a much shorter half-life of fucose than of methionine in normal liver is nearly cancelled in regenerating liver: the ratio of the respective half-lives is shifted from 0.17 in normal liver to 0.60 during liver regeneration. The ratio of sialic acid and methionine half-lives is only slightly elevated from 0.43 to 0.61. The ratio of mannose and methionine half-lives remains nearly unchanged. These results indicate a drastic reduction of the intramolecular fucose turnover in regenerating liver. Sialic acid is concerned to a lesser extent. Mannose turns over like the polypeptide in both normal and regenerating liver (Kreisel et al. 1984a). At the end of liver regeneration, the ratio of fucose and methionine half-livers approximates the normal value. This means that the normal state of intramolecular fucose turnover is restored (Kreisel et al. 1984b).

Table 4. Ratios of half-lives of carbohydrates and methionine of dipeptidylaminopeptidase IV in normal and regenerating liver and at the end of regeneration

Carbohydrate	$t_{1/2}$ Carbohydrate / $t_{1/2}$ Methionine		
	Normal liver	Regenerating liver	End of regeneration
L-Fucose	0.17	0.60	0.25
N-Acetylneuraminic acid	0.43	0.61	
D-Mannose	0.80	0.71	

5 Possible Mechanism and Function of the Intramolecular Turnover Heterogeneity

The intramolecular turnover of terminal carbohydrates requires the action of glycosidases and glycosyltransferases. Although a partial deglycosylation and reglycosylation of the membrane glycoprotein in the plasma membrane itself cannot be excluded, it seems to be much more probable that at least one of these processes occurs at intracellular sites. The fact that an intramolecular turnover of DPP IV can be proven for each of the hepatocyte plasma membrane domains (sinusoidal, bile canalicular, and lateral membrane) supports the latter possibility. An intracellular partial deglycosylation and/or reglycosylation of plasma membrane DPP IV together with the observed shorter half-lives of the terminal carbohydrates suggest a recycling of this glycoprotein back to the cellular surface. As mentioned before, such a recycling meanwhile has been demonstrated for several surface membrane components. Taking into account the fate of an internalized and recycled plasma membrane glycoprotein we propose the following model of the involvement of terminal carbohydrates in the turnover of plasma membrane glycoproteins of the liver (Fig. 3).

After internalization the glycoprotein comes into contact with lysosomes where terminal carbohydrates can be split off by glycosidases leading to partial deglycosylation. During further recycling the glycoprotein is fully reglycosylated in the Golgi complex. The completely restored glycoprotein is inserted again into the plasma membrane.

This model would be consistent with a higher recycling rate of DPP IV in the sinusoidal and bile canalicular membrane domain of the liver in comparison to the

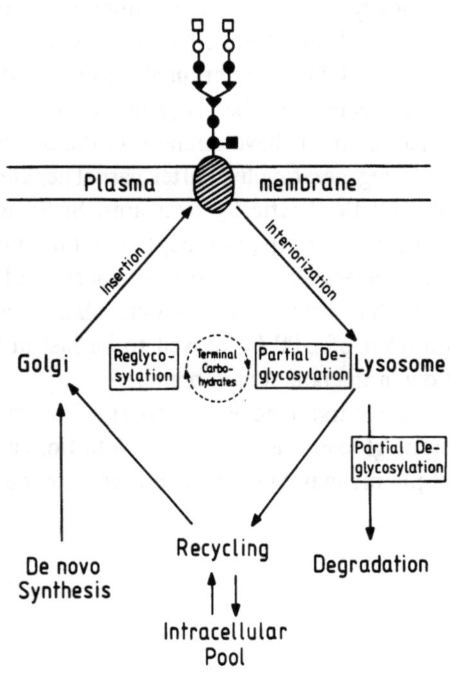

Fig. 3. Hypothetical model for the involvement of the intramolecular turnover of terminal carbohydrates in plasma membrane glycoprotein recycling and regulation of degradation. ● N-acetylglucosamine, ▲ mannose, ○ galactose, □ sialic acid, ■ fucose

lateral membrane. According to this hypothesis, the diminution of the intramolecular turnover heterogeneity of DPP IV in regenerating liver could reflect the diminution of plasma membrane glycoprotein recycling under this condition. A disturbed membrane glycoprotein recycling could be deduced from the depletion of cell surface asialoglycoprotein receptor in regenerating liver despite an unchanged total receptor binding capacity in the whole cell (Howard et al. 1981). The normalization of the intramolecular fucose turnover after the end of regeneration could indicate the normalization of the recycling.

Additionally one could speculate on a connection between the intramolecular fucose turnover and the regulation of degradation of membrane glycoproteins. Mammalian endo-N-acetyl-glucosaminidase, an enzyme crucial for final degradation of complex type liver glycoproteins, cannot cleave the bond between the two innermost N-acetyl-glucosamine residues of the carbohydrate chain if the glycoprotein is core-fucosylated (a fucose is bound to the innermost N-acetyl-glucosamine adjacent to the asparagine in the polypeptide chain) (Tachibana et al. 1981). If this is of importance in vivo also a partial deglycosylation (partial defucosylation) could be a decisive step in the initiation of degradation of membrane glycoproteins. The reduced intramolecular fucose turnover of DDP IV in regenerating liver could reflect the reduced degradation rate of this enzyme. Both processes are normalized after the end of regeneration.

6 Summary

The glycoproteins of the hepatocyte plasma membrane turn over with a marked intermolecular heterogeneity of different and individual half-lives and turnover rates. Additionally some plasma membrane glycoproteins exhibit an intramolecular heterogeneity of turnover, first demonstrated for dipeptidylaminopeptidase IV (DPP IV, EC 3.4.14.5). The terminal carbohydrates (fucose, sialic acid, galactose) turn over much faster than the polypeptide moiety, the core-carbohydrates (mannose, N-acetyl-glucosamine) behave similarly to the polypeptide.

In regenerating liver after partial hepatectomy the intramolecular turnover of fucose of DPP IV drastically decreases. Sialic acid is concerned to a lesser extent. Mannose turns over like the polypeptide in both normal and regenerating liver. After the end of the regeneration phase the normal state of intramolecular fucose turnover is restored.

In the different hepatocyte plasma membrane domains the intramolecular fucose turnover of DPP IV proved to be fast in the sinusoidal and canalicular membrane and slow in the lateral membrane.

The intramolecular turnover of terminal carbohydrates may be involved in membrane glycoprotein recycling. Additionally a relation to the regulation of degradation of plasma membrane glycoproteins can be discussed.

References

Ashwell G, Harford J (1982) Carbohydrate specific receptors of the liver. Ann Rev Biochem 51: 531–554

Baumann H, Hou E, Jahreis GP (1983) Preferential degradation of the terminal carbohydrate moiety of membrane glycoproteins in rat hepatoma cells and after transfer to the membranes of mouse fibroblasts. J Cell Biol 96:139–150

Blobel G (1980) Intracellular protein topogenesis. Proc Natl Acad Sci USA 77:1496–1500

Bock KW, Siekevitz P, Palade GE (1970) Localization and turnover studies of membrane nicotinamide adenine dinucleotide glycohydrolase in rat liver. J Biol Chem 246:188–195

Büchsel R, Berger D, Reutter R (1980) Routes of fucoproteins in plasma membrane domains. FEBS Lett 113:95–98

Büchsel R, Kreisel W, Heussner R, Hofmann W, Reutter W, Gerok W (in preparation)

Conlon JM, Sheehan L (1983) Conversion of substance P to C-terminal fragments in human plasma. Regul Pept 7:335–345

Doljanski F (1982) Cell surface shedding. In: Horowitz MJ (ed) The glycoconjugates, vol 4. Academic, New York, p 155

Doyle D, Cook J, Baumann H (1982) Regulation of membrane proteins in hepatocytes. In: Arias JM, Schachter D, Poppoer H, Shafritz DA (eds) The liver. Biology and pathobiology. Raven, New York, p 185

Dunn WA, Hubbard AL (1984) Receptor-mediated endocytosis of epidermal growth factor by hepatocytes in the perfused liver: ligand and receptor dynamics. J Cell Biol 98:2148–2159

Elovson J (1980) Biogenesis of plasma membrane glycoproteins II. Tracer kinetic studies of two rat liver plasma membrane glycoproteins in vivo. J Biol Chem 255:5816–5825

Evans WH (1980) A biochemical dissection of the functional polarity of the plasma membrane of the hepatocyte. Biochim Biophys Acta 604:27–64

Feller AC, Parawaresch MR (1981) Specificity and polymorphism of diaminopeptidase IV in normal and neoplastic T_μ-lymphocytes. J Cancer Res Clin Oncol 101:59–63

Hanskl C, Reutter W (1984) Turnover of the protein and carbohydrate moieties of a 105 kD glycoprotein of plasma membranes from mouse liver determined by immunoprecipitation. Eur J Cell Biol 33:123–126

Heymann E, Mentlein R (1984) Beeinflußt Dipeptidylpeptidase IV Blutdruck und Gerinnung? Klin Wochenschr 62:2–10

Hino M, Fuyamada H, Nagatsu T, Kurokawa S, Okuyama S (1976) Glycyl prolyl β-naphthylamidase activities in serum, liver and kidney of rats in chronic carbon tetrachloride intoxication. Clin Chim Acta 67:103–105

Howard DJ, Stockert RJ, Morell AG (1981) Asialoglycoprotein receptors in hepatic regeneration. J Biol Chem 257:2856–2858

Hubbard AL (1978) Turnover of mammalian polypeptides. In: Silverstein SC (ed) Transport of macromolecules in cellular systems. Dahlem Konferenzen, Berlin, p 1

Hughes RC (1976) Membrane glycoproteins. A review of structure and function. Butterworth, London

Ichinose M, Maeda R, Fukuda T, Watanabe B, Ishimaru T, Izumi M, Miyake S, Takamori M (1982) Partial purification and characterization of glycylprolyl dipeptidyl aminopeptidase in porcine pancreas. Biochim Biophys Acta 719:527–531

Imai K, Hama T, Kato T (1983) Purification and properties of rat brain dipeptidyl aminopeptidase IV. J Biochem (Tokyo) 93:431–437

Kenny J, Booth AG, George SG, Ingram J, Kershaw D, Wood EJ, Young AR (1976) Dipeptidylpeptidase IV, a kidney brush border serine peptidase. Biochem J 157:169–182

Kreisel W, Volk BA, Büchsel R, Reutter W (1980) Different half-lives of carbohydrate and protein moieties of a 110,000 Dalton glycoprotein isolated from plasma membrane of rat liver. Proc Natl Acad Sci USA 77:1828–1831

Kreisel W, Heussner R, Volk BA, Büchsel R, Reutter W, Gerok W (1982) Identification of the 110,000 M_r glycoprotein isolated from rat liver plasma membrane as dipeptidylaminopeptidase IV. FEBS Lett 147:85–88

Kreisel W, Büchsel R, Volk BA, Reutter W, Gerok W (1983) Tunorver of liver plasma membrane glycoproteins. In: Popper H, Reutter W, Köttgen E, Gudat F (eds) Structural carbohydrates in the liver. MTP Press, Boston, p 51

Kreisel W, Reutter W, Gerok W (1984a) Modification of the intramolecular turnover of terminal carbohydrates of dipeptidylaminopeptidase IV isolated from rat-liver plasma membrane during liver regeneration. Eur J Biochem 138:435–438

Kreisel W, Reutter W, Gerok W (1984b) Reversibility of the effect of liver regeneration on the intramolecular turnover of L-fucose of rat liver plasma membrane dipeptidylaminopeptidase IV (submitted)

Püschel G, Mentlein R, Heymann E (1982) Isolation and characterization of dipeptidyl peptidase IV from human placenta. Eur J Biochem 126:359–365

Scornik OA (1984) Role of protein degradation in the regulation of cellular protein content and amino acid pools. Fed Proc 43:1283–1288

Scornik OA, Botbol V (1976) Role of changes in protein degradation in the growth of regenerating livers. J Biol Chem 251:2891–2897

Skovbjerg H (1981) Immunoelectrophoretic studies on human small intestine brush border proteins – the longitudinal distribution of peptidases and disaccharidases. Clin Chim Acta 112: 205–212

Steinman RM, Mellmann JS, Muller WA, Cohn ZA (1983) Endocytosis and the recycling of plasma membrane. J Cell Biol 96:1–27

Struck DK, Lennarz WJ (1980) The function of saccharide lipids in synthesis of glycoproteins. In: Lennarz WJ (ed) The biochemistry of glycoproteins and proteoglycans. Plenum, New York, p 35

Tachibana Y, Yamashita K, Kobata A (1982) Substrate specificity of mammalian endo-N-acetyl-glucosaminidase: study with the enzyme of rat liver. Arch Biochem Biophys 214:199–210

Tanabe T, Pricer WE, Ashwell G (1979) Subcellular membrane topology and turnover of a rat hepatic binding protein specific for asialoglycoproteins. J Biol Chem 254:1038–1043

Tauber R, Reutter W (1978) Protein degradation in the plasma membrane of regenerating liver and Morris hepatomas. Eur J Biochem 83:35–42

Tauber R, Park CS, Reuter W (1983) Intramolecular heterogeneity of degradation in plasma membrane glycoproteins: Evidence for a general characteristic. Proc Natl Acad Sci USA 80:4026–4029

Vischer P, Reutter W (1980) Different turnover of fucose residues in plasma membranes of rat liver and Morris hepatoma 7777. Biochem J 190:51–55

Volk BA, Kreisel W, Köttgen E, Gerok W, Reutter W (1983) Heterogenous turnover of terminal and core carbohydrates within the carbohydrate chain of dipeptidylaminopeptidase IV isolated from rat liver plasma membrane. FEBS Lett 163:150–152

Wisher MH, Evans WH (1975) Functional polarity of the rat hepatocyte surface membrane. Biochem J 146:375–388

Appendix 1
Approaches to Establishing Ecto Orientation of Enzymes

The unequivocal identification of an enzyme as an ectoenzyme often presents difficulties, due to the relative permeability of the plasma membrane to various substrates and reaction products. As will be evident from the contributions to this book, progress in this area has therefore been confined to relatively few types of enzyme. It is to be hoped, however, that further investigations on other enzymes will continue to reveal rich metabolic activity at the cell surface due to their orientation with catalytic sites exposed to the extracellular space. It would therefore seem opportune to consider briefly the approaches currently employed to identify ectoenzymes in order to encourage and assist researchers embraking upon this topic.

The original criteria for establishing the "ectoness" of an enzyme as laid down by DePierre and Karnovsky (1973, 1974a,b) dealt with the case of intact cells such as blood cells or cells in culture free of contaminating membranous material. This represents an ideal case, which unfortunately does not pertain in most tissues in vivo, but which undoubtedly results in the least equivocal quantitative results. If more recent approaches are also included, the most useful criteria may be formulated as follows for such intact cell preparations (see also Chap. 1):

- Whole cells are able to attack non-penetrating substrates.
- Reaction products are located outside the cell.
- The enzymatic reaction can be inhibited by non-penetrating inhibitors of either a covalent (e.g. diazotized sulphanilic acid) or noncovalent nature (e.g. lectins, antibodies).
- The enzyme must not be released into the incubation medium.
- Mild proteolysis leads to loss of enzyme activity *without release of cytoplasmic components*.

In all these cases it is critical to ensure the maintenance of cell integrity throughout the experiments. This may be monitored by studying the uptake of vital dyes such as Trypan Blue into the cells or by measuring the release of a cytoplasmic marker such as lactate dehydrogenase into the incubation medium. A useful extension of these methods is to study the effect of controlled disruption of cell integrity on the expression of enzyme activity; in the case of a specifically "ecto" enzyme even cell damage should not lead to an increase in enzyme activity. In the case of an enzyme also occurring within the cell, comparison of substrate utilisation by intact cells and disrupted cell preparations or leaky cells (e.g., after detergent treatment) will give information on the relative amounts of enzyme on the cell surface and in intracellular compartments. In the case of substantial amounts of intracellular activity, experiments should also be performed on incubation medium after removal of cells to rule out the possibility of substrate utilisation by cytoplasmic enzymes released into the medium during incubation.

The rigorous application of the above methods should lead to unequivocal identification of an ectoenzyme in intact cell preparations. The situation is more complicated, however, in complex tissues in which the localisation of enzyme activity on the cell surface must be established in situ. Clearly, experiments with tissue slices maintained in oxygenated media in vitro cannot yield unequivocal results due to the inevitable presence of damaged cells at the cut surfaces. Comparison of substrate utilisation by slices of different thicknesses, i.e. where the ratio of intact cells to damaged surfaces is varied, may give a tentative impression of how much of an enzyme may be present on undamaged cell surfaces, but in this case it must be clearly established that cells in the centre of the slice are adequately oxygenated to prevent hypoxic disruption of the permeability barrier represented by the plasma membranes.

Somewhat different problems are encountered when an ectoenzyme must be established in a subcellular fraction such as the synaptosomal fraction (intact nerve endings) from nervous tissue. In this case all the criteria may be applied as for intact cells but contamination of the fractions with membrane fragments or other, intracellular components must be rigorously excluded by use of marker substances and examination of the fractions using electron microscopy.

The most direct method for studying ectoenzymes in situ is electron microscopy, in particular the use of cytochemical or immunocytochemical procedures. This approach can allow clear identification of reaction product or bound antibody at cell surfaces and is applicable to complex tissues, intact cells and subcellular fractions. It should be stressed, however, that electron microscopy gives an important but essentially qualitative view of enzyme distribution. With enzymes occurring at various subcellular sites, as is indeed often the case, studies on both intact tissues and on cells maintained in culture using both morphological and biochemical methods must be combined.

References

DePierre JW, Karnovsky ML (1973) Plasma membranes of mammalian cells: a review of methods for their characterization and isolation. J Cell Biol 56:275–303

DePierre JW, Karnovsky ML (1974a) Ectoenzymes of the guinea pig leukocyte. I. Evidence for an ecto-adenosine monophosphatase, adenosine triphosphatase, and p-nitrophenylphosphatase. J Biol Chem 249:7111–7120

DePierre JW, Karnovsky ML (1974b) Ectoenzymes of the guinea pig polymorphonuclear leukocyte. II. Properties and subability as markers for the plasma membrane. J Biol Chem 249:7121–7129

Appendix 2
Summary Table

In this table we have compiled information provided by the contributing authors concerning the distribution, localisation, and possible functions of the ectoenzymes considered in this volume. We have confined ourselves to these general aspects in order to assist the general reader, rather than giving detailed information of, for instance, molecular properties which can in most cases be found in the main body of the book. The information given here necessarily represents the state of the field at the time of writing and should be no means be considered the last word on the subject. In particular, in compiling the "possible function" column we have included possible roles that are still matters of debate as well as well established functions. It is hoped that this table will thus provide a useful overview of the state of the art.

Enzyme	Substrate(s)	Product(s)	Occurrence	Localisation	Possible functions
5'Nucleotidase (EC 3.1.3.5)	Nucleoside 5' monophosphates	Corresponding nucleosides	Fairly ubiquitous, high levels in heart, liver, lymphocytes, nervous tissue	Mostly as ectoenzyme, sometimes, as in hepatocytes and retinal Müller cells being localised on special domains of the plasma membrane. In some cells also as soluble form differing from the ectoenzyme in substrate preference and sensitivity to inhibition by nucleotides	Purine and pyrimidine recycling; production of humoral agent, adenosine; membrane-bound anchor for the intracellular microfilament system
Nucleoside triphosphatase (ATPase; EC 3.6.1.15)	ATP and other nucleoside triphosphates	Corresponding nucleoside diphosphates	Widespread	The ecto form, characterised by its sensitivity to Ca and Mg is apparently exclusively in an ecto position	Purine and pyrimidine recycling; inactivation of direct, receptor-mediated effects of ATP on nerve cells; source of ADP, AMP, adenosine, having effects on platelet aggregation, synaptic transmission, vasodilatation
Nucleoside diphosphatase (EC 3.6.1.6)	Nucleoside diphosphates	Corresponding nucleoside monophosphates	Probably widespread	Probably only as ectoenyzme	See nucleoside triphosphatase; would also terminate ADP effects on platelet aggregation
Nucleoside diphsophate kinase (EC 2.7.4.6)	Nucleoside di- and tri-phosphates	Corresponding tri- and di-phosphates	On human lymphoid cells, endothelial and smooth muscle cells. Other locations unknown	Not known	Not known
Protein kinase	Cell surface proteins or exogenous proteins (e.g. phosvitin) and nucleoside triphosphate	Phosphoproteins and nucleoside diphosphates	Probably widespread	Only one isoenyzme expressed on cell surface although several protein kinases are known to occur intracellularly	Not known

Endopeptidase-24.11 (enkephalinase; EC 3.4.24.11)	Several peptides including enkephalins, substance P, cholecystokinin, bradykinin. Not proteins	Peptides	Fairly widespread. Richest source is the kidney proximal tubule cell. Also found in nervous tissue, lymph nodes, intestine, glands, chondrocytes	Only as ectoenzyme	Inactivates regulatory peptides
Dipeptidylamino-peptidase IV (EC 3.4.14.5)	Substance P, fibrinogen; also probably other peptides with gly-pro on the N-terminal	Glycyl-proline and other fragments	Fairly ubiquitous. Highest in hepatocyte plasma membranes and in kidney and intestine brush border membranes	Mostly membrane-bound with small quantities of soluble enzyme in liver and serum	Regulation of regional blood flow and blood clotting
Leucine amino-peptidase (EC 3.4.11.2)	Aminoacyl peptide	Amino acid and oligopeptide	Possibly ubiquitous	On cell surface and in cytoplasm	Modulation of gene activity in lymphocytes
Acetylcholin-esterase (EC 3.1.1.7)	Acetylcholine or peptides (e.g. Substance P)	Acetate plus choline or peptide fragments	Widespread. Mostly in nervous system and muscle but also in erythrocytes, placenta, platelets, chromaffin cells	Intracellular, membrane bound (ectoenzyme) and as exoenzyme in extracellular matrix	Inactivation of acetylcholine and possibly peptides. Direct molatory function independent of esterase activity
UDP-glucuron-osyltransferase (EC 2.4.1.17)	Protein plus UDP-glucuronate	Protein-β-glucuronoside	Sponge cells	On cell surface and in the Golgi apparatus	Modulation of functional activity of cell aggregation receptor
β-Glucuronidase (EC 3.2.1.31)	β-D-glucuronoside	Glucuronic acid	On sponge cells. Possibly ubiquitous	On cell surface and in the Golgi apparatus	Modulation of functional activity of cell aggregation receptor
Galactosylfucosyl- and sialyltrans-ferases	Nucleotide-sugar plus protein	Protein-sugar	Possibly unbiqutous	On cell surface and in the Golgi apparatus	

Subject Index

MIX
Papier aus verantwortungsvollen Quellen
Paper from responsible sources
FSC® C105338

If you have any concerns about our products,
you can contact us on
ProductSafety@springernature.com

In case Publisher is established outside the EU,
the EU authorized representative is:
Springer Nature Customer Service Center GmbH
Europaplatz 3, 69115 Heidelberg, Germany

Printed by Libri Plureos GmbH
in Hamburg, Germany